T0181627

Bernard Jancewicz

Directed Quantities
in Electrodynamics

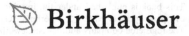 Birkhäuser

Bernard Jancewicz
Institute of Theoretical Physics
University of Wrocław
Wrocław, Poland

ISBN 978-3-030-90473-9   ISBN 978-3-030-90471-5   (eBook)
https://doi.org/10.1007/978-3-030-90471-5

Mathematics Subject Classification: 15A75, 53Z05, 78A25, 78A40, 83A05

This book is published under the imprint Birkhäuser, www.birkhauser-science.com, by the registered company Springer Nature Switzerland AG
The registered company address is: Gewerbestrasse 11, 6330 Cham, Switzerland

Preface

The first extensive description of electromagnetism using the notion of a *field*—
introduced by Michael Faraday—was in *A Treatise on Electricity and Magnetism*
by James Clerk Maxwell, published in 1873. Since then, Maxwell's equations have
constituted the core of electrodynamics—the classical theory of the electromagnetic
field. The traditional mathematical tools for the presentation of Maxwell's theory
were created by Josiah Willard Gibbs in his vector calculus, a blend of algebraic
and geometric notions. A typical example is the noncommutative vector product.
Such a product was introduced by Hermann Guenter Grassmann in about 1840 and
called the exterior product. However, the result of this product was not a vector,
but a *bivector*—an element of an exterior algebra of *multivectors*, later called a
Grassmann algebra. It consists of scalars, vectors, bivectors, trivectors (generally
k-vectors) and their linear combinations.

Grassmann did not develop fully algebraic structures corresponding to geometric
notions, because he lacked the idea of duality, that is, a notion of linear form
over given linear space. Duality allows one to replace the scalar product with one
factor fixed by a linear form. (The reader will see this in the example of the wave
vector replaced by a linear form of wavefronts.) Moreover, the scalar product with
two arbitrary factors can be replaced by a symmetric bilinear form. An object
dual to a $k$vector is $k$form, which is an exterior form of kth order. A set of such
forms constitutes an exterior algebra, also called a *Grassmann algebra*. Exterior
forms, depending on points in space, called *differential forms*, are appropriate to
integrate over curves, surfaces or three-dimensional regions, without referring to a
scalar product. They were introduced by Gregorio Ricci-Cubastro in about 1880
and developed by Elie Joseph Cartan at the turn of the nineteenth and twentieth
centuries.

The presentation of electrodynamics with a wide use of differential forms
has become quite popular; see Refs. [1–9, 11–14, 52]. Some authors address
electromagnetism through differential forms in relation to electrical engineering
[15–21, 23, 25]. Such a formulation ensures a deep synthesis of formulae and
simplifies many deductions. Usually, geometric imaging does not follow the formal

mathematical definitions of the notions, which could be done by analogy with vectors and can be easily depicted.

Other mathematical structures have been applied to physics, namely Clifford algebras, see [22, 24, 26, 27], where basic ingredients are multivectors as a generalization of vectors. The present book is intended to keep a balance between multivectors and exterior forms as two mathematical tools. As both physicists and engineers know, a vector can be shown as a directed segment which has a *direction* (a straight line with an orientation) and the *magnitude* (length). The linear form, as the dual object, also has a direction and magnitude. In addition to vectors and linear forms, we may introduce also *multivectors* and their duals, namely *exterior forms* of higher grade (known also as *multilinear forms* when acting on vectors). Here, I call these objects *directed quantities* because each of them has a direction and magnitude. Schouten [46] was probably the first author who considered the directions of multivectors and exterior forms. They were attractively illustrated in the book of Misner, Thorne, and Wheeler [40]. Similar pictures can be seen in publications of Burke [6–8] and Schutz [47], in an article by Warnick et al. [54]. Hehl and Obukhov illustrate them richly in their book [20]. However, the most careful presentation of the directed quantities in many domains of physics is contained in the book by Tonti [51] under the name "physical variables". The discussion of the directed nature of physical quantities is not covered in such depth here as in Tonti's book. Nevertheless, the present book is focused on the elaboration of the graphical presentation of directed quantities in mathematics, thereby deepening the understanding of their properties and the role played in physics. I have used as many diagrams as possible with the hope that this helps with visualization.

Not all publications using differential forms mention *odd forms*. These were introduced by Weyl [55] and developed by Schouten [46], who called them *W-p-vectors* (for Weyl). De Rham [10] called them *odd forms*, as did Ingarden with Jamiołkowski [28] after him. The term *twisted forms* was used by Frankel [17] and Burke [7, 8]. Edelen [15], and later Frankel [18], used the term *pseudo-forms*.

As it is shown in Chap. 1, sixteen directed quantities are connected with the three-dimensional vector space. We demonstrate in Chap. 2, that when a scalar product is present, the whole variety of directed quantities becomes unnecessary. There then appears a natural way to replace them with vectors, pseudovectors, scalars or pseudoscalars—in this way one can return to the description known from the time of Gibbs and employed in traditional textbooks.

A natural question arises: why do we introduce sixteen directed quantities when four of them can also do the job in three dimensions with the scalar product? In addition to the above mentioned simplicity of deductions, there are two extra reasons for this, which are discussed in Chap. 4. First, the authors of textbooks on electrodynamics, using exterior forms, have not stressed sufficiently strongly that to some extent one can proceed without the scalar product, i.e. without the metric. Van Dantzig [52] and Post [44] could be mentioned among authors treating a metric-free approach to electromagnetism, and more recent is the book by Hehl and Obukhov [20]. One may pose the question: how far can one go in electrodynamics without the metric? My answer is that we can only define physical quantities and formulate principal equations (Maxwell's equations, Lorentz force, continuity equation, gauge transformation, continuity conditions on interfaces and electromagnetic stress tensor). The part of it that can be formulated in a scalar product-independent way has been called *premetric electrodynamics*. The best advocates of this attitude are Hehl and Obukhov [20, 23], who also build up an axiomatic structure of electromagnetism.

It should be mentioned that the present book is an expanded version of my book [30], the Polish title of which should be translated as *Directed Quantities in Electrodynamics*. This is not a textbook on classical electrodynamics, and accordingly, its material is neither standard nor complete. It is intended for readers who have gained at least a basic knowledge of electrodynamics and special relativity. It also demands competence in the advanced calculus, linear algebra and differential equations. At advanced undergraduate and graduate levels in departments of mathematics and physics, the book will be useful for those lecturers and students who want to deepen their understanding of fundamental aspects of classical field theory and electrodynamics. The inclusion of exercises of different levels of complexity also contributes to the possibility of using the book as supplementary reading in combination with the more standard and complete textbooks.

I want to express my gratitude to Zbigniew Oziewicz who, when living in Wrocław, gave many lectures and seminars about differential geometry. They have revealed the beauty of this branch of mathematics and its usefulness when applied to physics. I am also grateful to Friedrich Hehl for plenty of valuable discussions on the subjects present in this book. However, not all his suggestions have been fulfilled. Thanks are due to Arkadiusz Jadczyk for cooperation when preparing the Appendix.

Wrocław, Poland   Bernard Jancewicz
April 2021

Introduction

By directed quantities I mean multivectors and, dual to them, exterior forms. They may be divided into two groups. The quantities of the first group, namely ordinary (even) multivectors, have internal orientations, and the ordinary (even) forms have external orientations, whereas the quantities of the second group are defined conversely: odd multivectors have external orientations odd forms have internal orientations. A detailed discussion of the adjectives "internal" and "external" is contained in the book by Tonti [51]. I can illustrate this distinction with the example of vectors: an even (ordinary) vector has an orientation chosen on its straight segment; an odd (pseudo-) vector has its orientation as a directed circle around the straight segment. The names "even" and "odd" are not good because they have too many meanings and connotations. For instance, the expression "even form" may equally well denote a form of even grade. If you call the odd objects pseudo-forms or pseudo-multivectors, there is no natural word to describe objects that are not "pseudo". The name "twisted quantities" is not appropriate for all quantities of the second group: for instance, an odd vector has its orientation marked as a directed ring surrounding a straight segment, and hence this can be seen as twisted direction, but the odd two-form has its orientation placed on a straight line, so this does not fit the intuitive meaning of the word "twist". Nevertheless, I accept the opinion of Burke in [8, p. 183], that "the language is forced on us by history", and I shall use the adjective "nontwisted" for the first group, and "twisted" for the second group of quantities.

Sixteen directed quantities of the three-dimensional vector space are introduced in Chap. 1 with careful definitions of their directions and magnitudes. In this mathematical chapter, Sect. 1.4 is an exception because it gives examples of physical quantities, showing that all the sixteen quantities have at least one physical counterpart. What is striking is that each of the four physical quantities describing the electromagnetic field, **E**—electric field strength, **D**—electric induction, **H**—magnetic field strength and **B**—magnetic induction, is an exterior form of a different directed nature. Some examples of physical quantities are also placed in other sections of the two mathematical chapters when the mathematical tools necessary for their introduction become available.

Chapter 2 describes linear spaces of directed quantities, their transformation properties and the status of the scalar product. Frankly speaking, the whole variety of directed quantities becomes unnecessary when a scalar product is given in the linear space of vectors and, therefore, also a metric defined by it. There then appears a natural way to replace them with vectors, pseudovectors, scalars or pseudoscalars— in this way one can return to the description known from the times of Gibbs and employed in traditional textbooks.

Chapter 3 is devoted to algebra and analysis of the directed quantities: exterior multiplication, contractions, linear operators as tensors, differential forms, exterior derivative and integration. A most striking feature is visible in Sect. 3.5, namely that the differential equations can be written in the same shape in an arbitrary system of curvilinear coordinates. This property goes under the name of general covariance. A well-recognized fact is that the calculus of differential forms was developed specifically to unify and simplify the integration theory. We present it rather briefly, only to provide tools for applications in electromagnetism.

Relevant physical quantities, the principal relations between them, and differential equations are presented in Chap. 4, in metric-free manner. However, in order to write down the solutions of previously introduced equations, a metric is necessary, which enters the constitutive relations $\mathbf{D} = \varepsilon(\mathbf{E})$, $\mathbf{B} = \mu(\mathbf{H})$. Then the Coulomb field, the Biot-Savart law, the field of rectilinear current, and plane waves can be obtained. The Laplace and d'Alembert equations also belong to this part of electrodynamics. It is useful to mention a lesson drawn from these considerations: the *constitutive relations* are needed to get solutions to Maxwell's equations. The authors who have stressed this role are, among others, van Dantzig [52], Post [44] and Edelen [15], who wrote on page 372: "...wave properties of solutions arise from the electromagnetic constitutive relations and the resulting differential relations between quantities...".

The second reason is that in anisotropic media, it is worth introducing another metric, dependent on the electric permittivity or magnetic permeability. Then the prescription for reducing forms to vectors or pseudovectors can be arranged so that the medium looks like an isotropic one. Thus, the typical solutions of the electrostatic and magnetostatic problems can be easily transposed from those known in isotropic media. Chapter 4 is devoted to this, so the Coulomb field and Biot-Savart law are discussed there.

Electromagnetic waves are treated in Chap. 5, and not only plane ones. An attempt is made to find waves with planar wave fronts, but not constant fields on them. For this purpose, cylindrical and spherical coordinates are used and so-called semiplane waves are introduced. Plane waves are considered in an anisotropic medium. In this case, the traditional wave vector does not describe the propagation direction of the wave: this role is taken by a Poynting two-form, whereas the wave vector is replaced by a one-form of phase density, well suited to surfaces of constant phase. So-called eigenwaves are useful in such a medium, as the phase velocity has a definite value for them.

Considerations of interfaces between two anisotropic media are placed in Chap. 6. This starts with the mathematical Sect. 6.1, which was not included in earlier chapters because it is not necessary for understanding Chaps. 4 and 5. Directed quantities of two-dimensional space (i.e. a plane) are introduced—only twelve of them are possible—and a transition from two- to three-dimensional directed quantities is discussed. The continuity conditions for the electromagnetic field on interfaces are derived. They follow from (integral) Maxwell's equations only, and therefore they belong to the premetric electrodynamics. These conditions help to obtain electromagnetic fields generated by infinite plates with uniform charges or currents, and fields of ideal electric and magnetic capacitors. The electromagnetic stress tensor is introduced. Reflection and refraction of plane waves are conclusions from the interfacial conditions.

The described transition from three-dimensional space to a plane helps us to move one step higher and predict what kinds of directed quantities can be present in four dimensions. There are twenty of them—they are introduced in Chap. 7. In the existing literature, the prolongation of the exterior forms from three dimensions to four was considered to be automatic, whereas—as I show— a distinguished time-like vector is necessary for this purpose, and without it, the directions of the prolonged forms cannot be defined. This knowledge has been used in Sects. 7.2 and 7.3 to present physical quantities appropriate in the space-time and the four-dimensional formulation of electrodynamics. After introducing a *Faraday two-form* $\mathbf{F} = \mathbf{B} + \mathbf{E} \wedge \mathbf{d}t$ and a *Gauss two-form* $\mathbf{G} = \mathbf{D} - \mathbf{H} \wedge \mathbf{d}t$, it is sufficient to write down the two Maxwell equations in very compact shape. The use of differential forms and exterior derivative ensures that they are invariant under arbitrary coordinate and frame transformations and independent of the metric of spacetime see [20]. Since the metric of spacetime serves as the gravitational potential in general relativity, it is useful to know that there is a gravity-free way of formulating Maxwell's equations. Accordingly, the equations are valid in a flat space-time and can be transferred to a curved one.

The plane electromagnetic wave in vacuum is discussed for four-dimensional space-time. The distinguished role of its phase velocity is used to find the Minkowski scalar product. A reason is shown why a metric tensor with one minus and three pluses is better than the alternative with opposite signs on the diagonal. The constitutive relation between $\mathbf{F}$ and $\mathbf{G}$ is found. It contains a Hodge map with the metric included. This relation allows one to express two Maxwell equations by the coordinates of the Faraday field only. The homogeneous equation is generally covariant, whereas the inhomogeneous one (with the four-current on the right-hand side) is only Lorentz covariant.

Contents

Chapter 1
Directed Quantities

Section 1.1 is devoted to considering the non-obvious scalar product in two-dimensional vector space, which nevertheless implies its own perpendicularity condition, magnitudes for vectors and "circles" looking like ellipses. This shows that a metric is non-unique for a given vector space and there is a need to consider vector spaces devoid of the metric.

The primary directed quantities are: vectors, bivectors, and trivectors. A vector arises in the process of abstraction from a segment with an orientation. Similarly, a *bivector* emerges from a flat figure with an orientation, and a *trivector* from a solid body with an orientation. Such objects are called *multivectors*. Connecting multivectors with segments, figures, and bodies has the advantage of easing their depiction in figures: see Ref. [27] for applications in classical mechanics and Ref. [29] for applications in electromagnetism. Multivectors are introduced in Sect. 1.2 by enumerating their relevant features, namely directions and relative magnitudes. The direction consists of two parts: attitude and orientation.

Objects dual to multivectors are *exterior forms*. They are called *differential forms* when they depend on the position in space. Differential forms are popular in theoretical physics, but writers rarely offer pictures. Nice exceptions are Refs. [6, 7, 40] where exterior forms are shown as specific "slicers". This term will become clear in Sect. 1.3, devoted to definitions of exterior forms with their directions and relative magnitudes.

Many physical quantities can be defined without invoking any metric. This is performed in Sect. 1.4.

© Springer Nature Switzerland AG 2021
B. Jancewicz, *Directed Quantities in Electrodynamics*,
https://doi.org/10.1007/978-3-030-90471-5_1

1.1 Linear Space of Vectors Without a Scalar Product

We know the linear space of vectors from the linear algebra courses. It is a set of elements endowed with two operations: addition and multiplication by scalars, that is, by real numbers. The two operations satisfy a series of axioms, which will not be covered here.

Each vector space has its dimension. Let us choose the two-dimensional space $\mathbf{R}^2$ to explain this point. A geometric model of such a space is the plane with a distinguished point O, corresponding to the zero vector. To any other point A there is a corresponding directed segment OA, beginning at O and ending at A (see Fig. 1.1 left). This directed segment with an arrow at point A is the *geometric image* of the vector. The arrow can also be placed somewhere between the ends (Fig. 1.1 right).[1]

The vector has a direction and a magnitude as its relevant features. *Direction* consists of a straight line (passing through the zero point on the plane), which after Lounesto [37, 38] will be called the *attitude*[2] of the vector, and of an arrow on the line, which is called the *orientation*. Two vectors with the same attitude are called *parallel*. For a fixed attitude only two different orientations are possible—they are called *opposite*.

What is magnitude? Usually one says that it is the length of the directed segment, but to determine this, a scalar product and a norm is needed, and here a problem begins: Does only one scalar product exist for a given vector space, that is, for the plane in our case? The answer is: no.

In order to see this, let us recall what a *scalar product* is: It is the *bilinear form* that is *symmetric* and *positive definite*. This means that having denoted the value of the bilinear form on two vectors $\mathbf{v}, \mathbf{w}$ by $(\mathbf{v},\mathbf{w})$, the named properties can be written down:

1. *linearity with respect to the first factor:* $(\mathbf{u} + \mathbf{v}, \mathbf{w}) = (\mathbf{u}, \mathbf{w}) + (\mathbf{v}, \mathbf{w})$, $(\lambda\mathbf{v}, \mathbf{w}) = \lambda(\mathbf{v}, \mathbf{w})$, $\lambda \in \mathbf{R}$, and the same with respect to the second factor,
2. *symmetry:* $(\mathbf{v}, \mathbf{w}) = (\mathbf{w}, \mathbf{v})$,

Fig. 1.1 Two possibilities for imaging a vector

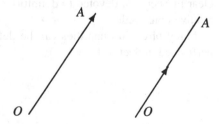

---

[1] This can be compared with Fig. 1.15 in which an orientation of a bivector is shown inside and on the boundary of its image.

[2] The attitude is another name for the straight line spanned by the considered vector.

3. *positivity:* $(\mathbf{v}, \mathbf{v}) \geq 0$,
4. *definiteness:* $(\mathbf{v}, \mathbf{v}) = 0 \Rightarrow \mathbf{v} = 0$.

The *length* or *norm* $|\mathbf{v}|$ is determined in known manner from the scalar product: $|\mathbf{v}|^2 = (\mathbf{v}, \mathbf{v})$ and also the *metric*, that is, *distance* $d(\mathbf{v}, \mathbf{u}) = |\mathbf{v} - \mathbf{u}|$.

If two linearly independent vectors are given $\mathbf{e}_1$, $\mathbf{e}_2$ on the plane, i.e. the *basis* in our vector space, then all vectors $\mathbf{v}$, $\mathbf{w} \in \mathbf{R}^2$ can be decomposed:

$$\mathbf{v} = v^1 \mathbf{e}_1 + v^2 \mathbf{e}_2, \quad \mathbf{w} = w^1 \mathbf{e}_1 + w^2 \mathbf{e}_2, \quad v^i, w^i \in \mathbf{R}.$$

Exercise 1.1 Check that the expression:

$$(\mathbf{v}, \mathbf{w}) = v^1 w^1 + v^2 w^2 \tag{1.1}$$

satisfies all properties from 1 to 4.

That is why expression (1.1) should be considered as the scalar product. It gives $(\mathbf{e}_1, \mathbf{e}_1) = 1 \cdot 1 + 0 \cdot 0 = 1$ and $(\mathbf{e}_2, \mathbf{e}_2) = 0 \cdot 0 + 1 \cdot 1 = 1$, which means that vectors $\mathbf{e}_i$ have unit length, and $(\mathbf{e}_1, \mathbf{e}_2) = 1 \cdot 0 + 0 \cdot 1 = 0$ which means that vector $\mathbf{e}_2$ is perpendicular to $\mathbf{e}_1$.

Let us consider a perspective view of a tiling on a bathroom floor as shown in Fig. 1.2. We know that the tiles are squares. We take one of them from Fig. 1.2, and reproduce it in Fig. 1.3a. It is, obviously, not square in the plane of the figure. We

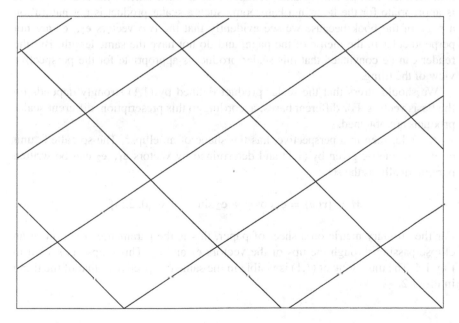

Fig. 1.2 A perspective view of a tiling

Fig. 1.3 (a) One tile from
Fig. 1.2. (b) Two sides as
basic vectors

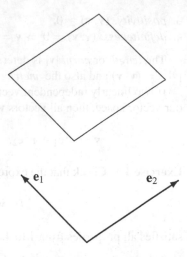

take two sides of it and form two basic vectors as in Fig. 1.3b. These vectors are not
perpendicular and not of equal length on the page of this book. We can, nevertheless
consider the metric of the tiling with two orthonormal vectors $\mathbf{e}_1$ and $\mathbf{e}_2$, on the page
of the book.

In this manner we see that for two arbitrary linearly independent vectors $\mathbf{e}_1$, $\mathbf{e}_2$,
as shown in Fig. 1.3, a scalar product can be chosen such that they become
orthonormal, that is, unitary and perpendicular to each other. This scalar product
is appropriate for the floor in a bathroom. Such a scalar product is not natural on
a page of the book because we see evidently that the two vectors $\mathbf{e}_1$, $\mathbf{e}_2$ are not
perpendicular in the metric of the page, and do not have the same length. But the
reader can be convinced that this scalar product is appropriate for the perspective
view of the tiling.

We should stress that the scalar product defined by (1.1) strongly depends on
the basis vectors. For different bases, according to this prescription, different scalar
products are obtained.

A circle, seen in a perspective, has the shape of an ellipse. The so called "unit
circle" for metric given by (1.1) and determined by vectors $\mathbf{e}_1$, $\mathbf{e}_2$ can be written
parametrically as the set

$$M = \{\mathbf{r}(\phi) = \mathbf{e}_1 \cos\phi + \mathbf{e}_2 \sin\phi; \ \phi \in \langle 0, 2\pi)\}.$$

For the ordinary metric on a sheet of paper, this is the parametric equation of an
ellipse passing through the tips of the vectors $\mathbf{e}_1$ and $\mathbf{e}_2$. This ellipse is shown in
Fig. 1.4. The unit circle of (1.1) is visible in the same perspective as that of the tiling
in Fig. 1.2.

Fig. 1.4 Ellipse generated by
two vectors from Fig. 1.3b

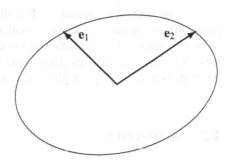

Exercise 1.2 The expression $\mathbf{v} = \frac{d\mathbf{r}(\phi)}{d\phi}$ is a vector tangent to the ellipse S. Check
that it satisfies the condition

$$(\mathbf{v}, \mathbf{r}(\phi)) = 0 \tag{1.2}$$

according to the scalar product (1.1).

We now leave the two-dimensional example, which allowed us to see that the
scalar product and the determined metric are not unique for a given vector space.
Therefore, there is a need to consider vector spaces devoid of a metric—to stress
this we shall call them *premetric vector spaces* (usually they are just called vector
spaces). One should be aware of the fact that not all known geometric notions
occur in such a space. Linearly dependent, linearly independent, parallel vectors,
and parallel planes exist in it. Angles, perpendicular vectors, circles, spheres, and
comparable length of nonparallel vectors are not present in such a space.

If the notion of length is not available, how can we interpret the magnitude of a
vector? The only possibility is *relative magnitude*, which is a measure to compare
only parallel vectors. If $\mathbf{a}$ and $\mathbf{b}$ are two parallel vectors, then a scalar λ exists such
that $\mathbf{b} = \lambda\mathbf{a}$, and then we say that $|\lambda|$ is *relative magnitude* of vector $\mathbf{b}$ *with respect
to* $\mathbf{a}$. In other words, the relative magnitude of a vector $\mathbf{b}$ parallel to $\mathbf{a}$ is its length in
units of $\mathbf{a}$. Two vectors $\mathbf{b}$, $\mathbf{c}$ of the same direction and the same relative magnitude
with respect to $\mathbf{a}$ are equal: $\mathbf{b} = \mathbf{c}$.

If we deal with premetric spaces, then there exists the need to introduce dual
objects known as *linear forms*, which to some extent replace the scalar product. The
possibility of distinguishing vectors from linear forms is important in those domains
of physics where formalisms of non-Euclidean spaces or spaces with variable metric
are introduced. The most spectacular place where such a distinction has a meaning
is gravitational theory, in which the metric itself becomes a dynamical variable.
Therefore, for a case of a premetric space it becomes important to establish the
algebraic nature of the physical quantities. In this book we shall learn that for
physical quantities, eight types of directed quantities should be introduced which
replace vectors and pseudovectors from three-dimensional space, as wells as another
eight quantities which replace scalars and pseudoscalars. This can be found in
Schouten's book [46] in which they appear under different names. After introducing

a scalar product and a metric with it, they all can be changed into vectors and pseudovectors or scalars and pseudoscalars.

In my opinion, the distinction of the directed quantities is not so difficult as to leave it to advanced lectures. They could be introduced for beginners—simply with the first definition of a given physical quantity.

1.2   Multivectors

Usually, when using the term *direction* for a vector in physics, it is not mentioned that this notion contains two ingredients: (1) the straight line and (2) the orientation attached to this line. For a fixed straight line, only two different orientations are possible. For instance, if the line is vertical, the possible orientations are up or down. If the line is horizontal and placed along the graze, the two orientations are forward or back. The orientation of the ordinary vector is marked by an arrow placed on the line.

However, I am going to introduce another orientation for a pseudovector, and hence it must be defined differently. This is a ring surrounding the straight line with an arrow placed on the ring, not on the line. This is illustrated in Figs. 1.5 and 1.6. In this manner, the direction of pseudovector consists of a straight line along with this new version of orientation.

Now, one could claim that the direction consists of straight line and (one or another kind) of an orientation. But a problem emerges for bivectors because the straight line must be replaced by a plane. Thus, a new word is needed to embrace various possibilities. I was not able to find such a word on my own. A Finnish mathematician Pertti Lounesto has found one and placed it in a modest publication [35] devoted to his computer program Clical, invented for Clifford algebras.[3]

We summarize now the relevant features of a vector:

1. *direction* which consists of: (a) *attitude*—straight line, (b) *orientation*—arrow on the line,
2. *relative magnitude*—length in units of some reference vector, defined separately for each attitude.

This description of the vector (it could also be called an *ordinary vector*) can be introduced in linear spaces of arbitrary dimension. The next quantity, namely a *twisted vector* or *pseudovector* has attitude and relative magnitude defined in the same way as for a vector, only the orientation is defined differently. If a segment

---

[3] A justification for why this word is appropriate: *Webster's Third New International Dictionary* published by Koeneman Verlag, Colonne 1993, places on page 141, among others, the following meaning for entry *attitude*: "the position or orientation of an aircraft in the air as seen by an observer stationary on the earth determined and expressed mathematically by the inclination of the axes of the aircraft to three fixed axes on the earth that form the reference frame." This meaning is close to what I mean.

Fig. 1.5 Two images of the
same orientation

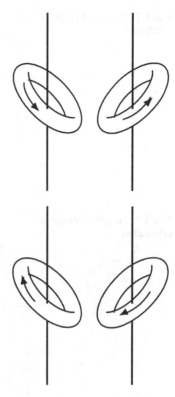

Fig. 1.6 Two images of
orientation opposite to the
previous one

is given, then the *orientation of a twisted vector* is taken as a ring with an arrow
surrounding the segment (see Fig. 1.5 where two of infinitely many situations
representing the same twisted vector are shown). One may agree that there are
only two distinct orientations for a fixed attitude—the other possible orientation is
shown in Fig. 1.6. These two distinct orientations are called *opposite*. Such pictorial
illustration of the twisted vector can be given only in the three-dimensional space.

Let us summarize: A twisted vector has the following features:

1. *direction*, which consists of: (a) *attitude*—straight line, (b) *orientation*—arrow
   on a ring surrounding the line;
2. *relative magnitude*, which is the length in units of some reference vector (no
   matter whether ordinary or twisted).

For this directed quantity, the orientation is *external* to the attitude.

From now on, when depicting the orientation of a twisted vector, an ellipse
with an arrow will be drawn instead of the thick ring as in Figs. 1.7 and 1.8, or a
parallelogram with an arrow as in Figs. 1.9 and 1.10. The segment with orientation
depicted in such a way is a *geometric image* of the twisted vector.

It is worthwhile to discuss the behaviour of twisted vectors under reflections.
They differ from ordinary vectors in this respect. This difference is shown in

Fig. 1.7 Simplified image of orientation

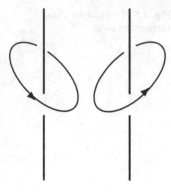

Fig. 1.8 Simplified image of orientation

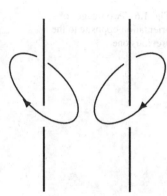

Fig. 1.9 Another image of orientation

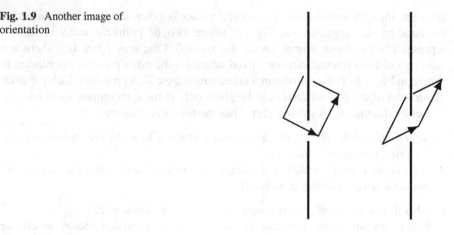

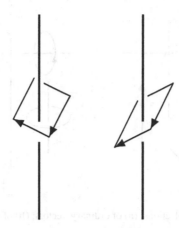

Fig. 1.10   Another image of orientation

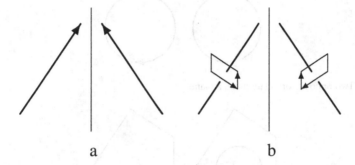

a   b

Fig. 1.11   Behaviour under reflections: (**a**) of an ordinary vector; (**b**) of a twisted vector

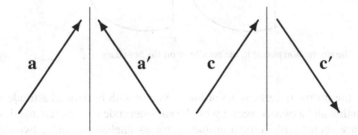

Fig. 1.12   Behaviour under reflections: of a polar vector **a**; of an axial vector **c**

Fig. 1.11 for reflection in the vertical plane. In the traditional presentation of physics, pseudovectors are drawn identically to vectors, and only their different behaviour under reflections is stated separately. Figure 1.12 is an example of such a procedure—normal behaviour of a vector **a** is depicted on the left-hand side, and the opposite behaviour of a pseudovector **c** on the right-hand side.

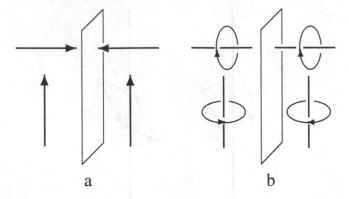

Fig. 1.13 Behaviour under reflections: (**a**) of ordinary vectors; (**b**) of twisted vectors

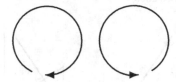

Fig. 1.14 Two opposite orientations on a plane

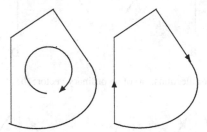

Fig. 1.15 The orientation placed in the middle or on the boundary

When the mirror is vertical, an ordinary vector with horizontal attitude reverses its orientation and a twisted vector with the same attitude is left unchanged, whereas an ordinary vector with vertical attitude remains unchanged and a twisted vector with the same attitude reverses its orientation. This is shown in Fig. 1.13.

A *bivector*, also called an *ordinary bivector*, is the next quantity. It has a *plane* as its attitude and *curved arrow* on the plane as its orientation. It should be obvious that for a fixed attitude only two distinct orientations are possible; they are called *opposite*. After closing the curved arrow to an oval the orientation is clockwise or counterclockwise—see Fig. 1.14. The *relative magnitude* of bivectors is based on comparing areas. In this manner we can depict bivectors as plane figures with curved arrows on them (see Fig. 1.15 left). The orientation can be also shown as an arrow placed on the boundary of the figure, see Fig. 1.15 right. The shape of the figure is not important, only the orientation and the area are relevant.

We now describe more closely how to determine the relative magnitude of a bivector. For this purpose one should fill up the given bivector with multiples of a (assumed smaller) reference bivector. The given bivector may have an arbitrary shape, and the reference bivector should be a parallelogram. The parallelogram can be defined without a scalar product because the notion of parallelity does not demand any scalar product. The parallelogram has a very important property that multiple copies can tile the plane without empty spots. When filling the given bivector with copies of the reference parallelogram, an integer number is not always obtained, hence fractions of the reference parallelogram are needed. A fraction of the parallelogram can be obtained by dividing all its sides onto n equal segments and drawing parallel lines through the ends of the small segments. In this manner one gets n^2 identical smaller parallelograms. Since they are identical, their areas are n^2 times smaller than the reference parallelogram and this relation is independent of any scalar product.

Let us summarize the relevant features of the bivector:

1. *direction:* (a) *attitude*—plane, (b) *orientation*—curved arrow on the plane,
2. *relative magnitude*—area in the units of some reference figure.

The orientation of the bivector is internal with respect to its attitude.

An ordinary bivector can be obtained from two ordinary vectors **a** and **b** as follows: we choose one of vectors as first, let it be **a**. Then by a parallel translation we juxtapose the beginning of **b** to the tip of **a** and draw two more parallel segments to obtain a parallelogram (see Fig. 1.16 left). The parallelogram is the figure that will be the image of the bivector **B** we are looking for. Vectors **a** and **b**, lying on its boundary, determine the orientation of **B**. The obtained bivector **B** is called the *exterior product of vectors* **a** *and* **b**, and this product is denoted by a wedge:

$$\mathbf{B} = \mathbf{a} \wedge \mathbf{b}. \tag{1.3}$$

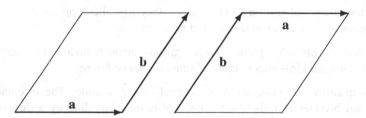

Fig. 1.16 Two exterior products **a** ∧ **b** and **b** ∧ **a**

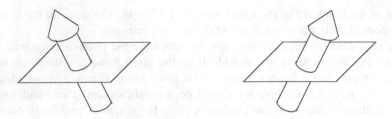

Fig. 1.17 Two images of the same orientation of a twisted bivector

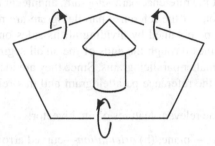

Fig. 1.18 The orientation placed in the middle or on the boundary

Such a representation of the bivector (called also *factorization onto an exterior product*) is, of course, not unique. After juxtaposing the vectors **b** and **a** in the other order (see Fig. 1.16, right), one can read off the identity

$$a \wedge b = - b \wedge a$$

meaning that the exterior product of vectors is *anticommutative*.

Another directed quantity is the *twisted bivector*, which differs from the ordinary bivector only in the definition of its orientation. Its orientation is chosen as an arrow piercing the plane denoting the attitude. Figure 1.17 shows two situations corresponding to the same orientation. It should be obvious that only two distinct orientations are possible for a fixed attitude—they are called *opposite*.

We summarize features of the twisted bivector:

1. *direction:* (a) *attitude*—plane, (b) *orientation*—arrow nonparallel to the plane;
2. *relative magnitude*—area in units of some reference figure.

For this quantity, the orientation is external to the attitude. The orientation of the twisted bivector can also be marked on its boundary line by a directed ring surrounding the line (see Fig. 1.18).

A twisted bivector **B** can be obtained from a vector **a** and twisted vector **c** by attaching **c** to the tip of **a** in the manner depicted in Fig. 1.19. An orientation of **B** is taken from how the direction of the curved arrow surrounding **c** pierces the plane of the parallelogram. The tip point of **a** touching both ends of **c** yields the same

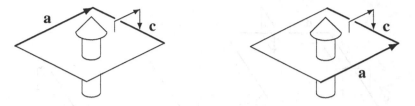

Fig. 1.19 Exterior product $\mathbf{a} \wedge \mathbf{c}$ of nontwisted vector $\mathbf{a}$ with twisted vector $\mathbf{c}$

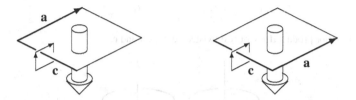

Fig. 1.20 Exterior product $\mathbf{c} \wedge \mathbf{a}$ of twisted vector $\mathbf{c}$ with nontwisted vector $\mathbf{a}$

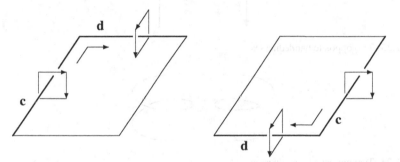

Fig. 1.21 Exterior product $\mathbf{c} \wedge \mathbf{d}$ of two twisted vectors $\mathbf{c}$ and $\mathbf{d}$

orientation of $\mathbf{a} \wedge \mathbf{c}$. The result is called the *exterior product* of quantities $\mathbf{a}$ and $\mathbf{c}$ and is also written with the wedge sign: $\mathbf{B} = \mathbf{a} \wedge \mathbf{c}$. The exterior product of $\mathbf{c}$ and $\mathbf{a}$ in the opposite order is shown in Fig. 1.20. Now an arbitrary end of $\mathbf{c}$ is attached to $\mathbf{a}$ to the other end od $\mathbf{a}$. By comparing Figs. 1.19 and 1.20 one notices that this product is *anticommutative*: $\mathbf{a} \wedge \mathbf{c} = -\mathbf{c} \wedge \mathbf{a}$.

The product of two twisted vectors is still missing, so we define it now. In order to obtain the *exterior product of two twisted vectors* $\mathbf{c}$ and $\mathbf{d}$, we juxtapose them such that the directed ring passing through the junction from one vector to another does not change its orientation—this can be done in two ways (see Fig. 1.21). Then we draw two other parallel segments to obtain a parallelogram. This parallelogram is taken as the exterior product $\mathbf{c} \wedge \mathbf{d}$ with the orientation from $\mathbf{c}$ to $\mathbf{d}$. The above mentioned two possibilities of juxtaposing twisted vectors give the same orientation, so the prescription is unique. Figure 1.22 shows, that the product $\mathbf{d} \wedge \mathbf{c}$ taken in the opposite order gives the opposite orientation, thus the exterior product of twisted vectors is *anticommutative*: $\mathbf{c} \wedge \mathbf{d} = -\mathbf{d} \wedge \mathbf{c}$.

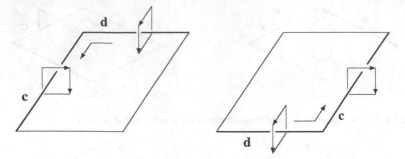

Fig. 1.22 Exterior product **d** ∧ **c** of two twisted vectors **d** and **c**

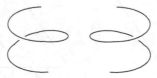

Fig. 1.23 Two opposite handednesses

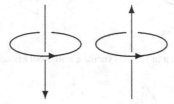

Fig. 1.24 Two opposite handednesses

For the three-dimensional objects a natural attitude is just the three-dimensional space, and an orientation is the composition of a round motion with a progressive motion nonparallel to the plane of the round one. We draw this orientation in the form of two intertwined arrows: one round, the other straight, as in Fig. 1.23. The orientation is considered to be the same when two arrows are rotated simultaneously. Therefore, only two distinct orientations are possible (see Fig. 1.23)—they are called *opposite*. Two kinds of screw correspond to this two orientations: left-handed and right-handed, or two kinds of helices shown in Fig. 1.24. For these reasons, the three-dimensional orientation is called *chirality* or *handedness*, which may be *right* or *left*.

The face of an analogue clock or watch is a natural example of a three-dimensional orientation, since it gives simultaneously a straight arrow—the direction of looking at the face; and a round arrow—the movement of the pointers of the face. We can probably agree that with such a convention the majority of analogue clocks on the Earth have the right handedness.

We are now ready to define a *trivector* or *ordinary trivector* as a geometrical object with the following features:

1. *direction:* (a) *attitude*—three-dimensional space, (b) *orientation*—handedness;
2. *relative magnitude*—volume in units of some reference body.

There is only one attitude possible for trivectors in the three-dimensional space, so all trivectors are parallel there. No continuous transformation of right-handed to left-handed orientation exists within this space—this can be accomplished only by passing to the fourth dimension.

A geometric image of the trivector is a solid body with two intertwined arrows or a fragment of a helix inside (see Fig. 1.25). One may also shift the round arrow in the direction shown by the straight one and place it on the boundary of the solid body, as is shown in Fig. 1.26. Both Figs. 1.25 and 1.26 are images of the same trivector with the right-handedness.

The ordinary trivector can be built of a vector **c** and a bivector **B** as follows. The tip point of **c** is juxtaposed to the boundary of **B** and shifted parallel around the boundary in order to obtain a titled cylinder with the base **B**, see Fig. 1.27.

In this manner, all relevant features of the ordinary trivector **T** are obtained: its orientation is the curved arrow of bivector **B** composed with the straight arrow of

Fig. 1.25 Orientation of non-twisted trivector, put inside it

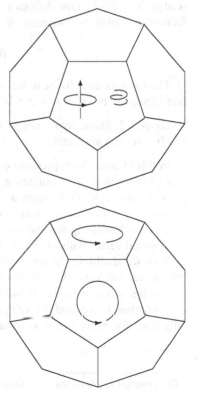

Fig. 1.26 Orientation of nontwisted trivector, put on its boundary

Fig. 1.27 The exterior
product $c \wedge B$. of nontwisted
quantities

Fig. 1.28 The exterior
product $c \wedge (a \wedge b)$ of
nontwisted vectors

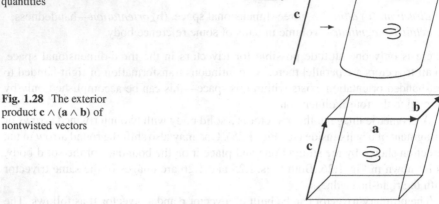

vector c; in the example depicted in Fig. 1.27, the handedness is right. Its attitude is
the only one possible in the three-dimensional space, and the magnitude is naturally
given as the volume of the obtained cylinder. The described operation ascribing
trivector T to the factors c and B is called the *exterior product* and denoted by the
wedge: $T = c \wedge B$. One defines also the exterior product in the opposite order of
factors, assuming its *commutativity*:[4]

$$B \wedge c = c \wedge B. \tag{1.4}$$

The trivector can also be represented as the exterior product of three vectors after
factorizing the bivector $B = a \wedge b$, which gives $T = c \wedge (a \wedge b)$ (see Fig. 1.28).

Exercise 1.2 Show, by drawing, that the three trivectors $c \wedge (a \wedge b)$, $a \wedge (b \wedge c)$
and $b \wedge (c \wedge a)$ are equal.

In this manner the expression $c \wedge (a \wedge b)$ is symmetric under a cyclic permutation of the factors. The equality $a \wedge (b \wedge c) = c \wedge (a \wedge b)$ proven in Exercise 1.2
in connection with (1.4) gives $a \wedge (b \wedge c) = (a \wedge b) \wedge c$, which means that the
exterior product of vectors is *associative*.

It is worth noticing that the trivector can also be obtained as the exterior product
of a twisted vector with a twisted bivector. This operation is defined in Fig. 1.29.

Now we shall introduce the next directed quantity, namely a *twisted trivector*.
As before, we leave the attitude and magnitude of the trivector fixed; only the
orientation will be different. It would be natural to assume that the dimension of the
straight arrow is one, and that of the curved arrow two. Hitherto we have considered
only cases for which the dimensions of the orientations of the ordinary and twisted

---

[4] This assumption will be important later when showing associativity of the exterior product.

Fig. 1.29 The exterior
product **c** ∧ **B** of twisted
quantities

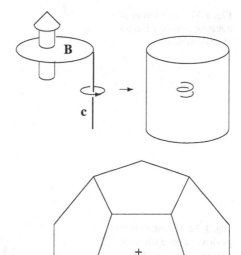

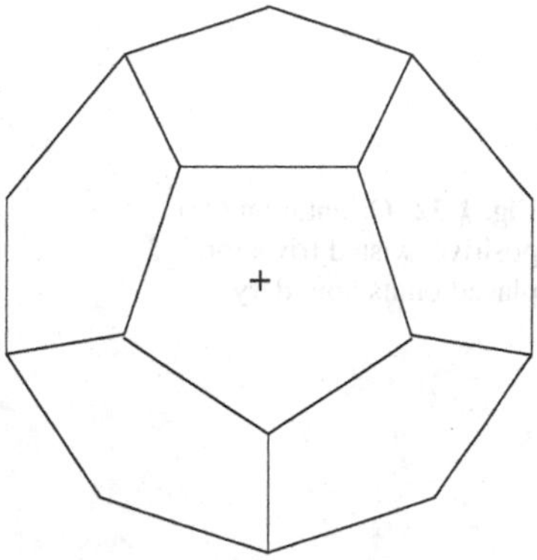

Fig. 1.30 Orientation of a
positive twisted trivector,
placed inside it

quantities of the same grade have a sum equal to three. Let us apply this observation
to the present situation. The ordinary trivector has a three-dimensional orientation.
Hence the twisted trivector should have a zero-dimensional orientation. A point is
the zero-dimensional object, and a zero-dimensional directed quantity is a scalar.
Thus we should take the orientation of the scalar. What is this? The orientation of a
scalar is the sign plus or minus. In this manner we arrive at the conclusion that the
orientation of the twisted trivector is a sign.

We can now summarize the features of *twisted trivectors*:

1. *direction*: (a) *attitude*—three-dimensional space, (b) *orientation*—sign;
2. *relative magnitude*—volume in units of some reference body.

Are we allowed to call this orientation external with respect to attitude? In my
opinion yes, because the sign is somehow outside the three-dimensional space.

A geometric image of the twisted trivector is a solid body with a sign inside
(see Figs. 1.30 and 1.31). If we want to mark the orientation on the boundary, that
is the surface of the body, the boundary should be treated as composed of twisted
bivectors. In this manner, the positive twisted trivector should have outgoing arrows
(Fig. 1.32) and the negative one, incoming arrows (Fig. 1.33).

The twisted trivector can be obtained as the exterior product of a twisted bivector
with a vector (Fig. 1.34 depicts two possible signs as orientations of the result) or
as the exterior product of a bivector with a twisted vector—two possible signs are
shown in Fig. 1.35.

Fig. 1.31 Orientation of a
negative twisted trivector
placed inside it

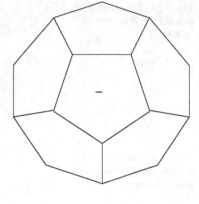

Fig. 1.32 Orientation of a
positive twisted trivector,
placed on its boundary

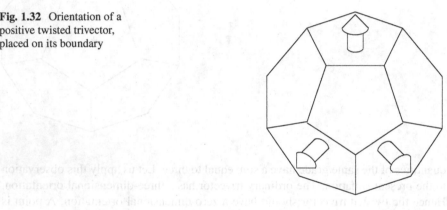

Fig. 1.33 Orientation of
negative twisted trivector put
on its boundary

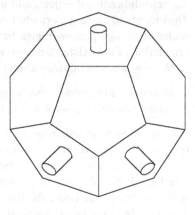

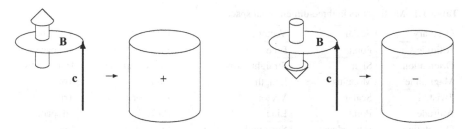

Fig. 1.34 Exterior product of non-twisted vector **c** with twisted bivector **B**

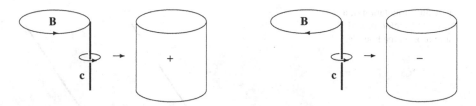

Fig. 1.35 Exterior product of twisted vector **c** with non-twisted bivector **B**

Exercise 1.3 Show on drawings that the associativity $\mathbf{a} \wedge (\mathbf{b} \wedge \mathbf{c}) = (\mathbf{a} \wedge \mathbf{b}) \wedge \mathbf{c}$ is satisfied when the factors **a**, **b**, **c** are ordinary or twisted vectors in any combinations.

By applying the established rules to construct twisted quantities from ordinary ones, we define a twisted scalar by first recalling the features of an ordinary scalar, which can also be called a *non-twisted scalar*.

1. *direction*: (a) *attitude*—point, (b) *orientation*—sign;
2. *magnitude*—absolute value or modulus.

We now list of features of the *twisted scalar*:

1. *direction*: (a) *attitude*—point, (b) *orientation*—handedness;
2. *magnitude*—absolute value or modulus.

As a summary of the considerations of this section we present Table 1.1. It collects the relevant features of ordinary scalars, vectors, bivectors, and trivectors, comprising the ordinary multivectors, as well as twisted scalars, vectors, bivectors, and trivectors, called jointly twisted multivectors. Twisted multivectors have an orientation external with respect to their attitude.

Table 1.1   Multivectors in three-dimensional space

Ordinary	Scalar	Vector	Bivector	Trivector
Attitude	Point	Line	Plane	3d space
Orientation	Sign	Straight arrow	Curved arrow	Handedness
Magnitude	Modulus	Length	Area	Volume
Twisted	Scalar	Vector	Bivector	Trivector
Attitude	Point	Line	Plane	3d space
Orientation	Handedness	Surrounding arrow	Straight arrow	Sign
Magnitude	Modulus	Length	Area	Volume

Fig. 1.36 Orientation of a
nontwisted vector, placed
inside it or on the boundary

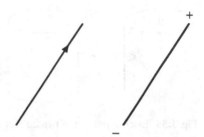

Each multivector has its *grade*, namely k-vectors have grade k, while scalars have grade zero. In other words—scalars are multivectors of zeroth grade. We shall also say that k-vector is a *directed quantity of grade k*.

In the case of bivectors and trivectors, we mentioned two ways of marking their orientation, namely inside and on the boundary of the geometric image. In the case of vectors, the geometric image of the vector is a segment with an orientation and its boundary is a pair of points. If one wants to mark the orientation of an ordinary vector on its boundary, one has to ascribe the zero-dimensional orientation, i.e., a sign, to the boundary points. We arrange that the initial point has a minus sign and the end point has a plus sign as suits the convention accepted in electrostatic for the electric dipole (see Fig. 1.36). For a twisted vector, one has to ascribe a three-dimensional orientation to the boundary points: one end should be right-handed, the other left-handed, according to the following rule: the round motion is taken from the ring surrounding the segment and the progressive motion is directed outside (see Fig. 1.37) where the left-handed motion is at the top, and the right-handed one at the bottom.

Notice a general rule: the orientations of multivectors can be placed on the attitude somewhere in the middle or on the boundary. In addition to the above two figures one can mention Figs. 1.15, 1.18, 1.25, 1.26, 1.30, 1.31, 1.32, and 1.33.

Fig. 1.37 Orientation of a twisted vector, placed in the middle or on the boundary

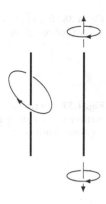

1.3 Exterior Forms

The next eight types of directed quantities are dual to the previous ones.

A *linear form, one-form* or *exterior form of first grade*[5] is the linear mapping of vectors into scalars: $\mathbf{f} : \mathbf{r} \to \mathbf{f}[[\mathbf{r}]] \in \mathbf{R}$.[6] In this sense it is an object dual to vectors. The value of the form $\mathbf{f}$ on a vector $\mathbf{r}$ will be denoted by the double square bracket; ordinary brackets will be used when forms depending on a point in space will be considered.

It should be known from algebra that a *kernel* of the mapping $\mathbf{f}$, i.e. the set $M = \{\mathbf{r} : \mathbf{f}[[\mathbf{r}]] = 0\}$ in $\mathbf{R}^3$ is a linear subspace of dimension 2; thus M is a plane. Hence with each linear form a plane can be associated that passes through the origin of the coordinates. One can also find another plane parallel to M on which the form $\mathbf{f}$ assumes the value 1. This pair of planes or, equivalently, the layer between them, constitutes a geometric image of the linear form. Instead of numbering the planes one can place an arrow joining them that shows the direction in which the form increases (see Fig. 1.38). The arrow needs not be perpendicular to the planes; in fact the notion of perpendicularity is absent, because a metric is missing. Such a visualization of the linear form was considered in the book [40].

Conversely, if we have such a pair of planes, we can ascribe to each vector $\mathbf{r}$ a number as follows: we put the beginning of a vector $\mathbf{r}$ on plane number zero, then draw additional layers at equal distances (the same as between the two original planes), count the layers cut by $\mathbf{r}$; this is the value $\mathbf{f}[[\mathbf{r}]]$. If $\mathbf{r}$ does not end on any plane, the number $\mathbf{f}[[\mathbf{r}]]$ is not an integer. For the situation depicted in Fig. 1.39, $\mathbf{f}[[\mathbf{r}]]$ is 3.5. If $\mathbf{r}$ cuts layers in the direction of decreasing numbers, the value $\mathbf{f}[[\mathbf{r}]]$ is negative. In this manner, the linear form becomes a kind of "slicer" that shreds vectors onto pieces—the number of pieces is the value of the form on a given vector. This prescription along with similar pictures can be found in the book [40].

---

[5] Frequently in literature such form is alternatively called *exterior form of degree one*.

[6] The notation $\mathbf{f}[[\mathbf{r}]]$ is used for the value of one-form $\mathbf{f}$ on vector $\mathbf{r}$.

Fig. 1.38 Image of a
nontwisted one-form

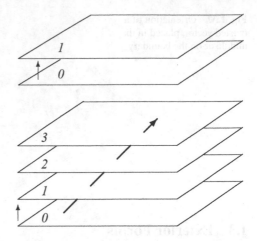

Fig. 1.39 Ascribing a
number to a vector by a
nontwisted one-form

For another form $\mathbf{f}' = 2\mathbf{f}$ which is twice as large as $\mathbf{f}$, i.e. $\mathbf{f}'[[\mathbf{r}]] = 2\,\mathbf{f}[[\mathbf{r}]]$ for each $\mathbf{r}$, the planes must be spaced closer by a factor of two. In this manner we arrive at the conclusion that the *magnitude* of the linear form must be related to the linear density of the planes, or in other words, the *(relative) magnitude* of a form $\mathbf{f}'$ with respect to $\mathbf{f}$ is the inverse ratio of the distances between neighbouring planes of $\mathbf{f}'$ to those of $\mathbf{f}$. If two linear forms are given, then a common reference vector can be used to measure their relative magnitudes by counting how many planes of a given form are cut by this vector. In this manner an appropriate unit for measuring the magnitudes in the family of parallel forms is the inverse of the length of a reference vector.

In this way we arrive at the following list of features of the ordinary linear form:

1. *direction*: (a) *attitude*—plane, (b) *orientation*—arrow nonparallel to the plane;
2. *relative magnitude*—linear density in inverse units of some reference vector.

We should remark that for this directed quantity the orientation is outer with respect to the attitude.

We shall need in the sequel forms changing from point to point. For this purpose a way of placing the geometric image of a form in the surrounding of a given point is necessary. The pair of planes is not convenient for this purpose. Therefore, I propose a simplified picture, namely an oblique cone with an elliptic base (see Fig. 1.40). The planar base of the cone lies on the zero plane of the form, and the vertex determines the plane with number 1, since one can draw a plane parallel to the base and passing through the vertex. This second plane is not necessary in the image; we marked it by broken lines.

A *twisted one-form* $\mathbf{g}$ is a linear mapping of twisted vectors into scalars. In this sense it is an object dual to twisted vectors. It can be represented geometrically as a layer, i.e., a pair of parallel planes with a curved arrow on one of them (see Fig. 1.41). The value $\mathbf{g}[[\mathbf{r}]]$ is the number of such layers pierced by the twisted vector $\mathbf{x}$, its sign established according to whether the curved arrow surrounding $\mathbf{r}$

Fig. 1.40 Simplified image
of a nontwisted one-form

Fig. 1.41 Image of a twisted
one-form

Fig. 1.42 Ascribing a
number to the twisted vector
by the twisted one-form

Fig. 1.43 Simplified image
of a twisted one-form

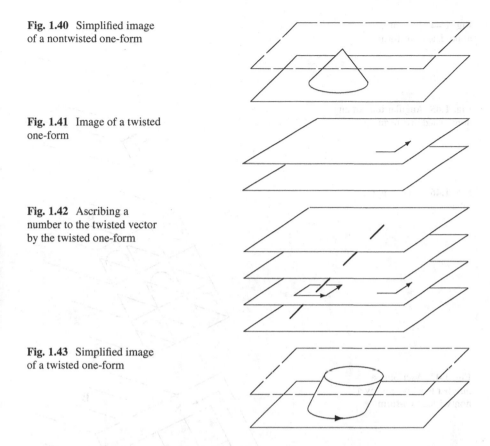

has the same orientation or not as the curved arrow on one of the planes of **g** (see Fig. 1.42).

We give now the features of the twisted one-form:

1. *direction*: (a) *attitude*—plane; (b) *orientation*—curved arrow on the plane,
2. *relative magnitude*—linear density.

The orientation of the twisted one-form is inner with respect to its attitude.

A simplified geometric image of the twisted one-form is an oblique cylinder, on the base of which the ellipse has a curved arrow—this gives the orientation of the twisted form (see Fig. 1.43). Two bases of the cylinder determine two planes of the form, with the distance one. The second plane is not necessary; we mark it with broken lines.

The next directed quantity is an *ordinary two-form* or *ordinary exterior form of second grade*, defined as the linear mapping of bivectors into scalars. In this sense, it is dual to ordinary bivectors. It can be represented geometrically as an infinite tube with a curved arrow surrounding its axis (see Fig. 1.44).

Fig. 1.44 Tube of a
nontwisted two-form

Fig. 1.45 Angular tube of an
nontwisted two-form

Fig. 1.46 Repeating the
tubes

Fig. 1.47 Ascribing a
number to a bivector by a
nontwisted two-form

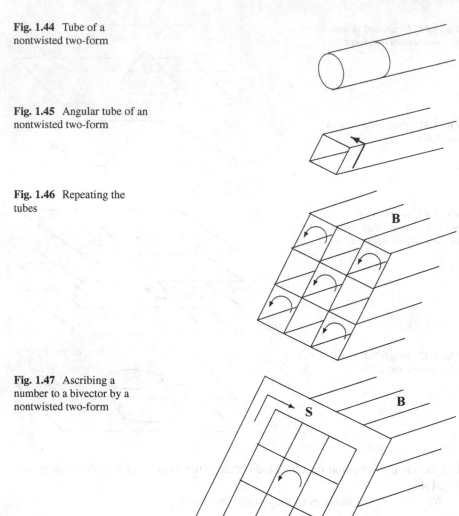

In order to determine the value of the two-form on a given bivector, we must
multitply the tubes. In the first step, we change the round tube into a prism with
the same axis, the same cross-section and the same sense of the arrow surrounding
(see Fig. 1.45). In the second step, we repeat the prisms in parallel, filling the
space tightly with them (see Fig. 1.46). Now the value of the two-form **B** on the
bivector **S** is the number of prisms passing through the bivector with the sign plus if
orientations of **S** and **B** coincide, and minus if they are opposite (see Fig 1.47).

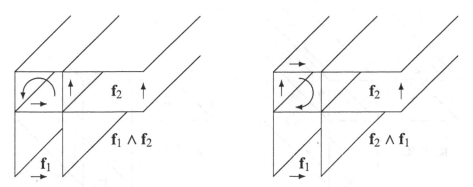

Fig. 1.48 Two possible exterior products of nontwisted one-forms

Now we have the list of relevant features of the ordinary two-form:

1. *direction*: (a) *attitude*—straight line; (b) *orientation*—curved arrow surrounding the line,
2. *relative magnitude*—surface density in inverse units of some reference figure.

The orientation of the two-form is outer with respect to its attitude. The two-form gives the value zero for all bivectors parallel to it.

One can build a two-form **B** from two linear forms $\mathbf{f}_1$ and $\mathbf{f}_2$, as their intersection as shown in Fig. 1.48, left. The attitude of **B** is a straight line—the intersection of the attitudes (planes) of $\mathbf{f}_1$ and $\mathbf{f}_2$—and the orientation is obtained by juxtaposing the arrows of both forms, the second after the first. The result is called the *exterior product of one-forms* $\mathbf{f}_1$ and $\mathbf{f}_2$ and is denoted by

$$\mathbf{B} = \mathbf{f}_1 \wedge \mathbf{f}_2 . \tag{1.5}$$

This product is anticommutative, which is visible upon comparing the left and the right sides of Fig. 1.48. The result of the product is parallel to its factors. The representation of a two-form **B** in formula (1.5), also called a *factorization in the exterior product*, is not unique.

The ordinary two-form **B** can also be obtained as an exterior product of two twisted one-forms $\mathbf{g}_1$ and $\mathbf{g}_2$. In order to determine the orientation of the product $\mathbf{B} = \mathbf{g}_1 \wedge \mathbf{g}_2$, we should draw the orientations of the twisted forms as broken arrows and places them close to the edge of the intersection of the planes so that the last segment of the (broken) arrow of $\mathbf{g}_1$ is parallel to the edge and to the first segment of the (broken) arrow of $\mathbf{g}_2$. These parallel segments should have opposite orientations. Then they cancel each other and the remaining "free", nonparallel fragments of the broken arrows constitute the orientation that we were seeking (see Fig. 1.49, left). To find the orientation of the reversed product, $\mathbf{g}_2 \wedge \mathbf{g}_1$ the last segment of the arrow of $\mathbf{g}_2$ must be parallel to the first segment of the arrow taken from $\mathbf{g}_1$ (see Fig. 1.49, right). Now, the "free" fragments of the arrows give an orientation opposite to the

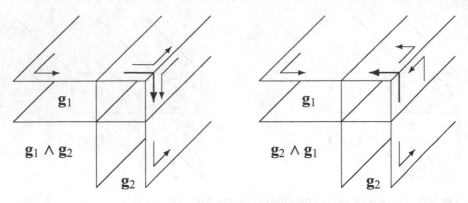

Fig. 1.49 Two possible exterior products of twisted one-forms

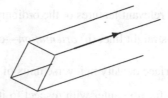

Fig. 1.50 Angular tube of a twisted two-form

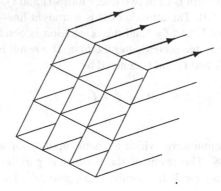

Fig. 1.51 Repeating the tubes

previous one, which proves that the exterior product of the twisted one-forms is *anticommutative*. The result of this product is also parallel to its factors.

A *twisted two-form* is a linear mapping of twisted bivectors into scalars. Its geometric image is very similar to that of an ordinary two-form. The only difference is that the orientation is marked by straight arrows on edges (Fig. 1.50). We repeat the prisms in parallel, filling the space tightly with them (see Fig. 1.51). Then the value of a twisted two-form **D** on a twisted bivector **P** is the number of tubes passing

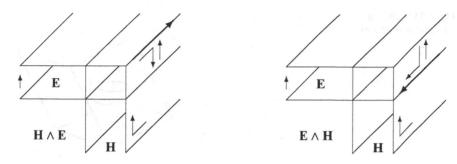

Fig. 1.52 Two possible exterior products of twisted one-forms

through **P** with a plus sign if the orientations of **P** and **D** are the same, and a minus if opposite. Hence the twisted two-form has the following features:

1. *direction*: (a) *attitude*—straight line, (b) *orientation*—arrow on the line;
2. *relative magnitude*—surface density.

The orientation of the twisted two-form is inner with respect to its attitude.

The twisted two-form can be obtained as the *exterior product* of an ordinary one-form **E** with a twisted one-form **H**, as shown in Fig. 1.52. The left-hand picture shows the product **R** = **H** ∧ **E**, in which the twisted one-form is the first factor. In order to find the orientation of **R**, one must move the broken arrow, representing the orientation of **H**, in such a manner that its second segment cancels with the arrow representing the orientation of **E**. Then the first segment defines the orientation of the product **R**. The right-hand picture shows the product **S** = **E** ∧ **H** with the twisted one-form as the second factor. Now, one should move the broken arrow of **H** in such way that its first segment cancels with the arrow of **E**. Then the remaining second segment represents the orientation of the product **S**. The two pictures demonstrate that this product is *anticommutative*: **E** ∧ **H** = − **H** ∧ **E**.

An *ordinary exterior form of third grade*, or in brief, an *ordinary three-form*, is a linear mapping of the trivectors into scalars. In this sense it is a dual object to trivector. It can be represented geometrically as a cell with a three-dimensional orientation, that is the fragment of a helix (see Fig. 1.53). We multiply the cells, tightly filling the space (see Fig. 1.54). A value of the three-form $\mathcal{R}$ on a trivector **T** is the number of cells occupied by the trivector with a plus sign if the orientations of **T** and $\mathcal{R}$ are the same, or minus if opposite.

An intriguing question is, what is an attitude of the three-form? As we could have noticed from the previous examples, the attitudes of the dual quantities are complementary to those of the primary quantities. Hence we should assume, that an attitude of the three-form is a point. In this manner we arrive at the list of features of the three-form:

1. *direction*: (a) *attitude*—point, (b) *orientation*—handedness;
2. *relative magnitude*—spatial density in inverse units of some reference figure.

Fig. 1.53 Cell of a
nontwisted three-form

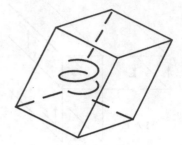

Fig. 1.54 Repeating the cells

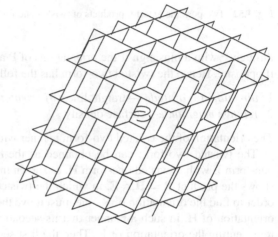

Fig. 1.55 The exterior
product of nontwisted forms

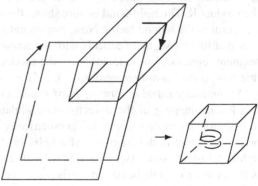

The nontwisted three-form can be made as the *exterior product* from a nontwisted
one-form and a nontwisted two-form. Figure 1.55 shows how one obtains the needed
features of the three-form from the intersection of the factors. The nontwisted three-
form can also be obtained as the exterior product of a twisted one-form with a
twisted two-form, as is visible in Fig. 1.56.

The next directed quantity is a *twisted three-form* for which the attitude and
magnitude are the same as for the ordinary three-form, only orientation should be

Fig. 1.56 The exerior
product of twisted forms

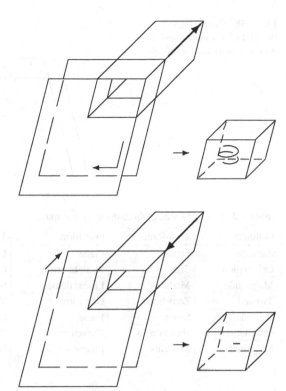

Fig. 1.57 The exterior
product of a nontwisted
one-form with a twisted
two-form

complementary so it should be the sign. In this manner we define the twisted three-form as a geometric object with the following features:

1. *direction*: (a) *attitude*—point, (b) *orientation*—sign;
2. *relative magnitude*—spatial density in inverse units of some reference body.

The twisted three-form can be represented as the exterior product of a twisted two-form with an ordinary one-form, as shown in Fig. 1.57, or as the exterior product of an ordinary two-form with a twisted one-form (see Fig. 1.58). In both illustrations, the orientations of the factors are opposite, hence the obtained twisted three form has a negative sign as its orientation.

Two types of directed quantities are still missing, namely dual to ordinary and twisted scalars. Their names are *ordinary* and *twisted zero-forms*. A *zero-form* is an object dual to scalars, that is, a linear mapping of scalars into scalars. It might seem that such a mapping is simply a multiplication by another scalar. But, according to our rules, such an object should have a complementary attitude to that of the scalar, thus its attitude should be three-dimensional. And this is the difference justifying the introduction of zero-forms. Now the features of an ordinary zero-form:

1. *direction*: (a) *attitude*—space, (b) *orientation*—sign;
2. *magnitude*—modulus.

Fig. 1.58 The exterior
product of a twisted one-form
with a nontwisted two-form

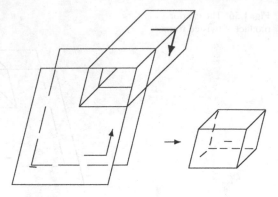

Table 1.2 Exterior forms in three-dimensional space

Ordinary	Zero-form	One-form	Two-form	Three-form
Attitude	Space	Plane	Line	Point
Orientation	Sign	Straight arrow	Curved arrow	Handedness
Magnitude	Modulus	Linear density	Surface density	Spatial density
Twisted	Zero-form	One-form	Two-form	Three-form
Attitude	Space	Plane	Line	Point
Orientation	Handedness	Curved arrow	Straight arrow	Sign
Magnitude	Modulus	Linear density	Surface density	Spatial density

In a similar way, we find the list of features of a *twisted zero-form* as an object dual
to the twisted scalar:

1. *direction*: (a) *attitude*—space, (b) *orientation*—handedness;
2. *magnitude*—modulus.

As a summary of the present section we give Table 1.2 containing features
of ordinary and twisted forms. It is worth to notice that ordinary forms have
orientations outer with respect to attitudes.

1.4   Premetric Physical Quantities

We now present physical quantities with their categorizations as directed quantities.

As the scalar quantities in 3-dimensional space we assume time t, energy $\mathcal{E}$,
power P, electric charge Q, electric current I, magnetic flux Φ_m.

The most natural vector physical quantity is *displacement vector* **l** of the same
nature as the *radius vector* **r**, which may also be considered as a displacement with
respect to the origin of coordinates. Hence the *velocity* $\mathbf{v} = d\mathbf{r}/dt$ as the derivative
of **r** over the scalar variable t is also a vector. The same is true for *acceleration*
$\mathbf{a} = d\mathbf{v}/dt$ and *electric dipole moment* $\mathbf{d} = q\mathbf{l}$.

A one-form occurs naturally in the description of the plane waves. We recall that a *wave* is a disturbance expanding in space and periodic in time. The disturbance is characterized by a displacement ξ of some physical quantity x from its stationary value x_0: $\xi = x - x_0$. The displacement depends on position $\mathbf{r}$ and time t: $\xi = \xi(\mathbf{r}, t)$. The locus of points with the same displacement is called *wavefront*. There can be different wavefronts with the same displacement. When the wavefronts are planes, the disturbance is called a *plane wave*. The distance between two neighbouring wavefronts of the same displacement is called the *wavelength* λ. The family of planes with the same displacements can be viewed as the geometric image of a physical quantity known as the *wave vector* κ with magnitude $1/\lambda$. The physical quantity κ in its directed nature is not a vector but rather a one-form, known in the literature as the *wave covector* or *wave one-form*. It can be interpreted as a linear density of the wave.

When the displacement function can be expressed by a trigonometric a function as follows:

$$\xi(\mathbf{r}, t) = A \cos(\mathbf{k} \cdot \mathbf{r} - \omega t),$$

it becomes function periodic in space. The *phase* $\phi = \mathbf{k} \cdot \mathbf{r} - \omega t$ of the cosine is the linear function of position. Its period 2π is related to the period λ in space by the relation $k\lambda = 2\pi$, where $k = |\mathbf{k}|$, hence $k = 2\pi/\lambda$. Unfortunately, $\mathbf{k}$ is called wave vector, as well. It is also a one-form in its directed nature. The one-form $\mathbf{k}$ counts layers between planes differing by one radian of phase. The one-form κ counts the layers between planes of the same fragment of the wave (i.e., a ridge or valley). The two named one-forms are proportional:[7]

$$\mathbf{k} = 2\pi\kappa. \tag{1.6}$$

We notice from these two examples that one-forms count a linear density of some scalar quantity distributed over space.

The situation with momentum and force is not easy. If we consider mass m as a scalar, then momentum $\mathbf{p} = m\mathbf{v}$ and force $\mathbf{F} = d\mathbf{p}/dt$ are vectors. But, if potential energy U is a scalar, then its relation with force in the traditional language has the form $dU = -\mathbf{F} \cdot d\mathbf{r}$, where the dot denotes the scalar product. It is clear from this that force is a linear map of the infinitesimal vector $d\mathbf{r}$ onto the infinitesimal scalar dU. Hence we should treat force as a one-form, and the rewrite the previous formula as $dU = -\mathbf{F}[[d\mathbf{r}]]$, where $\mathbf{F}[[d\mathbf{r}]]$ is to be read as the value of the one-form $\mathbf{F}$ on the vector $d\mathbf{r}$. In this manner, it may be legitimate to call a one-form a *potential energy density*.

---

[7] The difference between κ and $\mathbf{k}$ is analogous to the distinction between frequency ν and circular (or angular) frequency $\omega = 2\pi\nu$. The linear form $\mathbf{k}$ could be called a *phase density* whereas κ a *wave density*.

This, in turn, through Newton's equation $\mathbf{F} = d\mathbf{p}/dt$, implies that the momentum should also be a one-form and such a position is taken in some mathematical approaches to electromagnetism (see [52]), and to classical mechanics (see [53]). We shall do the same in the present book. Invoking the previous observation, the reader could ask: Is the *momentum one-form* a linear density of some scalar quantity? To answer this, let us mention a formula from the old theory of quanta, $\mathbf{p} \cdot d\mathbf{r} = dA$, which says that the scalar product on the left-hand side is an increment of action A. Thus, the answer is thatthe one-form of momentum is an *action density*.

We should mention that an important mechanical quantity, angular momentum $\mathbf{M} = \mathbf{r} \times \mathbf{p}$, cannot be represented in a metric-independent manner, because it is built of two vectorial quantities: the radius vector $\mathbf{r}$ and the momentum $\mathbf{p}$. The conversion of the one-form $\mathbf{p}$ into the vector $\mathbf{p}$ demands a scalar product, i.e., a metric, as will be explained in Sect. 2.3.

Another one-form is the *electric field strength* $\mathbf{E}$, since we consider it to be a linear map of the infinitesimal vector $d\mathbf{r}$ into the infinitesimal potential difference: $-dV = \mathbf{E}[[d\mathbf{r}]]$. The physical unit in SI $[\mathbf{E}] = \text{Vm}^{-1}$ fits this interpretation. The expression $\mathbf{F} = q\mathbf{E}$ for the force acting on a test charge q put in an electric field confirms our classification of the force as a one-form.

As a bivector quantity we can take the *area velocity* $\mathbf{S} = \mathbf{r} \wedge \mathbf{v}$. We know that the velocity $\mathbf{v}$ is fastened at the tip of the radius vector $\mathbf{r}$, so their order in the exterior product is natural, as shown in Fig. 1.59.

The best physical example of a bivector is a flat electric circuit—its attitude is the plane of the circuit, its magnitude is just the encompassed area and its orientation is given by the sense of the current. This bivector could be called the *directed area* $\mathbf{s}$ of the circuit. A connected bivectorial quantity is then the *magnetic moment* $\mathbf{m} = I\mathbf{s}$ of the circuit, where I is the current. The units in SI $[\mathbf{m}] = \text{Am}^2$ fit this interpretation.

The *magnetic induction* $\mathbf{B}$ is an example of a two-form quantity, since it can be treated as a linear map of the directed area bivector $d\mathbf{s}$ into the magnetic flux $d\Phi = \mathbf{B} \cdot d\mathbf{s}$ treated as a scalar. This is consistent with the SI dimension: $[\mathbf{B}] = \text{Wbm}^{-2}$. Stokes' theorem $\int \mathbf{B} \cdot d\mathbf{s} = \oint \mathbf{A} \cdot d\mathbf{l}$, in which $\mathbf{A}$ is the so called *vector potential*, says that the product $\mathbf{A} \cdot d\mathbf{l}$ is also the magnetic flux. Thus $\mathbf{A}$ is a one-form rather than a vector. The traditional name "vector potential" is not appropriate for the genuine directed nature of $\mathbf{A}$. Therefore, I would prefer the term *directed potential*. We should write Stokes' theorem in the form $\int \mathbf{B}[[d\mathbf{s}]] = \oint \mathbf{A}[[d\mathbf{l}]]$.

Now some examples of twisted quantities. The area $\mathbf{S}$ of a surface, through which a flow is measured is the first one. The side of the surface from which a substance

Fig. 1.59 Natural order of
factors in the product $\mathbf{r} \wedge \mathbf{v}$

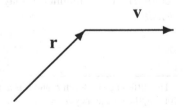

Fig. 1.60 Operational
definition of the electric
induction **D**

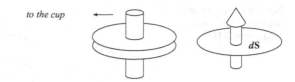

to the cup

dS

(mass, energy, electric charge etc.) passes is important. Hence the orientation of **S** should be marked as an arrow that is not parallel to the surface. This is the situation depicted in Fig. 1.17, hence we claim that the *area of a flow* is a twisted bivector quantity. Accordingly, *the flux density* **j** (or the *current density* in case of the electric current flowing) must be a twisted two-form quantity. It corresponds to the linear map: $dI = \mathbf{j} \cdot d\mathbf{S}$ of the area element $d\mathbf{S}$ into the electric current dI. The physical unit of the current density is $[\mathbf{j}] = \mathrm{Am}^{-2}$, which is consistent with this interpretation.

The *electric induction* **D** has a similar nature. We make use of a prescription of its measurement quoted from page 68 of Weyssenhoff's book [56]. This represents an *operational definition* of **D**.

> Take two identical discs each made of very thin sheet metal, and each with an isolated handle. Place one disc on top of the other, holding them by the handles, electrically discharge them and then place them in the presence of the field. As you separate the discs, the charges induced on them (one positive, the other negative) are also separated. Now measure one of them with the aid of a Faraday cup. It turns out that for a small enough disc the charge is proportional to its area.[8]

Hehl and Obukhov [20, page 136], claim that this prescription of measuring **D** was devised by Maxwell himself and is known by the name *Maxwellian double plates*.

One will agree that the *disc area* $d\mathbf{S}$ is a twisted bivector since its magnitude is the area, its attitude is the plane, and its orientation is given by an arrow showing which disc of the two is to be connected with the Faraday cup (see Fig. 1.60). Because of the proportionality relation $dQ = \mathbf{D}[[d\mathbf{S}]]$, we ascertain that as a linear map of twisted bivectors into scalars, the *electric induction* is a twisted two-form. The SI units $[\mathbf{D}] = \mathrm{Cm}^{-2}$ fit this interpretation. All planes $d\mathbf{S}$, for which $\mathbf{D} = 0$, intersect in a line that is the one-dimensional attitude of the two-form **D**. Notice that **D** is of the same directed nature as the electric current density. This is connected with the fact that $\dot{\mathbf{D}}$ is called the *displacement current*. If we agree that **D** is a two-form, we should accept that its geometric images are tubes, not arrows. This picture of electric induction is in agreement with Faraday's concept of "tubes of inductive force".

On another page, 347, of Weyssehhoff's book [56], one may find an operational definition of the *magnetic field strength*. Figure 1.61, taken from [56], is an illustration of this:

> Take a very small wireless solenoid prepared from a superconducting material. Close the circuit in a region of space where the magnetic field vanishes. Afterwards, introduce the

---

[8] This prescription is based on a similar one given in [23, p. 75].

Fig. 1.61 Operational
definition of the magnetic
field strength **H**

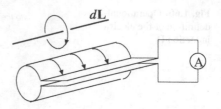

circuit into a region in the field where we want to measure **H**. A superconductor has the
property that the magnetic flux enclosed by it is always the same; a current will be induced
to compensate for this external field flux. Now measure the current dI flowing through the
superconducting solenoid. It turns out to be proportional to the solenoid length: $dI = \mathbf{H} \cdot d\mathbf{L}$.

A similar recipe for measuring **H** is described by Hehl and Obukhov [20, page
140].

The solenoid length in this measurement is apparently a twisted vector $d\mathbf{L}$; hence
the *magnetic field strength* **H** is a twisted one-form. The SI unit $[\mathbf{H}] = \text{Am}^{-1}$
confirms this interpretation. Figure 1.61 illustrates the discussed quantities. All
directions of $d\mathbf{L}$, for which $\mathbf{H} = 0$ lie in a plane determining the two-dimensional
attitude of the one-form **H**.

To the author's knowledge, the above-mentioned solenoid length is the first
example of a physical quantity with a true twisted vector nature. Many other
physical quantities hitherto called pseudovectors (like magnetic moment, magnetic
induction, magnetic field strength), have different natures, as shown previously.

We have summarized the quantities discussed up to now in this section in
Table 1.3. They are chosen as the ones that in traditional language are represented as
vectors or pseudovectors. To the list of their relevant features, typical examples of

Table 1.3 Typical examples of directed quantities in three-dimensional space

Primitive quantities	Ordinary vector	Twisted vector	Ordinary bivector	Twisted bivector
Attitude	Line	Line	Plane	Plane
Orientation	Straight arrow	Surrounding arrow	Curved arrow	Piercing arrow
Magnitude	Length l	Length **L**	Area s	Area **S**
Dual quantities	Ordinary one-form	Twisted one-form	Ordinary two-form	Twisted two-form
Attitude	Plane	Plane	Line	Line
Orientation	Piercing arrow	Curved arrow	Surrounding arrow	Straight arrow
Magnitude	Inverse length **E**	Inverse length **H**	Inverse area **B**	Inverse area **D**

geometric and electromagnetic field quantities are added. Each of the four quantities **E**, **D**, **H** and **B** is of a distinct directed nature. **E** and **B** are ordinary forms and they are called *intensity quantities* by Ingarden and Jamiołkowski [28] or *intensities of the electromagnetic field* by Frankel [17]; the two others, **D** and **H**, are twisted forms and they are called *magnitude quantities* in [28] or *quantities of the electromagnetic field* in [17].

Eight types of directed quantities remain, the attitudes of which—a point or 3-space—have only one possibility. Can we find a physical example for each of them?

As an example of a trivector quantity I propose an exterior product of three vectors: the velocity **v** of a particle, the acceleration **a** and the *jerk* $\frac{d\mathbf{a}}{dt}$. This is nonzero only for helical trajectories and has as an orientation the handedness of the helix.

A *volume* of something (a physical body or a region of space) is a twisted trivector quantity—it is natural to take the plus sign. This is the second example of a quantity with only one sign possible—the first one is the kinetic energy. Many authors using exterior forms treat the volume as an ordinary trivector, but this is so because they do not use twisted quantities. There is no reason to ascribe handedness to ordinary volume.

The integral of the electric induction over a closed surface ∂V surrounding a region V, according to Gauss' law, is equal to the electric charge contained in it:

$$\oint_{\partial V} \mathbf{D}[[d\mathbf{S}]] = Q. \tag{1.7}$$

Since the volume V of the region is a positive twisted trivector then, according to the rule illustrated in Figs. 1.30 and 1.31, the element $d\mathbf{S}$ of the boundary ∂V as a twisted bivector has the exterior orientation. Now we understand why traditionally the external orientation of the surface element is chosen for the integration over closed surfaces.

As for twisted three-forms, we can mention *spatial densities*, for instance energy density and electric charge density, since they are dual to the twisted trivector of the volume. The former occurs only with a plus sign, while the latter may assume two signs. The energy density of the electric field is given by the exterior product of the one-form **E** with the twisted two-form **D**:

$$w_e = \frac{1}{2}\mathbf{E} \wedge \mathbf{D} = \frac{1}{2}\mathbf{D} \wedge \mathbf{E},$$

for which Fig. 1.62 is an illustration. The energy density of the magnetic field is proportional to the exterior product of the twisted one-form **H** with the two-form **B**:

$$w_m = \frac{1}{2}\mathbf{B} \wedge \mathbf{H} = \frac{1}{2}\mathbf{H} \wedge \mathbf{B}$$

(Fig. 1.63). It is interesting that the situation depicted in Fig. 1.57 can never happen with **E** and **D** because the two electric field quantities cannot have opposite

Fig. 1.62 The exterior product $\mathbf{E} \wedge \mathbf{D}$

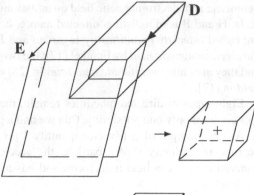

Fig. 1.63 The exterior product $\mathbf{H} \wedge \mathbf{B}$

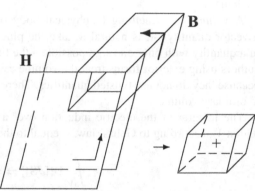

orientations, at least in static problems. The same is true for the two magnetic field quantities in connection with Fig. 1.58.

The *electric charge* is a scalar quantity. And what would *magnetic charge* be if it existed? In order to answer this question we recall the magnetic counterpart of equation (1.7) which we are used to write in the form $\oint_{\partial V} \mathbf{B} \cdot d\mathbf{s} = 0$. We now assume that the right hand side is not zero and ponder what directed quantity it is. It has been assumed until now that the integration element $d\mathbf{s}$ is the vector. We know that a vector is not appropriate for the surface element and we should only decide whether it is an ordinary or a twisted bivector. We should assume, as in (1.7), that it is a twisted bivector $d\mathbf{S}$ since we do not have any inner orientation available on the boundary ∂V of the volume V. In this manner we arrive at the formula

$$\oint_{\partial V} \mathbf{B}[[d\mathbf{S}]] = g \qquad (1.8)$$

expressing the magnetic charge g as the value of a two-form on a twisted bivector— we shall learn from the next mathematical sections that this is a twisted scalar. In this manner we conclude that the magnetic charge would be a twisted scalar. Up to now, it has not been discovered, but nevertheless theoreticians have discussed its possible properties, for instance Dirac, [12, 13], Schwinger [48], Rund [45],

Edelen [15] and Kaiser [32]. Only in the book of Edelen [15] on page 365 can one find arguments, based on behaviour under parity transformations, leading to the inference that magnetic charge should be a pseudoscalar quantity.

Let me present another argument leading to the conclusion that the magnetic charge should have handedness as its relevant feature. As we know from the history of physics, the magnetic charge was another name for the intensity of a single magnetic pole. What is a magnetic pole? It is merely one of the ends of a magnet or electromagnet which is a solenoid. The two ends of the solenoid differ by the sense of current flowing around the hole. An example of such a situation is shown in Fig. 1.64. The current on the boundary determines the round arrow. The straight arrow is just one showing direction outside the solenoid. The combination of these arrows (as in Fig. 1.23) gives the three-dimensional orientation, that is, the handedness of the magnetic pole. The south pole is shown in Fig. 1.64, hence we may claim that it is left-handed. One may also have another look at Fig. 1.37, repeated here as Fig. 1.65. The vertical line is the axis of solenoid, the directed circle in the middle is direction of flowing current, and the three-dimensional orientations at the ends characterize the poles—south pole at the top, north pole at the bottom.

Fig. 1.64 Three-dimensional orientation of one end of a solenoid

Fig. 1.65 Schematic images of the solenoid

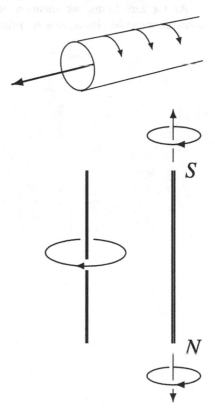

If one agrees that magnetic charge g is the pseudoscalar quantity, one may introduce a *magnetic dipole moment*, defined as $\mathbf{m} = g\mathbf{l}$, for a pair of magnetic charges $-g$ and g with $\mathbf{l}$ as the displacement vector from $-g$ to g. As the product of a pseudoscalar quantity with an ordinary vector, $\mathbf{m}$ is a second example of a true twisted vector.

We should mention another physical quantity that is to be considered as a twisted scalar. Particular media exist for which the electric field is accompanied by the magnetic field and vice versa (see [22]). Terms expressing these interrelations are of the form

$$\mathbf{D} = \alpha\mathbf{B}, \qquad \mathbf{H} = -\alpha\mathbf{E}. \tag{1.9}$$

The coefficient α is named the *axion parameter*, or *axion*. Since it relates the nontwisted quantities $\mathbf{B}, \mathbf{E}$ with the twisted ones $\mathbf{D}, \mathbf{H}$, it has to be a twisted scalar. A proposal for determining the presence of an axion in matter is contained in [41], whereas a report about finding it in chromium sesquioxide is in [24]. Publication [57] demonstrates the presence of axion electrodynamics in bismuth selenide, a topological insulator.

As for zero-forms, we mention only the electric scalar potential, which can sometimes can be viewed as a zero-form, but this distinction is not very important.

Chapter 2
Linear Spaces of Directed Quantities and Their Transformations

Directed quantities of the same type can be added and multiplied by scalars and, therefore, they constitute linear spaces—this is explained in Sect. 2.1. In each of the linear spaces, bases can be introduced and dimensions of the spaces determined. The coefficients of expanding a given quantity in the basis are called coordinates, not components, because the two notions can be ascribed to one vector and should be distinguished.[1]

Section 2.2 is devoted to explanation how transformation of ordinary vector basis influences transformations of bases in other linear spaces and transformations of coordinates for directed quantities.

By introducing the scalar product in the linear space of ordinary vectors it becomes Euclidean space. This product allows us to define magnitudes and orthogonality relation for vectors and other directed quantities, as presented in Sect. 2.3. Also, relations between two different scalar products in one linear space are considered. The scalar product also enables to replace the 16 types of directed quantities by only four: scalars, pseudoscalars, vectors and pseudovectors.

2.1 Linear Operations and Linear Spaces

For each type of directed quantity one can define the operations of addition and multiplication by scalars, called *linear operations*. In this manner the sets of quantities of the same type separately became linear spaces. We proceed now to define the linear operations.

Multiplication by a scalar can be defined once for all quantities. If **A** is an arbitrary directed quantity, then $\lambda\mathbf{A}$ has an attitude the same as that of **A**, a

---

[1] A coordinate of the vector is a scalar, whereas a component of the vector is a vector. One can speak about components of a vector, parallel or perpendicular to another vector.

© Springer Nature Switzerland AG 2021
B. Jancewicz, *Directed Quantities in Electrodynamics*,
https://doi.org/10.1007/978-3-030-90471-5_2

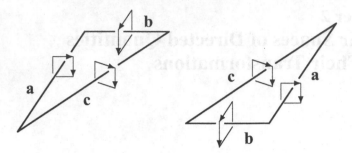

Fig. 2.1 Addition of two twisted vectors

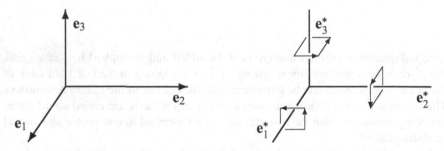

Fig. 2.2 An ordered basis and its twisted version

magnitude $|\lambda|$ times the magnitude of **A**, and orientation of **A** if $\lambda > 0$, or opposite if $\lambda < 0$.

We do not repeat the addition of ordinary (nontwisted) vectors. The addition of twisted vectors **a** and **b** is defined by juxtaposing their segments so that the directed ring from one segment after passing the junction has an orientation compatible with that of the other segment. Then joining the free ends by a straight line we obtain new segment **c**, which is the sum, and its orientation is obtained from the orientations of the summands by stretching their free ends up to the moment when **a** and **b** would be parallel to **c**. Figure 2.1 illustrates this procedure and shows that the addition does not depend on the way of juxtaposing, provided that the compatibility of orientations is maintained. One can check by construction that distributivity under multiplication by scalars is fulfilled, and similarly all other axioms of the linear space. In this manner we conclude that the set of twisted vectors is a linear space. What is its dimension? We shall now answer this question.

Let an *ordered basis* be given in the three-dimensional space of ordinary vectors, i.e., the basis $\{\mathbf{e}_1, \ \mathbf{e}_2, \ \mathbf{e}_3\}$ in which the order of the elements is essential. We can now create a basis $\{\mathbf{e}_1^*, \ \mathbf{e}_2^*, \ \mathbf{e}_3^*\}$ in the linear space of twisted vectors as follows. The twisted vector $\mathbf{e}_1^*$ has the attitude and magnitude the same as the vector $\mathbf{e}_1$, and the orientation as the rotation from $\mathbf{e}_2$ to $\mathbf{e}_3$. The definitions of twisted vectors $\mathbf{e}_2^*$ and $\mathbf{e}_3^*$ are similar to the cyclic permutation of the triple $\{\mathbf{e}_1, \ \mathbf{e}_2, \ \mathbf{e}_3\}$ (see Fig. 2.2). A reversal of the order of vectors $\mathbf{e}_2$ and $\mathbf{e}_3$ reverses the orientation of the twisted vector $\mathbf{e}_1^*$; hence we see why the order of the basis elements is essential for this definition.

Exercise 2.1 Show that a change in numbering the elements of a vector basis changes (besides numbering) the orientations of all elements of the twisted vector basis or none of them.

Since the introduced basis has three elements, we ascertain that the linear space of twisted vectors is three-dimensional.

Two bivectors $\mathbf{B}$ and $\mathbf{C}$ can be added after their factorization (1.3) with a common factor $\mathbf{a}$:

$$\mathbf{B} = \mathbf{a} \wedge \mathbf{b}, \qquad \mathbf{C} = \mathbf{a} \wedge \mathbf{c}$$

due to the distributivity of the exterior multiplication under addition:

$$\mathbf{B} + \mathbf{C} = \mathbf{a} \wedge \mathbf{b} + \mathbf{a} \wedge \mathbf{c} = \mathbf{a} \wedge (\mathbf{b} + \mathbf{c}). \tag{2.1}$$

In this formula the left-hand side is defined by the right-hand side (see Fig. 2.3). The possibility of a factorization of two bivectors $\mathbf{B}$, $\mathbf{C}$ with a common factor arises from the fact that in three-dimensional space two nonparallel planes intersect along a straight line. Vector $\mathbf{a}$ should be chosen on this line and the parallelograms $\mathbf{B}$ and $\mathbf{C}$ should be juxtaposed to each other such that the common edge has opposite orientation from the sides of the two figures.

Exercise 2.2 With the above defined addition of bivectors and the earlier defined multiplication by scalars, show that the axioms of a linear space are fulfilled.

With a given ordered vector basis $\{\mathbf{e}_1, \mathbf{e}_2, \mathbf{e}_3\}$, there exists a natural bivector basis $\{\mathbf{e}_1 \wedge \mathbf{e}_2, \mathbf{e}_2 \wedge \mathbf{e}_3, \mathbf{e}_3 \wedge \mathbf{e}_1\}$ shown in Fig. 2.4. A change of the order of basis vectors $\mathbf{e}_2$ and $\mathbf{e}_3$ changes not only the orientation of the bivector $\mathbf{e}_3 \wedge \mathbf{e}_3$, but also that of the two others. An arbitrary bivector $\mathbf{s}$ can be expressed by the formula

$$\mathbf{s} = s^{12}\mathbf{e}_1 \wedge \mathbf{e}_2 + s^{23}\mathbf{e}_2 \wedge \mathbf{e}_3 + s^{31}\mathbf{e}_3 \wedge \mathbf{e}_1 = \frac{1}{2}s^{ij}\mathbf{e}_i \wedge \mathbf{e}_j,$$

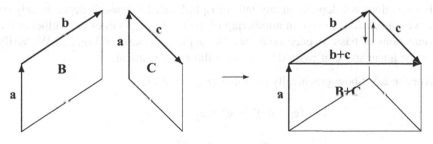

Fig. 2.3 Addition of bivectors, based on distributivity

Fig. 2.4 A nontwisted
bivector basis

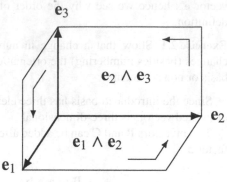

Fig. 2.5 A twisted bivector
basis

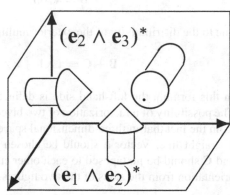

where the *Einstein convention* about summing over two repeating indices is applied
and the antisymmetry $s^{ij} = -s^{ji}$ is assumed. In this manner we see that the
linear space of bivectors is three-dimensional. If the underlying vector space is four-
dimensional, this linear space would have dimension six.

Addition of two twisted bivectors **B** and **C** is also based on formula (2.1), in
which now the common factor **a** is a twisted vector. The set of twisted bivectors also
constitutes a linear space. A natural basis $\{(\mathbf{e}_1 \wedge \mathbf{e}_2)^*, \ (\mathbf{e}_2 \wedge \mathbf{e}_3)^*, \ (\mathbf{e}_3 \wedge \mathbf{e}_1)^*\}$ can be
introduced as follows. Element $(\mathbf{e}_i \wedge \mathbf{e}_j)^*$ has attitude and magnitude of $\mathbf{e}_i \wedge \mathbf{e}_j$, but
orientation of $\mathbf{e}_k$ for $k \neq i, j$. Note that this rule of choosing orientations for basis
bivectors does not depend on any left- or right-handed screw—it depends only on
the vector basis. A change in numbering of the basis vectors does not influence the
orientations of basic twisted bivectors. We depict this basis in Fig. 2.5. We verify
that the linear space of twisted bivectors is three-dimensional.

Exercise 2.3 Show graphically the following identities:

$$(\mathbf{e}_i \wedge \mathbf{e}_j)^* = \mathbf{e}_i^* \wedge \mathbf{e}_j = \mathbf{e}_i \wedge \mathbf{e}_j^*,$$

$$\mathbf{e}_i \wedge \mathbf{e}_j = \mathbf{e}_i^* \wedge \mathbf{e}_j^*.$$

Exercise 2.4 Prove graphically the identities:

$$(\mathbf{e}_1^* \wedge \mathbf{e}_2) \wedge \mathbf{e}_3 = \mathbf{e}_1^* \wedge (\mathbf{e}_2 \wedge \mathbf{e}_3),$$

$$(\mathbf{e}_1 \wedge \mathbf{e}_2^*) \wedge \mathbf{e}_3 = \mathbf{e}_1 \wedge (\mathbf{e}_2^* \wedge \mathbf{e}_3),$$

$$(\mathbf{e}_1 \wedge \mathbf{e}_2) \wedge \mathbf{e}_3^* = \mathbf{e}_1 \wedge (\mathbf{e}_2 \wedge \mathbf{e}_3^*),$$

which denote the associativity of the exterior multiplication.

Exercise 2.5 Prove graphically the following identities:

$$\mathbf{e}_1^* \wedge \mathbf{e}_2 \wedge \mathbf{e}_3 = \mathbf{e}_1 \wedge \mathbf{e}_2^* \wedge \mathbf{e}_3 = \mathbf{e}_1 \wedge \mathbf{e}_2 \wedge \mathbf{e}_3^*.$$

What is the orientation (sign) of this twisted trivector?

This exercise shows that it is irrelevant of which factor the star is placed.

Two nontwisted linear forms **g** and **h** can be added according to a construction shown in Fig. 2.6. One should find two parallel straight lines being intersections of the planes: one line of planes with numbers 0 and 1, the other with numbers 1 and 0. Draw a plane passing through these two lines, and then the second plane parallel to the previously found and passing through the intersection line of the planes with numbers 0 and 0. The two new parallel planes, endowed by numbers 1 and 0, respectively, determine a new form **g** + **h**. Such defined addition satisfies all the needed axioms and in this manner the set of one-forms becomes a linear space.

Any basis vector $\mathbf{e}_i$ in the linear space of positions defines a straight line X^i passing through it, which is called the *coordinate axis*. A natural unit on this axis is determined by the length of $\mathbf{e}_i$. If we consider the position vector $\mathbf{r} = \sum_{i=1}^{3} x^i \mathbf{e}_i$, the coefficients x^i in this summation are *position coordinates*. For a linear form **g**,

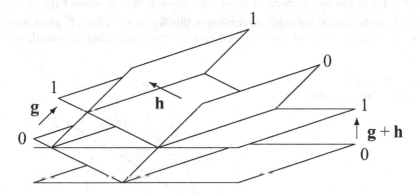

Fig. 2.6 Addition of nontwisted linear forms **g** and **h**

a distinguished plane, $\mathbf{g}[[\mathbf{r}]] = 1$, exists which is the set of points with coordinates satisfying the equation

$$\sum_{i=1}^{3} g_i x^i = 1. \tag{2.2}$$

This plane intersects the X^i axis at the point a_i fulfilling the condition (no summation over i): $g_i a^i = 1$, from which one obtains

$$a^i = \frac{1}{g_i}. \tag{2.3}$$

In this manner the distinguished plane (2.2) of the linear form $\mathbf{g}$ is characterized by three points

$$\{1/g_1, 0, 0\} \quad \{0, 1/g_2, 0\} \quad \{0, 0, 1/g_3\}, \tag{2.4}$$

(see Fig. 2.7).

For the sum $\mathbf{g} + \mathbf{h}$ of two linear forms, Eq. (2.2) takes the shape

$$\sum_{i=1}^{3} (g_i + h_i) x^i = 1, \tag{2.5}$$

from which the counterpart of (2.4) is

$$\{1/(g_1 + h_1), 0, 0\} \quad \{0, 1/(g_2 + h_2), 0\} \quad \{0, 0, 1/(g_3 + h_3)\}. \tag{2.6}$$

Given the basis $\{\mathbf{e}_1, \mathbf{e}_2, \mathbf{e}_3\}$ in the linear space of vectors we define a *dual basis* $\{\mathbf{f}^1, \mathbf{f}^2, \mathbf{f}^3\}$ in the linear space of one-forms through the condition $\mathbf{f}^i[[\mathbf{e}_j]] = \delta^i_j$, where δ^i_j is the *Kronecker delta*. According to this formula the form $\mathbf{f}^1$ gives zero on vectors $\mathbf{e}_2$ and $\mathbf{e}_3$, so these two vectors determine the plane being an attitude of the

Fig. 2.7 Plane $\mathbf{g}[[\mathbf{r}]] = 1$ intersecting three coordinate axes

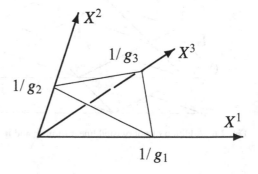

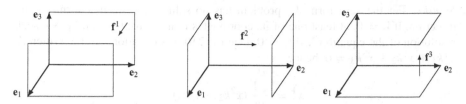

Fig. 2.8 Vectorial basis and its dual basis

form $\mathbf{f}^1$. The plane of the pair $\mathbf{e}_2$, $\mathbf{e}_3$ is thus the zero plane of $\mathbf{f}^1$. Since the condition $\mathbf{f}^1[[\mathbf{e}_1]] = 1$ should be fulfilled, the next plane of $\mathbf{f}^1$, parallel to the previous one passes through the tip of vector $\mathbf{e}_1$. In this manner the attitude of $\mathbf{f}^1$ is the plane of $\mathbf{e}_2$, $\mathbf{e}_3$, the orientation is that of $\mathbf{e}_1$, and the magnitude is equal to one in inverse units of the vector $\mathbf{e}_1$. The definitions of $\mathbf{f}^2$ and $\mathbf{f}^3$ are analogous. All the basic one-forms are shown in Fig. 2.8. Since the basis consists of three elements, the linear space of one-forms has the dimension three.

Let us see how the one-form $\mathbf{f}^1$ acts on an arbitrary vector $\mathbf{r} = x^1\mathbf{e}_1 + x^2\mathbf{e}_2 + x^3\mathbf{e}_3$. Due to the linearity $\mathbf{f}^1[[\mathbf{r}]] = x^1\mathbf{f}^1[[\mathbf{e}_1]] + x^2\mathbf{f}^1[[\mathbf{e}_2]] + x^3\mathbf{f}^1[[\mathbf{e}_3]] = x^1 \cdot 1 + x^2 \cdot 0 + x^3 \cdot 0 = x^1$. Similarly, for an arbitrary index j we obtain

$$\mathbf{f}^j[[\mathbf{r}]] = \mathbf{f}^j[[x^i\mathbf{e}_i]] = x^i\mathbf{f}^j[[\mathbf{e}_i]] = x^i\delta_i^j = x^j. \qquad (2.7)$$

We may interpret this equality by stating that the forms of the dual basis can be used to find the coordinates of a vector in the vector basis. This observation can be expressed more figuratively: one-forms of the dual basis are devices generating the coordinates of vectors.

Analogous calculations for a general linear form $\mathbf{k} = k_j\mathbf{f}^j$ yield

$$\mathbf{k}[[\mathbf{e}_i]] = k_j\mathbf{f}^j[[\mathbf{e}_i]] = k_j\delta_i^j = k_i. \qquad (2.8)$$

Hence we can say that the basis vectors can be used to find the coordinates of a form in the dual basis. And again in figurative language: basis vectors are devices to find the coordinates of forms.

The value of an arbitrary form $\mathbf{k}$ on an arbitrary vector $\mathbf{r}$ can be expressed by coordinates as follows:

$$\mathbf{k}[[\mathbf{r}]] = \mathbf{k}[[x^i\mathbf{e}_i]] = x^i\mathbf{k}[[\mathbf{e}_i]] = x^ik_j\mathbf{f}^j[[\mathbf{e}_i]] = x^ik_j\delta_i^j = k_ix^i. \qquad (2.9)$$

Exercise 2.6 Give a geometrical prescription for decomposition of a given linear form onto basis forms. Hint: use formula (2.8).

Problem 2.1 Let a linear form $\mathbf{k} = k_i\mathbf{f}^i$ be given. Find two vectors $\mathbf{r}_{(1)}$, $\mathbf{r}_{(2)}$ spanning the plane being the attitude of form $\mathbf{k}$.

Solution For the zero form the problem has no solution, since this form has no direction. If $\mathbf{k} \neq 0$, at least one of its coordinates in nonzero, let it be k_1. We seek a solution of the equation $\mathbf{k}[[\mathbf{r}]] = 0$. Due to (2.9) this equation can be written as $x^1 k_1 + x^2 k_2 + x^3 k_3 = 0$, hence

$$x^1 = -\frac{1}{k_1}(x^2 k_2 + x^3 k_3).$$

Two coordinates x^2 and x^3 are arbitrary in this equality, so we take them as parameters. Thus the two linearly independent vectors $\mathbf{r}_{(1)}$, $\mathbf{r}_{(2)}$ have for instance the following coordinates:

$$x^2_{(1)} = 1, \quad x^3_{(1)} = 0, \quad x^1_{(1)} = -\frac{k_2}{k_1};$$

$$x^2_{(2)} = 0, \quad x^3_{(2)} = 1, \quad x^1_{(2)} = -\frac{k_3}{k_1}.$$

Answer The two vectors are

$$\mathbf{r}_{(1)} = -\frac{k_2}{k_1}\mathbf{e}_1 + \mathbf{e}_2, \qquad \mathbf{r}_{(2)} = -\frac{k_3}{k_1}\mathbf{e}_1 + \mathbf{e}_3,$$

or after multiplying them by k_1:

$$\mathbf{r}'_{(1)} = -k_2\mathbf{e}_1 + k_1\mathbf{e}_2, \qquad \mathbf{r}'_{(2)} = -k_3\mathbf{e}_1 + k_1\mathbf{e}_3.$$

Two twisted one-forms can be added according to a prescription explained by Fig. 2.9. From two possible diagonals of the visible parallelogram we should choose the one from which the orientations of the two twisted one-forms are seen as the same.

A basis $\{\mathbf{f}^1_*, \mathbf{f}^2_*, \mathbf{f}^3_*\}$ in the linear space of twisted one-forms is introduced as follows. The attitude and magnitude of $\mathbf{f}^1_*$ is the same as for $\mathbf{f}^1$, and orientation is

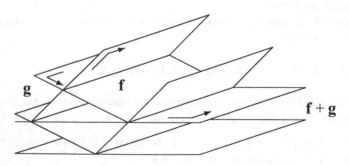

Fig. 2.9 Addition of twisted one-forms $\mathbf{f}$ and $\mathbf{g}$

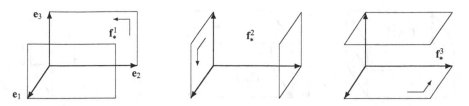

Fig. 2.10 Basis of twisted one-forms

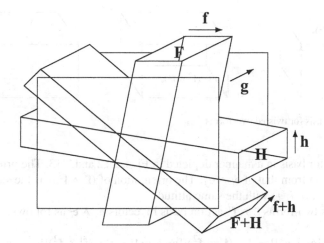

Fig. 2.11 The sum of two nontwisted two-forms **F** and **H**

taken from the bivector $\mathbf{e}_2 \wedge \mathbf{e}_3$ of the same attitude. Analogously, $\mathbf{f}_*^2$ and $\mathbf{f}_*^3$ are obtained by the cyclic permutation of indices. We depict this basis in Fig. 2.10.

A sum of two nontwisted two-forms **F** and **H** can be defined by means of their factorization (1.5) with a common factor **g**:

$$\mathbf{F} = \mathbf{g} \wedge \mathbf{f}, \qquad \mathbf{H} = \mathbf{g} \wedge \mathbf{h}$$

due to the distributivity of the exterior multiplication under addition:

$$\mathbf{F} + \mathbf{H} = \mathbf{g} \wedge \mathbf{f} + \mathbf{g} \wedge \mathbf{h} = \mathbf{g} \wedge (\mathbf{f} + \mathbf{h}).$$

This prescription is illustrated in Fig. 2.11.

The same prescription applies to the addition of two twisted two-forms with the exception that the common factor is a twisted one-form instead of a nontwisted one-form.

The sets of nontwisted and twisted two-forms are linear spaces and their dimensions are three. Natural bases can be introduced in them: $\{\mathbf{f}^1 \wedge \mathbf{f}^2,\ \mathbf{f}^2 \wedge \mathbf{f}^3,\ \mathbf{f}^3 \wedge \mathbf{f}^1\}$ for nontwisted ones, and $\{(\mathbf{f}^1 \wedge \mathbf{f}^2)_*,\ (\mathbf{f}^2 \wedge \mathbf{f}^3)_*,\ (\mathbf{f}^3 \wedge \mathbf{f}^1)_*\}$ for twisted ones, connected

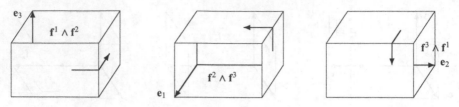

Fig. 2.12 A basis for nontwisted two-forms

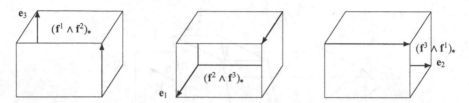

Fig. 2.13 A basis for twisted two-forms

with the vector basis in a manner depicted in Figs. 2.12 and 2.13. The orientation of $\mathbf{f}^i \wedge \mathbf{f}^j$ is taken from that of $\mathbf{e}_i \wedge \mathbf{e}_j$. The orientation of $(\mathbf{f}^i \wedge \mathbf{f}^j)_*$ is the same as that of vector $\mathbf{e}_k$ ($k \neq i, j$) with the same attitude.

The basis two-forms $\mathbf{f}^i \wedge \mathbf{f}^j$ act on basis bivectors $\mathbf{e}_k \wedge \mathbf{e}_l$ as follows:

$$(\mathbf{f}^1 \wedge \mathbf{f}^2)[[\mathbf{e}_1 \wedge \mathbf{e}_2]] = 1, \quad (\mathbf{f}^1 \wedge \mathbf{f}^2)[[\mathbf{e}_2 \wedge \mathbf{e}_3]] = 0, \quad (\mathbf{f}^1 \wedge \mathbf{f}^2)[[\mathbf{e}_3 \wedge \mathbf{e}_1]] = 0,$$

$$(\mathbf{f}^2 \wedge \mathbf{f}^3)[[\mathbf{e}_1 \wedge \mathbf{e}_2]] = 0, \quad (\mathbf{f}^2 \wedge \mathbf{f}^3)[[\mathbf{e}_2 \wedge \mathbf{e}_3]] = 1, \quad (\mathbf{f}^2 \wedge \mathbf{f}^3)[[\mathbf{e}_3 \wedge \mathbf{e}_1]] = 0,$$

$$(\mathbf{f}^3 \wedge \mathbf{f}^1)[[\mathbf{e}_1 \wedge \mathbf{e}_2]] = 0, \quad (\mathbf{f}^3 \wedge \mathbf{f}^1)[[\mathbf{e}_2 \wedge \mathbf{e}_3]] = 0, \quad (\mathbf{f}^3 \wedge \mathbf{f}^1)[[\mathbf{e}_3 \wedge \mathbf{e}_1]] = 1.$$

Can these formulas be expressed by Kronecker deltas? In order to answer this question, we should notice that the order of indices may be changed, for instance

$$(\mathbf{f}^1 \wedge \mathbf{f}^2)[[\mathbf{e}_2 \wedge \mathbf{e}_1]] = -1, \quad (\mathbf{f}^2 \wedge \mathbf{f}^1)[[\mathbf{e}_1 \wedge \mathbf{e}_2]] = -1, \quad (\mathbf{f}^2 \wedge \mathbf{f}^1)[[\mathbf{e}_2 \wedge \mathbf{e}_1]] = 1.$$

After some reflection we conclude that all possibilities are contained in the formula

$$(\mathbf{f}^i \wedge \mathbf{f}^j)[[\mathbf{e}_k \wedge \mathbf{e}_l]] = \delta_k^i \delta_l^j - \delta_l^i \delta_k^j. \tag{2.10}$$

This includes the fact that for $i = j$ or $k = l$ we obtain zero. This is the sought formula, expressing the action of basic two-forms on basic bivectors through Kronecker deltas.

Exercise 2.7 Prove that for arbitrary vectors $\mathbf{u}, \mathbf{v}$ the equality

$$(\mathbf{f}^i \wedge \mathbf{f}^j)[[\mathbf{u} \wedge \mathbf{v}]] = u^i v^j - u^j v^i$$

is fulfilled.

Exercise 2.8 Deduce formulas analogous to (2.7) and (2.8), namely

$$(\mathbf{f}^i \wedge \mathbf{f}^j)[[\mathbf{s}]] = s^{ij} \quad \text{and} \quad \mathbf{B}[[\mathbf{e}_i \wedge \mathbf{e}_j]] = B_{ij}.$$

We can comment on these results similarly to the formulas (2.7) and (2.8): the basis two-forms are devices to find the coordinates of bivectors, and basis bivectors are devices to find the coordinates of two-forms.

The value of an arbitrary two-form $\mathbf{B}$ on an arbitrary bivector $\mathbf{s}$ can be expressed by coordinates as follows

$$\mathbf{B}[[\mathbf{s}]] = \mathbf{B}\left[\left[\frac{1}{2}s^{ij}\mathbf{e}_i \wedge \mathbf{e}_j\right]\right] = \frac{1}{2}s^{ij}\mathbf{B}[[\mathbf{e}_i \wedge \mathbf{e}_j]]$$

$$= \frac{1}{2}B_{ij}s^{ij} = B_{12}s^{12} + B_{23}s^{23} + B_{31}s^{31}. \tag{2.11}$$

Exercise 2.9 Let $\mathbf{g}, \mathbf{h}$ be arbitrary one-forms, $\mathbf{u}, \mathbf{v}$—arbitrary vectors. Prove the formula:

$$(\mathbf{g} \wedge \mathbf{h})[[\mathbf{u} \wedge \mathbf{v}]] = \mathbf{g}[[\mathbf{u}]]\,\mathbf{h}[[\mathbf{v}]] - \mathbf{g}[[\mathbf{v}]]\,\mathbf{h}[[\mathbf{u}]]. \tag{2.12}$$

Problem 2.2 Let a nonzero two-form

$$\mathbf{B} = B_{12}\mathbf{f}^1 \wedge \mathbf{f}^2 + B_{23}\mathbf{f}^2 \wedge \mathbf{f}^3 + B_{31}\mathbf{f}^3 \wedge \mathbf{f}^1.$$

be given. For which bivectors $\mathbf{s}$ do we find the value $\mathbf{B}[[\mathbf{s}]]$ zero?

Solution Due to (2.11), the equation $\mathbf{B}[[\mathbf{s}]] = 0$ can be written as

$$B_{12}s^{12} + B_{23}s^{23} + B_{31}s^{31} = 0. \tag{2.13}$$

We have $\mathbf{B} \neq 0$, hence at least one of its coordinates is non zero: let it be B_{31}. Then we get

$$s^{31} = -\frac{B_{12}}{B_{31}}s^{12} - \frac{B_{23}}{B_{31}}s^{23}.$$

Two coordinates of the bivector, namely s^{12} and s^{23}, can be arbitrary. We change them into parameters α and β.

Answer The two-form **B** gives value zero on bivectors

$$\mathbf{s} = \alpha \mathbf{e}_1 \wedge \mathbf{e}_2 + \beta \mathbf{e}_2 \wedge \mathbf{e}_3 - \left(\frac{B_{12}}{B_{31}} \alpha + \frac{B_{23}}{B_{31}} \beta \right) \mathbf{e}_3 \wedge \mathbf{e}_1.$$

Since α and β are two arbitrary parameters, the set of such bivectors is a two-dimensional linear space.

Now a question arises: How can we characterize the directions of all such bivectors? A hint for an answer to this question is contained in the following.

Lemma 2.1 *Let a vector* **b** *have its coordinates taken from the two-form* **B**:

$$\mathbf{b} = B_{23}\mathbf{e}_1 + B_{31}\mathbf{e}_2 + B_{12}\mathbf{e}_3. \tag{2.14}$$

Then the bivector

$$\mathbf{s} = \mathbf{b} \wedge \mathbf{c}, \tag{2.15}$$

with an arbitrary vector **c**, *satisfies Eq. (2.13).*

Proof We calculate

$$\mathbf{s} = (B_{23}\mathbf{e}_1 + B_{31}\mathbf{e}_2 + B_{12}\mathbf{e}_3) \wedge (c^1 \mathbf{e}_1 + c^2 \mathbf{e}_2 + c^3 \mathbf{e}_3)$$

$$= (B_{23}c^2 - B_{31}c^1)\,\mathbf{e}_1 \wedge \mathbf{e}_2 + (B_{12}c^1 - B_{23}c^3)\,\mathbf{e}_3 \wedge \mathbf{e}_1 + (B_{31}c^3 - B_{12}c^2)\,\mathbf{e}_2 \wedge \mathbf{e}_3.$$

We find the coordinates of bivector **s**:

$$s^{12} = B_{23}c^2 - B_{31}c^1, \quad s^{23} = B_{31}c^3 - B_{12}c^2, \quad s^{31} = B_{12}c^1 - B_{23}c^3.$$

After inserting them into (2.13) we obtain zero.   ∎

The set of bivectors of the form (2.15) is a two-dimensional linear space, because it contains two arbitrary parameters (from the three components of the arbitrary vector **c**, one—parallel to **b**—is not present in **s**). Thus formula (2.15) describes all bivectors satisfying Eq. (2.13). All of them contain the common factor **b**, so their attitudes (recall: planes) have a common edge—this is the attitude of **b**. We know also from the action of two-forms on bivectors that for bivectors parallel to the two-form, the latter gives zero. Thus we have an answer to the question posed before the Lemma: vector (2.14) has the same attitude as that of the two-form **B**.

Exercise 2.10 Draw figures showing the identities:

$$(\mathbf{f}^i \wedge \mathbf{f}^j)_* = \mathbf{f}^i_* \wedge \mathbf{f}^j = \mathbf{f}^i \wedge \mathbf{f}^j_*,$$

$$\mathbf{f}^i \wedge \mathbf{f}^j = \mathbf{f}^i_* \wedge \mathbf{f}^j_*.$$

Exercise 2.11 Prove graphically the following associativity laws:

$$(\mathbf{f}^1_* \wedge \mathbf{f}^2) \wedge \mathbf{f}^3 = \mathbf{f}^1_* \wedge (\mathbf{f}^2 \wedge \mathbf{f}^3),$$

$$(\mathbf{f}^1 \wedge \mathbf{f}^2_*) \wedge \mathbf{f}^3 = \mathbf{f}^1 \wedge (\mathbf{f}^2_* \wedge \mathbf{f}^3),$$

$$(\mathbf{f}^1 \wedge \mathbf{f}^2) \wedge \mathbf{f}^3_* = \mathbf{f}^1 \wedge (\mathbf{f}^2 \wedge \mathbf{f}^3_*).$$

Exercise 2.12 Show graphically the identities:

$$\mathbf{f}^1_* \wedge \mathbf{f}^2 \wedge \mathbf{f}^3 = \mathbf{f}^1 \wedge \mathbf{f}^2_* \wedge \mathbf{f}^3 = \mathbf{f}^1 \wedge \mathbf{f}^2 \wedge \mathbf{f}^3_*.$$

What is the orientation of this twisted three-form?

Up to now we have discussed addition of directed quantities from Table 1.3 for which attitudes can vary. It turned out that, for each of the eight types of the quantities, their linear spaces are three-dimensional. And this means that after establishing the bases, three coordinates are sufficient. That's why physicist treated them as vectors, and some of them from time to time as pseudovectors.

Now we define addition of the other directed quantities. For each of the eight types only one attitude is possible; thus for adding two quantities of the same type it is sufficient to know whether they have the same orientation or not. When the orientations are the same we add their magnitudes, keeping the common orientation; when they are opposite, we subtract the smaller attitude from the greater one and choose the orientation of the greater. In this manner we obtain eight linear spaces, each of them one-dimensional.

The vector basis again determines bases in each linear space, this time one-element bases. So the basis trivector is $\mathbf{e}_{123} = \mathbf{e}_1 \wedge \mathbf{e}_2 \wedge \mathbf{e}_3$. Its geometric image is the parallelepiped shown in Fig. 2.14. The numbering of vector bases is relevant because it determines the three-dimensional orientation. It is well known as the name *orientation of the basis* in the vector space. Linear transformations of the basis $\{\mathbf{e}_1, \mathbf{e}_2, \mathbf{e}_3\}$ with a negative determinant change this orientation. Therefore,

Fig. 2.14 A basic nontwisted trivector

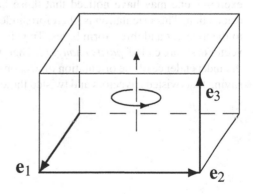

in mathematical textbooks one can find definitions of the orientation of space as an equivalence class of bases, and the equivalence relation is the existence of a linear transformation with positive determinant.

Exercise 2.13 Prove the identity:

$$\mathbf{e}_i \wedge \mathbf{e}_j \wedge \mathbf{e}_k = \epsilon_{ijk}\mathbf{e}_{123}, \tag{2.16}$$

where ϵ_{ijk} is the *totally antisymmetric (Levi-Civita) symbol*:

$$\epsilon^{ijk} = \epsilon_{ijk} = \begin{cases} 1 & \text{for even permutation of } ijk \text{ with respect to } 1, 2, 3, \\ -1 & \text{for odd permutation of } ijk, \\ 0 & \text{when two indices repeat.} \end{cases}$$

Exercise 2.14 Show that the exterior product of vector $\mathbf{a}$ with bivector $\mathbf{S}$ can be expressed by their coordinates as follows:

$$\mathbf{a} \wedge \mathbf{S} = \frac{1}{2}\,\epsilon_{ijk}a^i\,S^{jk}\,\mathbf{e}_{123}. \tag{2.17}$$

The vector basis also determines the basis element for twisted scalars and twisted zero-forms—it is number one with the handedness of the vector basis. The basis of Fig. 2.2 or 2.13, for instance, distinguishes right-handed unity as the basis element for twisted scalars. A reversal of any of the basis vectors changes the basis twisted scalar to the opposite one. In traditional language, this is expressed by saying that pseudoscalars change sign under reflections in the vector space.

The nontwisted one-form basis $\{\mathbf{f}^1, \mathbf{f}^2, \mathbf{f}^3\}$ (also determined, as we know, by the vector basis) defines the basis element for the nontwisted three-forms, namely $\mathbf{f}^{123} = \mathbf{f}^1 \wedge \mathbf{f}^2 \wedge \mathbf{f}^3$. After some thought we probably agree that its handedness is the same as that of the basis trivector.

Similarly $\mathbf{e}^*_{123} = \mathbf{e}^*_1 \wedge \mathbf{e}_2 \wedge \mathbf{e}_3$ is the basis element for twisted trivectors, and $\mathbf{f}^{123}_* = \mathbf{f}^1_* \wedge \mathbf{f}^2 \wedge \mathbf{f}^3$ for twisted three-forms. From Exercises 2.5 and 2.12, we know that it is not important on which factor the star is placed. When solving these exercises one may have noticed that those basis elements have the same positive orientation. This orientation is, therefore, independent of the vector basis—contrary to the trivector and three-form bases. Transformations changing the orientation of a vector basis are called *parity changing*. Therefore, in the spirit of Burke's article [8], the independence of the orientation of $\mathbf{e}^*_{123}$ on the vector basis could be expressed by saying that twisted trivectors and twisted three-forms are *parity-invariant* quantities.

2.2 Transformations of Bases

Let us assume that we have two vector bases $\{\mathbf{e}_1, \mathbf{e}_2, \mathbf{e}_3\}$ and $\{\mathbf{m}_1, \mathbf{m}_2, \mathbf{m}_3\}$, connected with each other through the matrix A with the elements $(A)_i{}^j$:

$$\mathbf{m}_i = (A)_i{}^j \mathbf{e}_j. \tag{2.18}$$

Again the Einstein summation convention is applied from 1 to 3 over repeating indices: one up, the other down. Formula (2.18) could be called a *transition from the first to the second vector basis*. The inverse matrix is present in the reverse transition:

$$\mathbf{e}_i = (A^{-1})_i{}^j \mathbf{m}_j. \tag{2.19}$$

To both vector bases there exist separate dual bases of one-forms, that is, $\{\mathbf{f}^j\}$ to $\{\mathbf{e}_i\}$,

$$\mathbf{f}^j[[\mathbf{e}_i]] = \delta_i^j,$$

and $\{\mathbf{h}^j\}$ to $\{\mathbf{m}_i\}$,

$$\mathbf{h}^j[[\mathbf{m}_i]] = \delta_i^j.$$

Of course, the dual bases also are connected by some matrix B:

$$\mathbf{h}^k = (B)^k{}_j \mathbf{f}^j \tag{2.20}$$

and

$$\mathbf{f}^k = (B^{-1})^k{}_j \mathbf{h}^j. \tag{2.21}$$

We want to find the matrix B. Let us take the value of $\mathbf{f}^k$ on $\mathbf{m}_i$, making use of (2.18):

$$\mathbf{f}^k[[\mathbf{m}_i]] = \mathbf{f}^k[[(A)_i{}^j \mathbf{e}_j]] = (A)_i{}^j \mathbf{f}^k[[\mathbf{e}_j]] = (A)_i{}^j \delta_j^k = (A)_i{}^k.$$

Calculate the same, using (2.21),

$$\mathbf{f}^k[[\mathbf{m}_i]] = (B^{-1})^k{}_j \mathbf{h}^j[[\mathbf{m}_i]] = (B^{-1})^k{}_j \delta_i^j = (B^{-1})^k{}_i.$$

Note the identity

$$(A)_i{}^k = (B^{-1})^k{}_i.$$

We cannot write the matrix identity $A = B^{-1}$, since the indices on both sides are not in the same place—the left one is interchanged with the right one. However, we are allowed to write

$$A = (B^{-1})^T,$$

where T denotes the transposed matrix. This equality is equivalent to the following

$$B = (A^{-1})^T. \tag{2.22}$$

A reciprocal matrix and transposed to A is called *contragredient to A*. The formula (2.22) should be interpreted as follows. If the vector basis transforms according to the matrix A, then the dual basis of forms transforms according to a matrix contragredient to A. Thus the formulas (2.20) and (2.21) should be rewritten in the form

$$\mathbf{h}^k = ((A^{-1})^T)^k{}_j \mathbf{f}^j = (A^{-1})^k{}_j \mathbf{f}^j; \qquad \mathbf{f}^k = (A^T)^k{}_j \mathbf{h}^j = (A)^k{}_j \mathbf{h}^j. \tag{2.23}$$

Each nontwisted vector $\mathbf{r}$ can be expanded into vectors of both considered bases:

$$\mathbf{r} = x^i \mathbf{e}_i = x'^j \mathbf{m}_j.$$

The linear relation (2.18) between the vector bases implies a linear relation between coordinates x^i and $x'j$ of the same vector in the two bases. This relation can be found due to formula (2.7) and its counterpart for the second basis:

$$x^j = \mathbf{f}^j[[\mathbf{r}]], \qquad x'^k = \mathbf{h}^k[[\mathbf{r}]]. \tag{2.24}$$

Now, by virtue of (2.20), we may write

$$\mathbf{h}^k[[\mathbf{r}]] = (B)^k{}_j \mathbf{f}^j[[\mathbf{r}]],$$

and due to (2.24),

$$x'^k = (B)^k{}_j x^j.$$

We use now (2.22):

$$x'^k = ((A^{-1})^T)^k{}_j x^j. \tag{2.25}$$

This result can be summarized in the sentence: if the vector basis transforms according to matrix A, the coordinates of a vector transform according to the matrix contragredient to A. In the traditional approach, a vector is defined as a collection of its coordinates appropriately behaving under transformations of the basis. Then the transformation law (2.25) is usually written in the form:

$$x'^k = \frac{\partial x'^k}{\partial x^j} x^j.$$

Such a collection of coordinates is called a *contravariant vector*.

An arbitrary one-form **E** can be expanded into one-forms dual to both bases considered:

$$\mathbf{E} = E_j \mathbf{f}^j = E'_i \mathbf{h}^i.$$

Let us find a relation between the coordinates E_j and E'_i of the same form in two bases. For this purpose, we use two counterparts of formula (2.8),

$$E_j = \mathbf{E}[[\mathbf{e}_j]], \qquad E'_i = \mathbf{E}[[\mathbf{m}_i]]$$

and Eq. (2.7):

$$\mathbf{E}[[\mathbf{m}_i]] = \mathbf{E}[[(A)_i{}^j \mathbf{e}_j]] = (A)_i{}^j \mathbf{E}[[\mathbf{e}_j]].$$

Thus, we obtain

$$E'_i = (A)_i{}^j E_j, \tag{2.26}$$

which can be summarized as follows: coordinates of a one-form transform according to the same matrix as the basis vectors. Therefore, the collection of coordinates of a one-form is called a *covariant vector* in the traditional approach. The relation (2.26) then assumes the shape

$$E'_i = \frac{\partial x^j}{\partial x'^i} E_j$$

Let us consider now how a change (2.7) of the nontwisted vector basis implies a change of twisted vector basis. Let the twisted vector basis $\{\mathbf{e}_j^*\}$ be related to the vector basis $\{\mathbf{e}_j\}$ in the manner described at the beginning of Sect. 2.1, and similarly basis $\{\mathbf{m}_i^*\}$ with $\{\mathbf{m}_i\}$. We are looking for a matrix C linking the twisted vector bases in the following linear relation:

$$\mathbf{m}_i^* = (C)_i{}^j \mathbf{e}_j^*. \tag{2.27}$$

Elements of the twisted vector basis have their attitudes parallel to the elements of the nontwisted vector basis with the same index: $\mathbf{e}_j^* \| \mathbf{e}_j$, $\mathbf{m}_i^* \| \mathbf{m}_i$, with equal magnitudes, hence matrix C should be proportional to A: $C = \lambda A$, where λ is a scalar coefficient possibly depending on A.

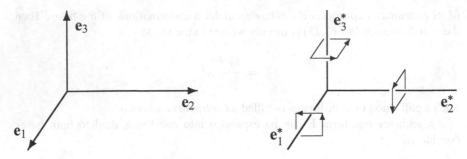

Fig. 2.15 An ordered nontwisted basis $\mathbf{e}_i$ and its twisted version

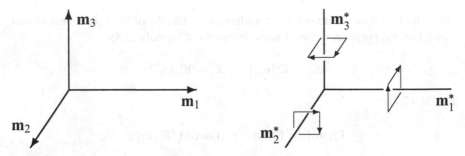

Fig. 2.16 An ordered nontwisted basis $\mathbf{m}_i$ and its twisted version

In order to find this coefficient, we consider a specific transformation of the vector basis consisting of renumbering only, for instance

$$\mathbf{m}_1 = \mathbf{e}_2, \quad \mathbf{m}_2 = \mathbf{e}_1, \quad \mathbf{m}_3 = \mathbf{e}_3. \tag{2.28}$$

In this case A has the form

$$A = \begin{pmatrix} 0 & 1 & 0 \\ 1 & 0 & 0 \\ 0 & 0 & 1 \end{pmatrix}. \tag{2.29}$$

This is the situation considered in Exercise 2.1. We repeat Fig. 2.2 as Fig. 2.15, illustrating the relation between the vector and twisted vector bases.

Figure 2.16 illustrates how the twisted vector basis $\{\mathbf{m}_i^*\}$ is obtained from the nontwisted vector basis $\{\mathbf{m}_i\}$.

In this case the change of twisted vector basis consists of the same renumbering as in (2.28) and, additionally, on the change of sign:

$$\mathbf{m}_1^* = -\mathbf{e}_2^*, \quad \mathbf{m}_2^* = -\mathbf{e}_1^*, \quad \mathbf{m}_3^* = -\mathbf{e}_3^*.$$

The expected coefficient is $\lambda = -1$. It so happens that the determinant of the matrix (2.29) is -1. Checking of the other possible cases of renumbering (there are six possibilities, the same number as the permutations of three elements) gives the result: the sign is minus only when the determinant of the matrix is negative.

In this manner, we arrive at the conclusion that the coefficient λ is equal to the sign of the determinant of A: $\lambda = \mathrm{sgn}(\det A)$, hence Eq. (2.27) can be written as

$$\mathbf{m}_i^* = \mathrm{sgn}(\det A)(A)_i{}^j \mathbf{e}_j^*. \tag{2.30}$$

Reasoning similar to that previously done for basis one-forms can be carried out for basic twisted one-forms and leads to the following counterpart of formula (2.23):

$$\mathbf{h}_*^k = \mathrm{sgn}[\det(A^{-1})^T]((A^{-1})^T)^k{}_j \mathbf{f}_*^j.$$

Since the determinant of the transposed matrix is the same as that of the original one and the sign of the determinant of the inverse matrix is the same as that of the original one, we can write

$$\mathbf{h}_*^k = \mathrm{sgn}(\det A)((A^{-1})^T)^k{}_j \mathbf{f}_*^j. \tag{2.31}$$

It remains to consider how the change of the vector basis influences changes of other multivector bases. Now, due to Eq. (2.18) we have

$$\mathbf{m}_i \wedge \mathbf{m}_j = [(A)_i{}^k \mathbf{e}_k] \wedge [(A)_j{}^l \mathbf{e}_l] = (A)_i{}^k (A)_j{}^l \mathbf{e}_k \wedge \mathbf{e}_l. \tag{2.32}$$

Similarly for the twisted bivector basis by virtue of (2.29) we have

$$\mathbf{m}_i \wedge \mathbf{m}_j^* = \mathrm{sgn}(\det A)(A)_i{}^k (A)_j{}^l \mathbf{e}_k \wedge \mathbf{e}_l^*. \tag{2.33}$$

For the single basis element for trivectors we obtain

$$\mathbf{m}_1 \wedge \mathbf{m}_2 \wedge \mathbf{m}_3 = (A)_1{}^i (A)_2{}^j (A)_3{}^k \mathbf{e}_i \wedge \mathbf{e}_j \wedge \mathbf{e}_k.$$

Use Exercise 2.13 to show that

$$\mathbf{m}_1 \wedge \mathbf{m}_2 \wedge \mathbf{m}_3 = (A)_1{}^i (A)_2{}^j (A)_3{}^k \epsilon_{ijk}\, \mathbf{e}_1 \wedge \mathbf{e}_2 \wedge \mathbf{e}_3.$$

The scalar coefficient on the right-hand side formally is the sum of 27 terms (3 values of i) $\times$ (3 values of j) $\times$ (3 values of k), but only the six terms with different indices i, j, k are nonzero. They can be written as $\epsilon_{ijk} = \sigma(i, j, k)$, where $\sigma(i, j, k)$ is called the sign of permutation of the numbers i, j, k with respect to 1, 2, 3 and is equal $+1$ for even permutations, and -1 for odd ones. Thus we have the coefficient

$$(A)_1^i (A)_2^j (A)_3^k \epsilon_{ijk} = \sum_{\sigma \in S_3} \sigma(i, j, k)(A)_1^i (A)_2^j (A)_3^k$$

with σ summed over all permutations of three numbers. We recognize in this expression the determinant of A. Thus we may write the following relation between the two basis trivectors:

$$\mathbf{m}_1 \wedge \mathbf{m}_2 \wedge \mathbf{m}_3 = \det A \, \mathbf{e}_1 \wedge \mathbf{e}_2 \wedge \mathbf{e}_3. \tag{2.34}$$

We similarly arrive at the relation between basic twisted trivectors,

$$\mathbf{m}_1 \wedge \mathbf{m}_2 \wedge \mathbf{m}_3^* = \mathrm{sgn}(\det A)\det A \, \mathbf{e}_1 \wedge \mathbf{e}_2 \wedge \mathbf{e}_3^*,$$

which can be written differently,

$$\mathbf{m}_1 \wedge \mathbf{m}_2 \wedge \mathbf{m}_3^* = |\det A| \, \mathbf{e}_1 \wedge \mathbf{e}_2 \wedge \mathbf{e}_3^*, \tag{2.35}$$

where the coefficient is the absolute value of the determinant. This result should not be a surprise for us, since at the end of Sect. 2.1 we noticed that the basic twisted trivectors always have positive orientation, so we cannot change this orientation by a change of vector basis.

We still need transformations of the three-form and twisted three-form bases. By virtue of (2.23), analogously to previous deductions, we obtain the following transformation of the basic nontwisted three-form:

$$\mathbf{h}^1 \wedge \mathbf{h}^2 \wedge \mathbf{h}^3 = (\det A)^{-1} \mathbf{f}^1 \wedge \mathbf{f}^2 \wedge \mathbf{f}^3 \tag{2.36}$$

and of the basic twisted three-form:

$$\mathbf{h}^1 \wedge \mathbf{h}^2 \wedge \mathbf{h}_*^3 = |\det A|^{-1} \mathbf{f}^1 \wedge \mathbf{f}^2 \wedge \mathbf{f}_*^3. \tag{2.37}$$

2.3 Euclidean Space and Reduction of the Number of Directed Quantities

When considering three-dimensional physical space, we have no problem deciding which segments are perpendicular and which nonparallel segments have the same length. Therefore, we can easily choose three equally long vectors and define the

Fig. 2.17 A bivector s
determined by the directed
angle α

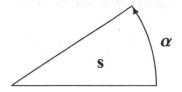

scalar product according to formula (1.1). The only problem is with picking out a unit vector.[2]

If to each pair of vectors **u, v** the *scalar product* **(u,v)** is given, the linear space of vectors is called *Euclidean space*. In such a case, there is also a *norm* $|\mathbf{v}|$ given as the square root of the scalar product of the vector with itself:

$$|\mathbf{v}|^2 = (\mathbf{v}, \mathbf{v}).$$

Now each vector has a *magnitude* or a *length* and we can compare the lengths of nonparallel vectors.

Euclidean space is also required to introduce an *angle* because in its definition the *unit circle* occurs, in which we put nonparallel vectors of the same length. A *directed angle* α can be considered as a bivector since the planar angle is always defined in a certain plane—it can be an attitude of the directed angle. Its orientation will also be known if we give a sense to the angle (for instance a rotation to one or the other side). The magnitude of the angle is proportional to the arc traced on the unit circle and is also proportional to area of the circle segment. Therefore, in spite of usually measuring the angle in radians as the arc length on the circumference, we may equally well measure it as the area s of the circle segment. The two bivectors defined in this manner are linked by the formula $\mathbf{s} = \frac{1}{2}\alpha$ (see Fig. 2.17).

Now we can introduce a further physical quantity, *angular velocity*, for planar motions $\omega = d\alpha/dt$, the derivative of α with respect to time—this is, obviously, also a bivector.

If, according to the prescription given in Sect. 1.1, we define the scalar product for an arbitrary basis $\{\mathbf{e}_1, \mathbf{e}_2, \mathbf{e}_3\}$ then, obviously, this basis is orthonormal for this scalar product. Now the question arises: what is the *sphere*, i.e., the set of vectors **r**

---

[2] By a unit vector I mean here a vector that has length one without any dimension. One can easily point out a vector with a length of one meter, one centimeter, one inch or whatever, but not with a dimensionless length one (one without any attribute). This is an abstract notion to which, nevertheless, physicists and mathematicians are accustomed.

Fig. 2.18 Metric spheres

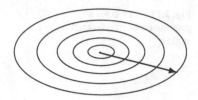

of equal magnitudes $|\mathbf{r}| = a$? As one may check, this set is given parametrically by the formula:

$$S = \{\mathbf{r}(\theta, \varphi) = a\mathbf{e}_1 \sin\theta \cos\varphi + a\mathbf{e}_2 \sin\theta \sin\varphi + a\mathbf{e}_3 \cos\theta :$$

$$0 \le \theta \le \pi, \ 0 \le \varphi \le 2\pi\}. \tag{2.38}$$

This is the surface of an ellipsoid.

A family of ellipsoids (2.38) taken for $a = 1, 2, 3$ and so on serves to find the length of a vector. Vectors with tips on the same ellipsoid have equal length. We display several ellipsoids in Fig. 2.18, which for the sake of simplicity is drawn in two dimensions, hence the ellipsoids are shown as ellipses. In this particular case the length of the chosen vector is 5 units. The existence of the prescription for finding lengths of vectors, and therefore distances between points, means that the scalar product determines a *metric* in the Euclidean space. The above-mentioned ellipsoids can be called *metric spheres*.

The partial derivative of $\mathbf{r}$ given in (2.38)

$$\frac{\partial \mathbf{r}}{\partial \theta} = a\mathbf{e}_1 \cos\theta \cos\varphi + a\mathbf{e}_2 \cos\theta \sin\varphi - a\mathbf{e}_3 \sin\theta,$$

is the vector tangent to the coordinate line of θ, whereas

$$\frac{\partial \mathbf{r}}{\partial \varphi} = -a\mathbf{e}_1 \sin\theta \sin\varphi + a\mathbf{e}_2 \sin\theta \cos\varphi.$$

is the vector tangent to the coordinate line of φ.

Exercise 2.15 Check that the two vectors are orthogonal to $\mathbf{r}$.

Hence the nontwisted bivector tangent to S can be written as

$$\mathbf{s} = \frac{\partial \mathbf{r}}{\partial \theta} \wedge \frac{\partial \mathbf{r}}{\partial \varphi} = a^2 \mathbf{e}_1 \wedge \mathbf{e}_2 \sin\theta \cos\theta \cos^2\varphi - a^2 \mathbf{e}_2 \wedge \mathbf{e}_1 \sin\theta \cos\theta \sin^2\varphi$$

$$+ a^2 \mathbf{e}_3 \wedge \mathbf{e}_1 \sin^2\theta \sin\varphi - a^2 \mathbf{e}_3 \wedge \mathbf{e}_2 \sin^2\theta \cos\varphi$$

$$= a^2 \sin\theta(\mathbf{e}_1 \wedge \mathbf{e}_2 \cos\theta + \mathbf{e}_2 \wedge \mathbf{e}_3 \sin\theta \cos\varphi + \mathbf{e}_3 \wedge \mathbf{e}_1 \sin\theta \sin\varphi).$$

Fig. 2.19 Nontwisted
bivector tangent to the metric
sphere

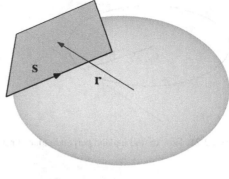

Fig. 2.20 Twisted bivector
tangent to the metric sphere

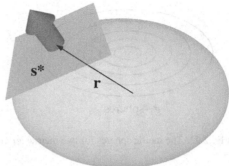

As the exterior product of two vectors perpendicular to $\mathbf{r}$, it is perpendicular to $\mathbf{r}$; see Fig. 2.19. It can easily be changed into the twisted bivector $\mathbf{s}^*$ just through the replacement of basic nontwisted bivectors by basic twisted bivectors:

$$\mathbf{s}^* = a^2 \sin\theta[(\mathbf{e}_1 \wedge \mathbf{e}_2)^* \cos\theta + (\mathbf{e}_2 \wedge \mathbf{e}_3)^* \sin\theta \cos\varphi + (\mathbf{e}_3 \wedge \mathbf{e}_1)^* \sin\theta \sin\varphi]; \tag{2.39}$$

see Fig. 2.20. We want to know whether the orientation of this bivector is external with respect to the ellipsoid surface. Let us take a point given by (2.38), e.g., for $\theta = 0$, we obtain then $\mathbf{r} = a\mathbf{e}_3$. We take now the twisted bivector (2.39) at this point—we obtain $\mathbf{s}^* = a^2(\mathbf{e}_1 \wedge \mathbf{e}_2)^*$. This has direction of one of the basic twisted bivectors. We know from Fig. 2.5 that its orientation is the same as that of $\mathbf{e}_3$, so the same as that of $\mathbf{r} = a\mathbf{e}_3$, hence we ascertain that the orientation of the twisted bivector (2.39) is external to the surface of the ellipsoid.

The ellipsoid (2.38) serves to find vectors perpendicular to a given vector $\mathbf{r}$. If we only draw the plane tangent to the ellipsoid at the point $\mathbf{r}$, then after shifting to the origin of coordinates we obtain the plane of all vectors perpendicular to $\mathbf{r}$. This is displayed in Fig. 2.21 in a two-dimensional simplification.

Let us consider two scalar products g_1 and g_2 in the same linear space such that g_1 is not proportional to g_2. Each of them has its own spheres, which for us are

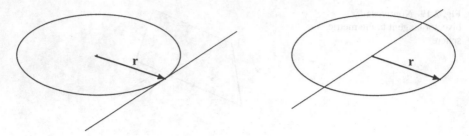

Fig. 2.21 How to find a plane perpendicular to **r** and passing through the origin

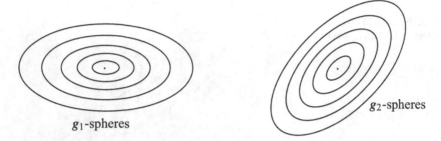

g_1-spheres

g_2-spheres

Fig. 2.22 The metric spheres for two metrics g_1 and g_2

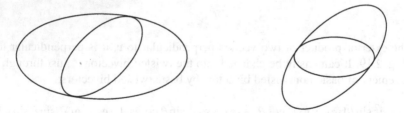

Fig. 2.23 Spheres of one metric tangent to spheres of the other

ellipsoids. We visualize them in Fig. 2.22 which for the sake of simplicity is made in two dimensions, hence the ellipsoids are shown as ellipses.

One can find pairs of ellipses, one being the g_1-sphere, the other g_2-sphere such that they are tangent to each other. We show two such pairs in Fig. 2.23. Now we draw two vectors c_1, c_2 from the origin to the point of tangency of the ellipses and two vectors b_1, b_2 tangent to both ellipses simultaneously (see Fig. 2.24).

From the properties of the scalar product the orthogonality $b_1 \perp c_1$ and $b_2 \perp c_2$ follows with respect to both scalar products g_i. Accordingly, $c_1 \perp c_2$ with respect to both g_i's. In the plane one can find only two attitudes for which vectors (like c_1 and c_2) have tips at points of tangency of the ellipses of the two families. Thus, in two dimensions, when g_1 and g_2 are not proportional, there exist only two attitudes perpendicular to each other with respect to g_1 and g_2 simultaneously. It is obvious

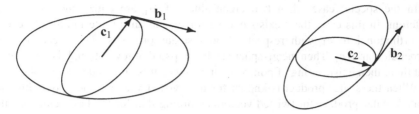

Fig. 2.24 Vectors from the origin to points of tangency and vectors tangent to both spheres

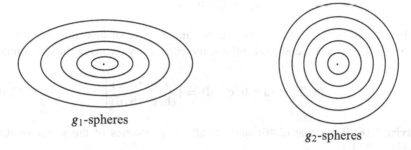

g_1-spheres

g_2-spheres

Fig. 2.25 The metric spheres for two metrics g_1 and g_2

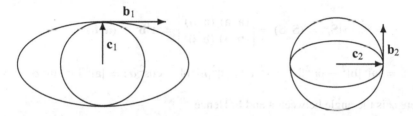

Fig. 2.26 Vectors from the origin to points of tangency and vectors tangent to both spheres

that in three dimensions, three such attitudes exist. We shall call them *principal attitudes* or *principal axes* of one scalar product with respect to the other.

If one of the scalar products—let it be g_2—is the ordinary scalar product of our space, its g_2-spheres are ordinary spheres (see Fig. 2.25) where the ellipsoids are represented by ellipses and the spheres by circles.

Now the counterpart of Fig. 2.24 looks as shown in Fig. 2.26, where we have an ellipsoid inscribed in a circle and a circle inscribed in an ellipsoid.

In three dimensions, three vectors c_1, c_2 and c_3 exist that are perpendicular to each other. We say that they determine the *principal attitudes* or *principal axes* of g_1 with respect to the ordinary scalar product. At the tip point of c_1, the two vectors c_2 and c_3 span a plane tangent to the ellipsoids and to the spheres simultaneously. The same is valid for the circular permutation of indices. This is the situation when the scalar product matrix $\mathcal{G}_1$, when put into diagonal form, has three different values on the diagonal.

In the special case when two eigenvalues of $\mathcal{G}_1$ are equal, the situation is different. In this case, there exists one vector c_1 and an infinite family of vectors c_α orthogonal to c_1 with respect to both scalar products. They scan a plane perpendicular to c_1. Then the g_1-sphere is the ellipsoid of revolution. It is not strange that there then exists a ring of points of tangency of this ellipsoid to a g_2-sphere.

When the scalar product is known for nontwisted vectors, one may define the related scalar product for twisted vectors assuming that for the basic elements the following identity occurs:

$$(e_i^*, e_j^*) = (e_i, e_j).$$

Also related is the scalar product in the linear space of bivectors, namely for $S = a \wedge b, T = c \wedge d$, we introduce the scalar product by means of the determinant

$$(S, T) = (a \wedge b, c \wedge d) = \begin{vmatrix} (a, c) & (a, d) \\ (b, c) & (b, d) \end{vmatrix}. \tag{2.40}$$

Exercise 2.16 Prove that (2.40) satisfies all the properties of the scalar product listed in Sect. 1.1.

The scalar product (2.40) serves to define the magnitude $|S|$ as follows

$$|S|^2 = (S, S) = \begin{vmatrix} (a, a) & (a, b) \\ (b, a) & (b, b) \end{vmatrix} = a^2 b^2 - (a, b)^2$$

$$= |a|^2 |b|^2 - |a|^2 |b|^2 \cos^2 \alpha = |a|^2 |b|^2 (1 - \cos^2 \alpha) = |a|^2 |b|^2 \sin^2 \alpha,$$

where α is the angle between a and b. Hence

$$|S| = |a| \, |b| \sin \alpha.$$

Accordingly, we obtain the known expression for the area of a parallelogram spanned by the vectors a, b. This is quite obviously a relation between the magnitude of a bivector and the magnitudes of vectors forming it in the exterior product.

We also assume that a related scalar product for one-forms can be introduced such that the dual basis one-forms are also orthonormal:

$$(f^i, f^j) = \delta^{ij}. \tag{2.41}$$

Then the scalar product is extended by linearity to arbitrary one-forms. Also, the *norm of a one-form* can be introduced as the square root of the scalar product of the form with itself. This norm determines the *magnitude of the one-form*.[3]

Let us now consider another scalar product $\mathbf{u}, \mathbf{v} \to g(\mathbf{u}, \mathbf{v})$ such that the basis $\{\mathbf{e}_i\}$ is no longer orthonormal. How can this product be expressed by coordinates of vectors? We have

$$g(\mathbf{u}, \mathbf{v}) = g(u^i \mathbf{e}_i, v^j \mathbf{e}_j) = u^i v^j g(\mathbf{e}_i, \mathbf{e}_j).$$

A matrix G is needed with elements

$$g_{ij} = g(\mathbf{e}_i, \mathbf{e}_j), \tag{2.42}$$

which helps us to write

$$g(\mathbf{u}, \mathbf{v}) = u^i g_{ij} v^j. \tag{2.43}$$

The matrix G bears the traditional name of a *metric tensor*.

Assume that $\{\mathbf{m}_i\}$ is an orthonormal basis for the scalar product g, i.e.,

$$g(\mathbf{m}_k, \mathbf{m}_l) = \delta_{kl}.$$

Let the bases $\{\mathbf{m}_k\}$ and $\{\mathbf{e}_j\}$ be linked through matrix A as in (2.19):

$$\mathbf{e}_j = (A^{-1})_j^{\;k} \mathbf{m}_k. \tag{2.44}$$

Then the elements of the metric tensor can be expressed as follows:

$$g_{ij} = g(\mathbf{e}_i, \mathbf{e}_j) = g((A^{-1})_i^{\;k} \mathbf{m}_k, (A^{-1})_j^{\;l} \mathbf{m}_l) = (A^{-1})_i^{\;k} (A^{-1})_j^{\;l} g(\mathbf{m}_k, \mathbf{m}_l)$$

$$= (A^{-1})_i^{\;k} \delta_{kl} (A^{-1})_j^{\;l} = (A^{-1})_i^{\;k} (A^{-1})_j^{\;k}.$$

This can be written in matrix form:

$$G = A^{-1}(A^{-1})^T. \tag{2.45}$$

We calculate the determinant,

$$\det G = \det(A^{-1}) \det[(A^{-1})^T] = [\det(A^{-1})]^2. \tag{2.46}$$

---

[3] We meet the same trouble as with unit vectors—it is difficult to pick out forms satisfying (2.41), since they should have dimensionless magnitudes equal to one.

It can be seen from this formula that det$G$ is positive. This can be proven also from the properties of the scalar product. Moreover,

$$|\det A| = (\det G)^{-1/2}. \tag{2.47}$$

Let $\{\mathbf{h}^j\}$ be basis dual to $\{\mathbf{m}_i\}$, i.e., $\mathbf{h}^j[[\mathbf{m}_i]] = \delta_i^j$. Since $\{\mathbf{m}_i\}$ is an orthonormal basis for g, then $\{\mathbf{h}^j\}$ is orthonormal for a scalar product $\tilde{g}$ appropriate for one-forms:

$$\tilde{g}(\mathbf{h}^i, \mathbf{h}^j) = \delta^{ij}. \tag{2.48}$$

What values does the scalar product $\tilde{g}$ assume on one-forms $\{\mathbf{f}^j\}$ of basis dual to $\{\mathbf{e}_i\}$? Now, due to (2.23), we have

$$\tilde{g}(\mathbf{f}^k, \mathbf{f}^l) = \tilde{g}((A)_i^{\ k}\mathbf{h}^i, (A)_j^{\ l}\mathbf{h}^j) = (A)_i^{\ k}(A)_j^{\ l}\tilde{g}(\mathbf{h}^i, \mathbf{h}^j)$$

$$= (A)_i^{\ k}\delta^{ij}(A)_j^{\ l} = (A)_i^{\ k}(A)_i^{\ l}.$$

If we introduce a *metric tensor* $\tilde{G}$ *for one-forms* with the elements $\tilde{g}^{kl} = \tilde{g}(\mathbf{f}^k, \mathbf{f}^l)$, the last result can be written in terms of matrices:

$$\tilde{G} = A^T A. \tag{2.49}$$

After comparing (2.45) and (2.49) we ascertain that the matrices G and $\tilde{G}$ are inverses of each other:

$$\tilde{G} = G^{-1}. \tag{2.50}$$

This means that the *metric tensor for one-forms is inverse (as the matrix) to the metric tensor for vectors*. We also write a counterpart of formula (2.43) for the one-forms:

$$\tilde{g}(\mathbf{E}, \mathbf{F}) = E_i \, \tilde{g}^{ij} F_j. \tag{2.51}$$

We need formulas linking two basic twisted trivectors and two basic twisted three-forms connected with the two vector bases. By virtue of (2.47), the formula (2.35) can be written as

$$\mathbf{m}_1 \wedge \mathbf{m}_2 \wedge \mathbf{m}_3^* = (\det G)^{-1/2} \, \mathbf{e}_1 \wedge \mathbf{e}_2 \wedge \mathbf{e}_3^*. \tag{2.52}$$

We similarly obtain from (2.37)

$$\mathbf{h}^1 \wedge \mathbf{h}^2 \wedge \mathbf{h}_*^3 = (\det G)^{1/2} \, \mathbf{f}^1 \wedge \mathbf{f}^2 \wedge \mathbf{f}_*^3. \tag{2.53}$$

Let us return to the scalar product of bivectors. We want to express it by the coordinates of the factors. We do this generally in a basis that is not orthonormal. We rewrite formula (2.40) for such a scalar product:

$$g(\mathbf{S}, \mathbf{T}) = g(\mathbf{a} \wedge \mathbf{b}, \mathbf{c} \wedge \mathbf{d}) = \begin{vmatrix} g(\mathbf{a}, \mathbf{c}) & g(\mathbf{a}, \mathbf{d}) \\ g(\mathbf{b}, \mathbf{c}) & g(\mathbf{b}, \mathbf{d}) \end{vmatrix}.$$

In particular, for the basis bivectors, this yields

$$g(\mathbf{e}_i \wedge \mathbf{e}_j, \mathbf{e}_k \wedge \mathbf{e}_l) = \begin{vmatrix} g(\mathbf{e}_i, \mathbf{e}_k) & g(\mathbf{e}_i, \mathbf{e}_l) \\ g(\mathbf{e}_j, \mathbf{e}_k) & g(\mathbf{e}_j, \mathbf{e}_l) \end{vmatrix} = g_{ik}g_{jl} - g_{il}g_{jk}.$$

Now for arbitrary bivectors $\mathbf{S}$ and $\mathbf{T}$

$$g(\mathbf{S}, \mathbf{T}) = g\left(\frac{1}{2}S^{ij}\mathbf{e}_i \wedge \mathbf{e}_j, \frac{1}{2}T^{kl}\mathbf{e}_k \wedge \mathbf{e}_l\right) = \frac{1}{4}S^{ij}T^{kl}(g_{ik}g_{jl} - g_{il}g_{jk})$$

$$= \frac{1}{4}(S^{ij}T^{kl}g_{ik}g_{jl} - S^{ij}T^{kl}g_{il}g_{jk}).$$

Rename the summation indices k and l in the second term:

$$g(\mathbf{S}, \mathbf{T}) = \frac{1}{4}(S^{ij}T^{kl}g_{ik}g_{jl} - S^{ij}T^{lk}g_{ik}g_{jl}).$$

Interchange the indices of T^{lk} in the second term, which changes its sign:

$$g(\mathbf{S}, \mathbf{T}) = \frac{1}{4}(S^{ij}T^{kl}g_{ik}g_{jl} + S^{ij}T^{kl}g_{ik}g_{jl}).$$

The two terms in the bracket are equal; hence at last we obtain the formula

$$g(\mathbf{S}, \mathbf{T}) = \frac{1}{2}S^{ij}T^{kl}g_{ik}g_{jl}. \tag{2.54}$$

Knowing the scalar product of two vectors, we may also introduce the product of a vector $\mathbf{c}$ with a bivector $\mathbf{S} = \mathbf{a} \wedge \mathbf{b}$, which is zero when $\mathbf{c}$ is perpendicular to $\mathbf{S}$. Now the expression

$$\mathbf{c} \cdot \mathbf{S} = \mathbf{c} \cdot (\mathbf{a} \wedge \mathbf{b}) = (\mathbf{c}, \mathbf{a})\mathbf{b} - (\mathbf{c}, \mathbf{b})\mathbf{a} = \mathbf{w} \tag{2.55}$$

is called the *inner product of a vector with a bivector*, whereas

$$\mathbf{S} \cdot \mathbf{c} = (\mathbf{a} \wedge \mathbf{b}) \cdot \mathbf{c} = \mathbf{a}(\mathbf{b}, \mathbf{c}) - \mathbf{b}(\mathbf{a}, \mathbf{c}) = -\mathbf{w} \tag{2.56}$$

is the *inner product of a bivector with a vector*. Since (2.56) is opposite to (2.55), we see that this product is anticommutative. Several facts are significant:

1. The two products are distributive with respect to the addition of two factors.
2. The operations give vectors lying in the plane of **S**.
3. The result **w** is perpendicular to **c**, since

$$(\mathbf{w}, \mathbf{c}) = (\mathbf{c}, \mathbf{a})(\mathbf{b}, \mathbf{c}) - (\mathbf{c}, \mathbf{b})(\mathbf{a}, \mathbf{c}) = 0.$$

4. The condition $\mathbf{c} \perp \mathbf{S}$ (understood as the pair of conditions $\mathbf{c} \perp \mathbf{a}$ and $\mathbf{c} \perp \mathbf{b}$) occurs if and only if $\mathbf{c} \cdot \mathbf{S} = 0$. In this fact a similarity to the scalar product of vectors is visible—its vanishing is equivalent to the orthogonality of factors. This observation allows to state that only the component $\mathbf{c}_{||}$ of $\mathbf{c}$ parallel to **S** makes a nonzero contribution to (2.55):

$$\mathbf{c} \cdot \mathbf{S} = \mathbf{c}_{||} \cdot \mathbf{S}. \tag{2.57}$$

5. Making use of some arbitrariness of factorization of **S**, choose **b** perpendicular to **c**; then (2.55) yields $\mathbf{w} = (\mathbf{c}, \mathbf{a})\mathbf{b}$. And, due to (2.57), we may write

$$\mathbf{w} = (\mathbf{c}_{||}, \mathbf{a})\mathbf{b}. \tag{2.58}$$

Now, choose the vector **a** parallel to $\mathbf{c}_{||}$, with the same orientation; then $\mathbf{a} \perp \mathbf{b}$ and $(\mathbf{c}_{||}, \mathbf{a}) \geq 0$; hence Eq. (2.58) says that **w** has direction of **b**. This situation is illustrated in Fig. 2.27.

6. Taking **a** and **b** as in item 5, we have $\mathbf{a} \perp \mathbf{b}$, i.e., $|\mathbf{a} \wedge \mathbf{b}| = |\mathbf{a}|\,|\mathbf{b}|$, and the magnitude of **w** is

$$|\mathbf{w}| = (\mathbf{c}_{||}, \mathbf{a})|\mathbf{b}| = |\mathbf{c}_{||}|\,|\mathbf{a}|\,|\mathbf{b}| = |\mathbf{c}_{||}|\,|\mathbf{a} \wedge \mathbf{b}| = |\mathbf{c}|\,|\mathbf{a} \wedge \mathbf{b}|\cos\theta,$$

Fig. 2.27 The inner product
c · **S** of vector **c** with
bivector **S**

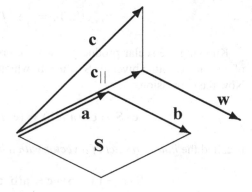

where θ is an angle between $\mathbf{c}$ and $\mathbf{a} \wedge \mathbf{b}$. This formula can be also written as

$$|\mathbf{c} \cdot \mathbf{S}| = |\mathbf{c}|\,|\mathbf{S}|\cos\theta. \tag{2.59}$$

This resembles the analogous formula for the scalar product of vectors.

In particular, for basis elements, we obtain from (2.55) and (2.56) (we use now shortened notation $\mathbf{e}_i \wedge \mathbf{e}_j = \mathbf{e}_{ij}$):

$$\mathbf{e}_i \cdot \mathbf{e}_{jk} = (\mathbf{e}_i, \mathbf{e}_j)\mathbf{e}_k - (\mathbf{e}_i, \mathbf{e}_k)\mathbf{e}_j,$$

$$\mathbf{e}_{jk} \cdot \mathbf{e}_i = \mathbf{e}_j(\mathbf{e}_k, \mathbf{e}_i) - \mathbf{e}_k(\mathbf{e}_j, \mathbf{e}_i).$$

If the basis $\{\mathbf{e}_i\}$ is not orthonormal for the considered scalar product, we are obliged to use the elements $g_{ij} = g(\mathbf{e}_j, \mathbf{e}_j)$ of the metric tensor:

$$\mathbf{e}_i \cdot \mathbf{e}_{jk} = g_{ij}\mathbf{e}_k - g_{ik}\mathbf{e}_j,$$

$$\mathbf{e}_{jk} \cdot \mathbf{e}_i = g_{ki}\mathbf{e}_j - g_{ji}\mathbf{e}_k.$$

Therefore, for arbitrary factors $\mathbf{c}$ and $\mathbf{S}$ we get

$$\mathbf{c} \cdot \mathbf{S} = c^i \mathbf{e}_i \cdot \left(\frac{1}{2}S^{jk}\mathbf{e}_{jk}\right) = \frac{1}{2}c^i S^{jk}(g_{ij}\mathbf{e}_k - g_{ik}\mathbf{e}_j) = \frac{1}{2}c^i(g_{ij}S^{jk}\mathbf{e}_k - g_{ik}S^{jk}\mathbf{e}_j).$$

We interchange the summation indices j, k in the second term in the bracket:

$$\mathbf{c} \cdot \mathbf{S} = \frac{1}{2}c^i(g_{ij}S^{jk}\mathbf{e}_k - g_{ij}S^{kj}\mathbf{e}_k) = \frac{1}{2}c^i g_{ij}(S^{jk} - S^{kj})\mathbf{e}_k.$$

The coordinates S^{jk} are antisymmetric under interchange of the indices, hence

$$\mathbf{c} \cdot \mathbf{S} = c^i g_{ij}S^{jk}\mathbf{e}_k.$$

This means that the coordinates of the inner product have the form

$$(\mathbf{c} \cdot \mathbf{S})^k = c^i g_{ij}S^{jk}. \tag{2.60}$$

One similarly arrives at the result

$$(\mathbf{S} \cdot \mathbf{c})^k = S^{kj}g_{ji}c^i. \tag{2.61}$$

We may similarly define the inner product of a one-form $\mathbf{E}$ with a two-form $\mathbf{D} = \mathbf{g} \wedge \mathbf{h}$ in two possible orders:

$$\mathbf{D} \cdot \mathbf{E} = (\mathbf{g} \wedge \mathbf{h}) \cdot \mathbf{E} = \mathbf{g}\,(\mathbf{h}, \mathbf{E}) - \mathbf{h}\,(\mathbf{g}, \mathbf{E}), \tag{2.62}$$

$$\mathbf{E} \cdot \mathbf{D} = \mathbf{E} \cdot (\mathbf{g} \wedge \mathbf{h}) = (\mathbf{E}, \mathbf{g})\,\mathbf{h} - (\mathbf{E}, \mathbf{h})\,\mathbf{g}. \tag{2.63}$$

This product is also anticommutative. It can be written in coordinates:

$$(\mathbf{D} \cdot \mathbf{E})_k = D_{kj}\,\tilde{g}^{ji}\,E_i, \tag{2.64}$$

$$(\mathbf{E} \cdot \mathbf{D})_k = E_i\,\tilde{g}^{ij}\,D_{jk}. \tag{2.65}$$

When we have at our disposal the scalar product in the linear space of vectors, we can easily define a mapping of the linear forms into vectors. Namely, for a linear form $\mathbf{E}$, there exists one and only one vector $\mathbf{E}$ such that

$$\mathbf{E}[[\mathbf{r}]] = g(\mathbf{E}, \mathbf{r}) \quad \text{for any vector } \mathbf{r}, \tag{2.66}$$

where $g(\cdot, \cdot)$ denotes the scalar product. Of course, $\mathbf{E}$ is perpendicular to the plane forming the attitude of $\mathbf{E}$; see Fig. 2.28. (We recall that the word "perpendicular" only makes sense in Euclidean spaces. To find a straight line perpendicular to a given plane we use ellipsoids shown in Figs. 2.19 and 2.20.) The vector $\mathbf{E}$ inherits its orientation from $\mathbf{E}$, but its attitude is orthogonal to that of $\mathbf{E}$. If we want to express (2.66) in coordinates corresponding to a non-orthonormal basis $\{\mathbf{e}_i\}$, we write

$$E_i x^i = E^j g_{ji} x^i.$$

Thus the vector coordinates of $\mathbf{E}$ are connected with its form coordinates by the formula $E_i = g_{ji} E^j$ or, using the symmetry of $\mathcal{G}$,

$$E_i = g_{ij} E^j. \tag{2.67}$$

Fig. 2.28 Relation of a nontwisted vector $\mathbf{E}$ to a nontwisted one-form $\mathbf{E}$

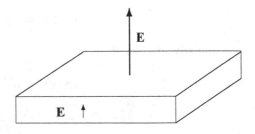

Fig. 2.29 Relation of a
twisted vector **H** to a twisted
one-form **H**

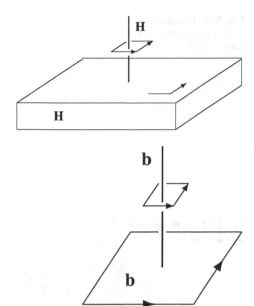

Fig. 2.30 Relation of a
twisted vector **b** to a
nontwisted bivector **b**

This relation is expressed by saying that "the metric tensor G serves to lower the
indices". After multiplying this equality by the inverse matrix $\tilde{G}$ we obtain the
inverse relation:

$$E^k = \tilde{g}^{ki} E_i,\tag{2.68}$$

which is characterized by the words: "the metric tensor $\tilde{G}$ serves to raise the
indices". The relation between the nontwisted one-form **E** and the nontwisted vector
E is illustrated in Fig. 2.28.

Formula (2.66) gives also a definition of the (nontwisted) wave vector **k** if the
phase function φ is known for a plane wave: $\varphi(\mathbf{r}) = \mathbf{k}[[\mathbf{r}]] = g(\mathbf{k}, \mathbf{r})$.[4]

One may also define a mapping of twisted one-forms into twisted vectors. For
a given twisted one-form **H** there exists one and only one twisted vector **H** such
that $\mathbf{H}[[\mathbf{v}]] = g(\mathbf{H}, \mathbf{v})$ for any twisted vector **v**. We illustrate this prescription
in Fig. 2.29. Twisted vector **H** inherits its orientation from **H** but its attitude is
perpendicular to that of **H**. The formulas connecting corresponding coordinates in a
non-orthonormal basis are the same: you can use (2.67) and (2.68) with H replacing
E there.

We map nontwisted bivectors **b** into perpendicular twisted vectors **b** (see
Fig. 2.30) whereas twisted bivectors **c** are mapped into perpendicular vectors **c** (see
Fig. 2.31), keeping their magnitudes and orientations.

---

[4] This prescription contains an element of convention, because there is no natural way to point out
a vector that has length measured in inverse meters or inverse centimeters.

Fig. 2.31 Relation of a
nontwisted vector **c** to a
nontwisted bivector **c**

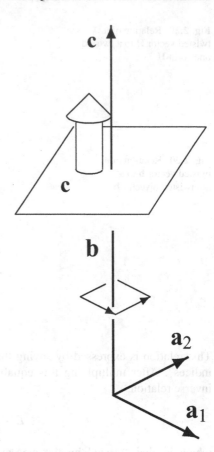

Fig. 2.32 Vector product
$b = a_1 \times a_2$

If we represent a bivector **b** as the exterior product of vectors $b = a_1 \wedge a_2$, then in a twisted vector **b**, corresponding to it, we recognize the vector product $b = a_1 \times a_2$, which is illustrated in Fig. 2.32. Very often, the given relation between vectors of the right-handed basis is improperly written as $e_1 \times e_2 = e_3$. This is inappropriate because it neglects the fact that the vector product of two vectors is a twisted vector or pseudovector. This relation should be written as $e_1 \times e_2 = e_3^*$.

On the other hand, if we represent a twisted bivector **c** as the exterior product of a vector **a** and a twisted vector **d**, i.e., $c = a \wedge d$, then the corresponding vector **c** would be the following vector product $c = a \times d$, which is shown in Fig. 2.33. Incidentally, the vector product does not always yield a twisted vector: sometimes it is a nontwisted vector if one of the factors is a twisted vector.

Since we can define magnitudes for two-forms, we may formulate a prescription for the change of a given two-form **B** onto a twisted vector **B**: take the direction of **B**, ascribe it to a twisted vector **B** and use the magnitude of |**B**| to define the length of the twisted vector **B**. This prescription is illustrated in Fig. 2.34. An analogous prescription for transforming twisted bivectors **D** into vectors **D** is depicted in Fig. 2.35.

Fig. 2.33 Vector product
$c = a \times d$

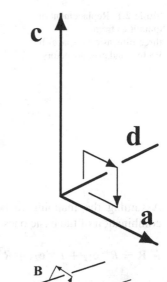

Fig. 2.34 Relation of a
twisted vector **B** to a
nontwisted two-form **B**

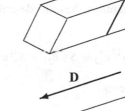

Fig. 2.35 Relation of a
nontwisted vector **D** to a
nontwisted two-form **D**

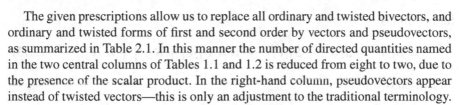

The given prescriptions allow us to replace all ordinary and twisted bivectors, and ordinary and twisted forms of first and second order by vectors and pseudovectors, as summarized in Table 2.1. In this manner the number of directed quantities named in the two central columns of Tables 1.1 and 1.2 is reduced from eight to two, due to the presence of the scalar product. In the right-hand column, pseudovectors appear instead of twisted vectors—this is only an adjustment to the traditional terminology.

The prescriptions can be expressed in formulas. For instance, in the case of an orthonormal basis, replace bivectors by pseudovectors as applied to three basic bivectors; compare Figs. 2.2 and 2.4. This is accomplished by the mapping:

$$\mathbf{e}_{12} \;\to\; \mathbf{e}_3^*, \qquad \mathbf{e}_{23} \;\to\; \mathbf{e}_1^*, \qquad \mathbf{e}_{31} \;\to\; \mathbf{e}_2^*.$$

Table 2.1 Replacement of
quantities from
three-dimensional spaces by
vectors and pseudovectors

Previous quantity	New quantity
Vector	Vector
Bivector	Pseudovector
Twisted vector	Pseudovector
Twisted bivector	Vector
One-form	Vector
Two-form	Pseudovector
Twisted one-form	Pseudovector
Twisted two-form	Vector

Assuming the mapping to be linear, we may write an arbitrary bivector as a combination of the basic ones:

$$\mathbf{R} = R^{12}\mathbf{e}_{12} + R^{23}\mathbf{e}_{23} + R^{31}\mathbf{e}_{31} \quad \rightarrow \quad \mathbf{R} = R^{12}\mathbf{e}_3^* + R^{23}\mathbf{e}_1^* + R^{31}\mathbf{e}_2^*, \qquad (2.69)$$

After the decomposition of other directed quantities into basic ones (also the shortened notation $\mathbf{f}^i \wedge \mathbf{f}^j = \mathbf{f}^{ij}$ is used, the star denoting the corresponding twisted quantity), the other described mappings are expressed as follows: for twisted bivectors

$$\mathbf{S} = S^{12}\mathbf{e}_{12}^* + S^{23}\mathbf{e}_{23}^* + S^{31}\mathbf{e}_{31}^* \quad \rightarrow \quad \mathbf{S} = S^{12}\mathbf{e}_3 + S^{23}\mathbf{e}_1 + S^{31}\mathbf{e}_2, \qquad (2.70)$$

for one-forms

$$\mathbf{E} = E_1\mathbf{f}^1 + E_2\mathbf{f}^2 + E_3\mathbf{f}^3 \quad \rightarrow \quad \mathbf{E} = E_1\mathbf{e}_1 + E_2\mathbf{e}_2 + E_3\mathbf{e}_3, \qquad (2.71)$$

for twisted one-forms

$$\mathbf{H} = H_1\mathbf{f}_*^1 + H_2\mathbf{f}_*^2 + H_3\mathbf{f}_*^3 \quad \rightarrow \quad \mathbf{H} = H_1\mathbf{e}_1^* + H_2\mathbf{e}_2^* + H_3\mathbf{e}_3^*, \qquad (2.72)$$

for two-forms

$$\mathbf{B} = B_{12}\mathbf{f}^{12} + B_{23}\mathbf{f}^{23} + B_{31}\mathbf{f}^{31} \quad \rightarrow \quad \mathbf{B} = B_{12}\mathbf{e}_3^* + B_{23}\mathbf{e}_1^* + B_{31}\mathbf{e}_2^*, \qquad (2.73)$$

for twisted two-forms

$$\mathbf{D} = D_{12}\mathbf{f}_*^{12} + D_{23}\mathbf{f}_*^{23} + D_{31}\mathbf{f}_*^{31} \rightarrow \mathbf{D} = D_{12}\mathbf{e}_3 + D_{23}\mathbf{e}_1 + D_{31}\mathbf{e}_2. \qquad (2.74)$$

As we see, the coordinates are retained; only the basic quantities are changed. Notice the similarity of (2.73) and (2.14). Remembering the observation after Lemma 2.1, we claim that the twisted vector $\mathbf{B}$ has the same direction as the two-form $\mathbf{B}$. A similar relation holds between the vector $\mathbf{D}$ and the twisted two-form $\mathbf{D}$. We return to some of these mappings in Chap. 3, where more mathematical tools are available.

Table 2.2 Replacement of quantities from one-dimensional spaces by scalars and pseudoscalars

Previous quantity	New quantity
Scalar	Scalar
Trivector	Pseudoscalar
Twisted scalar	Pseudoscalar
Twisted trivector	Scalar
Zero-form	Scalar
Three-form	Pseudoscalar
Twisted zero-form	Pseudoscalar
Twisted three-form	Scalar

A natural unit volume exists in Euclidean space, namely a cube with three edges made of the three basic orthonormal vectors (see Fig. 2.13). In this manner the magnitude of the trivector and twisted trivector is just the volume in units of the basic cube. The same cube serves to ascribe magnitudes to three-forms and twisted three-forms. Now we may formulate a prescription of how to change quantities from one-dimensional linear spaces into scalars or twisted scalars (pseudoscalars): for a trivector, take a twisted scalar of the same magnitude and handedness; for a twisted trivector take a scalar of the same magnitude and sign; for a three-form take a twisted scalar of the same magnitude and handedness; and for a twisted three-form take a scalar of the same magnitude and sign. Zero-forms are easily changed into scalars, and twisted zero-forms into twisted scalars, as has already been discussed.

After these replacements only two quantities remain: scalars and pseudoscalars. In traditional approaches, pseudoscalars are not endowed with handedness; a sign is ascribed to them in a similar way to scalars, only their different behaviour under reflections of basic vectors is underlined—they should change sign in this transformation. The last prescriptions are summarized in Table 2.2. In this manner the number of directed quantities named in the first and fourth columns of Tables 1.1 and 1.2 is reduced from eight to two, due to the presence of the scalar product. Generally, the 16 directed quantities present in the three-dimensional space are reduced to four: scalars, pseudoscalars, vectors, and pseudovectors.

For the sake of completeness I should also mention another reduction from 16 to 4, namely to four distinct grade multivectors. The difference is that instead of pseudovectors one uses bivectors and in place of pseudoscalars, trivectors. This is done for instance in Refs. [16–18] and [25]. The adherents of Clifford algebras are accustomed to call trivectors pseudoscalars, just because of this identification.

Chapter 3
Algebra and Analysis of Directed Quantities

Two opposite twisted scalars—right-handed and left-handed—are introduced and their exterior products with arbitrary multivectors are considered in Sect. 3.1. One of these twisted scalars serves to change nontwisted multivectors into twisted ones and vice versa. It also enables us to define the scalar product of nontwisted vectors with twisted vectors and of two twisted bivectors.

Having at our disposal the exterior product of two arbitrary nontwisted and twisted multivectors, due to distributivity, we extend it to the set of linear combinations of elements of various grades. In this manner, we obtain an extended Grassmann algebra of multivectors in Sect. 3.1.1. The word "extended" is applied, because the Grassmann algebras known in the literature are used for nontwisted quantities only. Analogously, an extended Grassmann algebra of exterior forms is introduced in Sect. 3.1.2.

Another kind of product can be introduced, in which one of the factors is a multivector and the second is an exterior form. This product is called a contraction. There are four possibilities for it, depending on the grades of the factors. They are introduced and considered in Sect. 3.2.

A particular linear operator from one vector space to another vector space has a simple shape—it is a composition of an exterior one-form **f** with multiplication by a vector **v**. This operator is called the tensor product and is denoted as $\mathbf{v} \otimes \mathbf{f}$. We show in Sect. 3.3 that an arbitrary linear operator between two given vector spaces is a combination of such tensor products. A lot of other linear operators represented in similar manner are also considered in this section.

A unit system is presented in Sect. 3.4 which enables us to identify the grade of a directed quantity by the physical unit needed to express it. The mechano-electrical fundamental units are units of length, time, action and electric charge. We present arguments that this is a system compatible with geometry.

When an exterior form depends on a point in space, it is called a differential form. Such notions are briefly described in Sect. 3.5.

© Springer Nature Switzerland AG 2021
B. Jancewicz, *Directed Quantities in Electrodynamics*,
https://doi.org/10.1007/978-3-030-90471-5_3

The exterior derivative of differential form is introduced in Sect. 3.6. It maps a k-form into a $(k + 1)$-form. It replaces, among others, the gradient, divergence and rotation known from vector analysis. The exterior derivatives of specific scalar-valued functions, namely of curvilinear coordinates, are computed. The calculus of exterior forms has been developed to unify and simplify integration of directed quantities. This is briefly described at the end of Sect. 3.6.

3.1   Grassmann Algebras

We return to the situation without the scalar product. Denote by M the set of all ordinary and twisted multivectors, including ordinary and twisted scalars. Each quantity has its grade, namely ordinary and twisted k-vectors have grade k. Ordinary and twisted scalars have grade 0. We are going to show that for each pair of quantities from M, the exterior product may be defined. The grades of the factors in such a product are added; however, when their sum is greater than three, their product is zero.

The exterior product of any quantity N from M with a scalar α is the same as the ordinary multiplication by scalars, considered in Sect. 2.1, and this product is commutative:

$$\alpha \wedge N = N \wedge \alpha = \alpha N.$$

We assume similarly that the exterior product with a twisted scalar λ is commutative:

$$\lambda \wedge N = N \wedge \lambda = \lambda N. \tag{3.1}$$

Such a product changes a nontwisted quantity into a twisted one and vice versa. The magnitudes of the factors are multiplied in this product and the directions are changed according to the rules shown in Figs. 3.1, 3.2, 3.3, and 3.4. In the multiplication (3.1) the factor N has the same attitude as that of the result, so we may claim that N and λN are parallel.

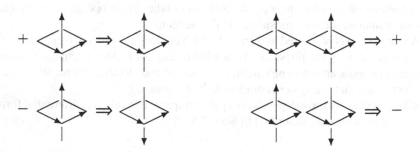

Fig. 3.1 Multiplication of nontwisted and twisted scalars by twisted scalars

Fig. 3.2 Multiplication of nontwisted and twisted vectors by twisted scalars

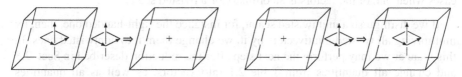

Fig. 3.3 Multiplication of nontwisted and twisted bivectors by twisted scalars

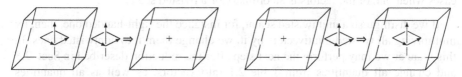

Fig. 3.4 Multiplication of nontwisted and twisted trivectors by twisted scalars

Table 3.1 Multiplication group of four ordinary and twisted scalars

	1	−1	r	l
1	1	−1	r	l
−1	−1	1	l	r
r	r	l	1	−1
l	l	r	−1	1

Let us denote by r the right-handed unit twisted scalar and by l the left-handed one. We see from the left column of Fig. 3.1 that $l = -r$, and a missing picture would show that $r = -l$. The right column of the same figure shows that $r^2 = 1$, and similarly we obtain $l^2 = 1$. In this way we notice that the ordinary and twisted unit scalars form an Abelian group—their multiplication table is shown in Table 3.1. This group is isomorphic to the *four-group of Klein*.

The exterior products of ordinary vectors with nontwisted and twisted multivectors other than scalars have been already defined in Figs. 1.16, 1.19, 1.20, 1.27 and 1.34. The missing exterior products of a vector with a nontwisted or twisted trivector are zero due to the rule of adding grades.

The exterior products of twisted vectors with nontwisted and twisted multivectors other than scalars have been already defined by Figs. 1.19, 1.20, 1.21, 1.22, 1.29 and 1.35. By virtue of those figures and Figs. 2.1, 2.2, 2.3, and 2.4, the *homogeneity* of the exterior product with respect to multiplication by an ordinary scalar λ,

$$\lambda(N_1 \wedge N_2) = (\lambda N_1) \wedge N_2 = N_1 \wedge (\lambda N_2),$$

can be proven for N_1, $N_2 \in \mathcal{M}$. The exterior product of two elements from $\mathcal{M}$ has its attitude spanned on attitudes of both factors. In short, the exterior product is parallel to its factors.

Exercise 3.1 Prove graphically the identity $r(\mathbf{a} \wedge \mathbf{b}) = (r\mathbf{a}) \wedge \mathbf{b} = \mathbf{a} \wedge (r\mathbf{b})$ for the unit right-handed twisted scalar r and arbitrarily chosen vectors $\mathbf{a}$ and $\mathbf{b}$.

The vector basis $\{\mathbf{e}_i\}$ distinguishes some unit twisted scalar 1^* with the handedness of the basis trivector $\mathbf{e}_{123} = \mathbf{e}_1 \wedge \mathbf{e}_2 \wedge \mathbf{e}_3$. We call 1^* the *orientation of the vector basis*.

Exercise 3.2 Show on pictures that the twisted vector basis $\{\mathbf{e}_i^*\}$ is connected with the nontwisted vector basis through the formula $\mathbf{e}_i^* = 1^*\mathbf{e}_i$.

Exercise 3.3 Prove that the exterior multiplication is associative. Consider also cases when one of the factors is an ordinary or a twisted scalar.

If we distinguish a unit twisted scalar, for instance the right-handed one, then, by multiplying all twisted multivectors by it, we change them into nontwisted ones. In this manner we may perform the next step after the reduction described in Sect. 2.3 and change all quantities from Table 2.1 into vectors, as well as all quantities from Table 2.2 into scalars. As a matter of fact, this occurs in introductory physics courses, when magnetic induction or magnetic field intensity are presented as a vector (segment with an arrow at the end). From time to time one has to mention the right-hand screw—this is just recalling the distinguished twisted unit scalar. The choice of the right-handed screw is purely a convention connected with the traditional definition of the vector product. This convention is absent in the present book—if a physical quantity with a one-dimensional attitude is connected with a rotation, its orientation is not a straight arrow but a curved one, surrounding the attitude (see Sect. 1.4).

We now have the means at our disposal to define *scalar products of nontwisted with twisted vectors*. Namely, if $\mathbf{a}$ is a nontwisted vector and $\mathbf{b}$ is a twisted one, we represent the latter as $\mathbf{b} = \lambda \mathbf{b}'$ for a twisted unit scalar λ, and then define

$$(\mathbf{a}, \mathbf{b}) = \lambda(\mathbf{a}, \mathbf{b}'); \quad (\mathbf{b}, \mathbf{a}) = \lambda(\mathbf{b}', \mathbf{a}).$$

In this manner we obtain a twisted scalar as the result. This scalar product is symmetric and linear in both factors.

The watchful reader has probably noticed that in Sect. 2.3 we did not define a scalar product of two twisted bivectors. We are able to do this now, namely for two twisted bivectors $\mathbf{B} = \mathbf{a} \wedge \mathbf{b}$, $\mathbf{C} = \mathbf{c} \wedge \mathbf{d}$, where $\mathbf{a}$ and $\mathbf{c}$ are nontwisted vectors and $\mathbf{b}$ and $\mathbf{d}$ are twisted ones. We introduce their *scalar product* through the determinant

$$(\mathbf{B}, \mathbf{C}) = (\mathbf{a} \wedge \mathbf{b}, \mathbf{c} \wedge \mathbf{d}) = \begin{vmatrix} (\mathbf{a}, \mathbf{c}) & (\mathbf{a}, \mathbf{d}) \\ (\mathbf{b}, \mathbf{c}) & (\mathbf{b}, \mathbf{d}) \end{vmatrix}.$$

We have ordinary scalars on the main diagonal and twisted scalars on the other one, but the whole determinant is an ordinary scalar.

Once multiplication by twisted scalars is defined, we are able to introduce a directed physical quantity, namely the *magnetic moment of two opposite magnetic charges* $\mathbf{m}_m = g\mathbf{l}$, where $\mathbf{l}$ is the vector joining the charges and g is the *magnetic charge* at the tip of $\mathbf{l}$. Notice that as a twisted vector, this quantity has a different directed nature from the *magnetic moment* $\mathbf{m}$ *of an electric circuit*, considered in Sect. 1.4. They have different dimensions in SI units: $[\mathbf{m}] = \mathrm{Am}^2$, $[\mathbf{m}_m] = \mathrm{Wb\,m} = \mathrm{Vsm}$, so they are distinct physical quantities. Their ratio has the dimension of the magnetic permeability μ:

$$\left[\frac{\mathbf{m}_m}{\mathbf{m}}\right] = \frac{\mathrm{Wb}}{\mathrm{Am}} = [\mu].$$

Also, a *magnetic current* $I_m = dg/dt$ can be introduced as the twisted scalar quantity along with the *electric moment of the magnetic circuit* $\mathbf{d}_m = I_m\mathbf{s}$, where $\mathbf{s}$ is the bivector of the directed area of the circuit. In this manner the latter quantity becomes the twisted bivector. Its dimension in SI units is $[\mathbf{d}_m] = \mathrm{Wb\,m}^2\mathrm{s}^{-1} = \mathrm{Vm}^2$, whereas that of the ordinary electric dipole moment is $[\mathbf{d}] = \mathrm{Cm}$. Their ratio has the dimension of the electric permittivity:

$$\left[\frac{\mathbf{d}}{\mathbf{d}_m}\right] = \frac{\mathrm{C}}{\mathrm{Vm}} = [\epsilon].$$

3.1.1 Extended Grassmann Algebra of Multivectors

Let $L(\mathcal{M})$ be a linear span of $\mathcal{M}$, that is, the set of all linear combinations of elements of various grades. Due to distributivity, we extend the exterior product on $L(\mathcal{M})$. In this manner we obtain an *algebra* (i.e., a set with the addition and multiplication of the elements), which will be called *extended Grassmann algebra of multivectors in* $\mathbf{R}^3$. The word "extended" is used here because the algebra of multivectors alone with the exterior product was long ago called a *Grassmann algebra*. Linear combinations of nontwisted multivectors span a linear subspace $L(\mathcal{M})_+$ and subalgebra of $L(\mathcal{M})$, which is just the traditional Grassmann algebra. On the other hand the linear combinations of twisted multivectors span only a linear subspace $L(\mathcal{M})_-$. The exterior product of two twisted quantities is a nontwisted quantity; the mixed exterior product (one quantity nontwisted, one twisted) gives a twisted quantity. In this manner, we obtain the so-called Z_2-*gradation* of our algebra, written usually as the *direct sum* $L(\mathcal{M}) = L(\mathcal{M})_+ \oplus L(\mathcal{M})_-$. The nontwisted quantities constitute the nontwisted part of the algebra.

Multiplication by two twisted unit scalars r, l gives two natural isomorphisms between two linear subspaces present in this sum. We may, for instance, take the right-handed unit r to convert twisted multivectors into nontwisted ones according to the prescriptions shown in the right-hand parts of Figs. 3.1, 3.2, 3.3, and 3.4, which may be summarized in the formula $rL(M)_- = L(M)_+$. This includes the famous rule of the right-hand screw hidden in the vector product and is used to draw nontwisted vector in place of the twisted one. We remind ourselves of Fig. 2.4 for the right-handed basis and the formula $\mathbf{e}_1 \times \mathbf{e}_2 = \mathbf{e}_3^*$ following from it. If one writes the nontwisted vectors on the right-hand side, one in fact multiplies the vector product by r:

$$r\mathbf{e}_1 \times \mathbf{e}_2 = \mathbf{e}_3, \quad r\mathbf{e}_2 \times \mathbf{e}_3 = \mathbf{e}_1, \quad r\mathbf{e}_3 \times \mathbf{e}_1 = \mathbf{e}_2,$$

A mapping in the other direction is also given as the multiplication by r: $rL(M)_+ = L(M)_-$. Thus the extended Grassmann algebra of multivectors can be written as the direct sum

$$L(M) = L(M)_+ \oplus rL(M)_+,$$

in which the first summand is the ordinary Grassmann algebra of multivectors connected with $\mathbf{R}^3$.

One may also introduce a general vector product. This means that for an arbitrary orthonormal (but possibly not right-handed) vector basis the relations between the basis elements of the type $\mathbf{e}_1 \times \mathbf{e}_2 = \mathbf{e}_3^*$ can be written with the use of the orientation 1^* of the vector basis:

$$\mathbf{e}_1 \times \mathbf{e}_2 = 1^*\mathbf{e}_3, \quad \mathbf{e}_2 \times \mathbf{e}_3 = 1^*\mathbf{e}_1, \quad \mathbf{e}_3 \times \mathbf{e}_1 = 1^*\mathbf{e}_2,$$

In a summary of these considerations we state that $L(M)$ is the extended Grassmann algebra over the field of real numbers The dimension of $L(M)$ is 16. *The exterior product of two elements from M is parallel to its factors.* The product of an element K of grade p with element N of grade q has the property

$$K \wedge N = (-1)^{pq} N \wedge K. \tag{3.2}$$

Thus, if one of the factors has even grade the product is commutative.

3.1.2 *Extended Grassmann Algebra of Exterior Forms*

Let $\mathcal{F}$ denote a set of all nontwisted and twisted forms. We assume that nontwisted and twisted k-forms have grade $-k$. We should now define an exterior product for all pairs of elements from $\mathcal{F}$, such that the grades of factors add up. If, however, the sum of grades is less than minus three, their product is zero.

We start from exterior multiplication by zero-forms. We define it similarly to multiplication by scalars. When q is an ordinary scalar, then for arbitrary element $\mathbf{E} \in \mathcal{F}$ the product $q\mathbf{E}$ has magnitude $|q|\,|\mathbf{E}|$, with direction the same as that of $\mathbf{E}$

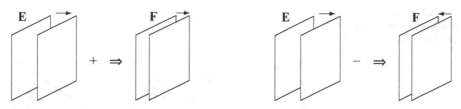

Fig. 3.5 Multiplication of a nontwisted one-form by a nontwisted scalar

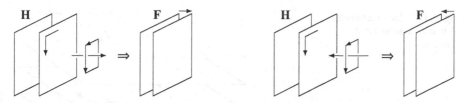

Fig. 3.6 Multiplication of a twisted one-form by a twisted scalar

for positive q or opposite for negative q. If we assume that q is the electric charge then it connects the one-form of the electric field strength **E** with the one-form of force through the formula $\mathbf{F} = q \wedge \mathbf{E}$. For fixed **E** one may obtain two possible orientations of **F** depending on sign of q (see Fig. 3.5).

The same prescription is valid for exterior multiplication of nontwisted zero-form μ in place of nontwisted scalar q. Therefore, Fig. 3.5 also illustrates the product $\mu\mathbf{E}$.

Multiplication by twisted scalars changes nontwisted into twisted quantities and vice versa. It can be considered as an example of magnetic charge g, which by (1.8) is a twisted scalar. For the twisted one-form **H** of the magnetic field strength the product $g\mathbf{H} = \mathbf{F}$ is a nontwisted one-form, which fits to the interpretation of **F** as the force. For fixed **H** two orientations of **F** are possible, depending on the handedness of g, as shown in Fig. 3.6. An exterior multiplication by a twisted zero-form changes nontwisted forms into twisted ones and vice versa. Figure 3.6 also illustrates the exterior product $\lambda\mathbf{H}$ for the twisted zero-form λ in place of g.

The exterior products of nontwisted one-forms with nontwisted and twisted forms other than zero-forms are defined in Figs. 1.48, 1.53, 1.56 and 1.58. The missing products of nontwisted one-forms with nontwisted or twisted three-forms give zero due to the rule of adding grades. The exterior products of twisted one-forms with nontwisted and twisted forms other than zero-forms are defined through Figs. 1.49, 1.53, 1.57 and 1.59.

In this manner we obtain another algebra $L(\mathcal{F})$, which could be called an *extended Grassmann algebra of exterior forms in* $\mathbf{R}^3$. The exterior product of elements from $\mathcal{F}$ has its attitude being an intersection of attitudes of both factors. We are allowed to say that the *product is parallel to its factors*.

Fig. 3.7 Value of a
nontwisted form **f** on a
twisted vector **b**

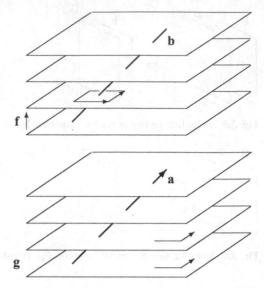

Fig. 3.8 Value of a twisted
form **g** on a nontwisted
vector **a**

Since each twisted quantity N can be represented as $N = rM$ for a nontwisted
quantity M, we may also define a value of a one-form **f** on a twisted vector **b** $= r$**a**
as

$$\mathbf{f}[[\mathbf{b}]] = \mathbf{f}[[r\mathbf{a}]] = r\mathbf{f}[[\mathbf{a}]], \tag{3.3}$$

which results in a twisted scalar. According to this prescription, the value of a
nontwisted one-form on a twisted vector of Fig. 3.7 is the right-handed twisted
scalar. In this figure, the segment representing **b** passes through four layers of **f**,
hence $\mathbf{f}[[\mathbf{a}]] = 4$, so we are willing to say that (3.3) is "right-handed four".

Similarly, the value of a twisted one-form **g** $= r$**f** (here r could equally well be a
twisted scalar or a twisted zero-form) on a vector **a** is defined by

$$\mathbf{g}[[\mathbf{a}]] = (r\mathbf{f})[[\mathbf{a}]] = r(\mathbf{f}[[\mathbf{a}]]).$$

Figure 3.8 shows how one finds the value of a twisted one-form on a nontwisted
vector, which is a twisted scalar. For this figure again the result is "right-handed
four".

Analogous prescriptions for finding values of forms can be formulated with
respect to $L(\mathcal{M})_2$ and $L(\mathcal{M})_3$.

3.2 Contractions

Still another kind of product can be introduced, namely the *contraction*. The rules are the same for twisted and nontwisted quantities, hence we omit these adjectives in the definitions. We define contractions of exterior forms with multivectors from both sides, and later contractions of multivectors with forms also from both sides. The definitions are inductive.

A *contraction of a one-form* **f** *with a vector* **v** is $\mathbf{v}|\mathbf{f} = \mathbf{f}[[\mathbf{v}]]$, where $\mathbf{f}[[\mathbf{v}]]$ is the value of **f** on **v**. We similarly define the contraction from the other side: $\mathbf{f}|\mathbf{v} = \mathbf{f}[[\mathbf{v}]]$.

Let **f, g** be one-forms. A *contraction of a two-form* $\mathbf{B} = \mathbf{f} \wedge \mathbf{g}$ *with a vector* **v** is

$$\mathbf{v}\rfloor\mathbf{B} = \mathbf{v}\rfloor(\mathbf{f} \wedge \mathbf{g}) = \mathbf{f}[[\mathbf{v}]]\mathbf{g} - \mathbf{g}[[\mathbf{v}]]\mathbf{f}, \tag{3.4}$$

The values $\mathbf{f}[[\mathbf{v}]]$ and $\mathbf{g}[[\mathbf{v}]]$ are scalars, so they commute with all elements of $L(\mathcal{F})$. We define also a contraction of **B** with **v** from the other side:

$$\mathbf{B}\lfloor\mathbf{v} = (\mathbf{f} \wedge \mathbf{g})\lfloor\mathbf{v} = \mathbf{fg}[[\mathbf{v}]] - \mathbf{gf}[[\mathbf{v}]] = \mathbf{g}[[\mathbf{v}]]\mathbf{f} - \mathbf{f}[[\mathbf{v}]]\mathbf{g}. \tag{3.5}$$

After comparing this with (3.4), we notice that this contraction is *anticommutative*:

$$\mathbf{B}\lfloor\mathbf{v} = -\mathbf{v}\rfloor\mathbf{B}. \tag{3.6}$$

Let us denote the result of (3.4) by **h**: $\mathbf{h} = \mathbf{v}\rfloor\mathbf{B}$. This is obviously a nontwisted or twisted one-form. It yields for the same vector **v** the value:

$$\mathbf{h}[[\mathbf{v}]] = \mathbf{f}[[\mathbf{v}]]\mathbf{g}[[\mathbf{v}]] - \mathbf{g}[[\mathbf{v}]]\mathbf{f}[[\mathbf{v}]] = 0.$$

This can be interpreted geometrically by saying that **h** is parallel to **v** or, in other words: *the contraction of a two-form with a vector is parallel to the vector.*

Making use of the arbitrariness in the factorization of a two-form into two one-forms $\mathbf{B} = \mathbf{f} \wedge \mathbf{g}$, we may choose one of the factors parallel to **v**—let it be **f**. Figure 3.9 shows how this is possible. Then $\mathbf{f}[[\mathbf{v}]] = 0$ and from (3.5) only

$$\mathbf{B}\lfloor\mathbf{v} = \mathbf{g}[[\mathbf{v}]]\mathbf{f} \tag{3.7}$$

Fig. 3.9 Contraction
$(\mathbf{f} \wedge \mathbf{g})\lfloor\mathbf{v}$ when $\mathbf{f}\|\mathbf{v}$

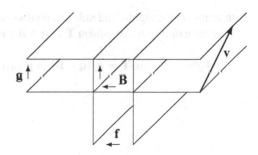

remains. If $\mathbf{g}[[\mathbf{v}]] \neq 0$ we can still use arbitrariness of the factorization and assume $\mathbf{g}[[\mathbf{v}]] = 1$. Then from (3.7) we get

$$\mathbf{B}\lfloor \mathbf{v} = \mathbf{f}.$$

This equality says that the contraction $\mathbf{B}\lfloor \mathbf{v}$ is parallel to $\mathbf{B}$ (see Fig. 3.9). By combining this with the previous underlined sentence we ascertain that the *contraction of a two-form with a vector* $\mathbf{v}$ *is parallel to both factors*. Thus, the (two-dimensional) attitude of the one-form $\mathbf{B}\lfloor \mathbf{v}$ is spanned on attitudes of the factors $\mathbf{B}$ and $\mathbf{v}$. If, however, $\mathbf{g}[[\mathbf{v}]] = 0$, the other factor $\mathbf{g}$ in $\mathbf{B} = \mathbf{f} \wedge \mathbf{g}$ is also parallel to $\mathbf{v}$. Hence the whole two-form $\mathbf{B}$ is parallel to $\mathbf{v}$ and then $\mathbf{B}\lfloor \mathbf{v} = 0$.

Exercise 3.4 Let $\mathbf{B} = \frac{1}{2} B_{lm} \mathbf{f}^l \wedge \mathbf{f}^m$, $\mathbf{v} = v^k \mathbf{e}_k$. Prove the identity

$$\mathbf{B}\lfloor \mathbf{v} = B_{jk} v^k \mathbf{f}^j. \tag{3.8}$$

Exercise 3.5 Check for the basis elements that

$$\mathbf{f}^{ij}\lfloor \mathbf{e}_k = \delta_k^j \mathbf{f}^i - \delta_k^i \mathbf{f}^j. \tag{3.9}$$

Exercise 3.6 Let $\mathbf{B} = \frac{1}{2} B_{kj} \mathbf{f}^k \wedge \mathbf{f}^j$ and $\mathbf{v} = v^i \mathbf{e}_i$. Prove the identity $\mathbf{v}\rfloor \mathbf{B} = v^i B_{ij} \mathbf{f}^j$.

An example of such a contraction is provided by the magnetic part $\mathbf{F}_m$ of the Lorentz force, which acts on an electric charge q moving in a magnetic field $\mathbf{B}$ with velocity $\mathbf{v}$: $\mathbf{F}_m = q\mathbf{B}\lfloor \mathbf{v}$. This contraction is also illustrated in Fig. 3.9, where again the factors of $\mathbf{B}$ are chosen such that $\mathbf{f}||\mathbf{v}$ and $\mathbf{g}[[\mathbf{v}]] = 1$. In this case the magnetic force is $\mathbf{F}_m = q\mathbf{B}\lfloor \mathbf{v} = q\mathbf{f}$. Thus, the one-form of the force is parallel to the velocity. By adding this to the above-mentioned electric force, we arrive at the formula for the full *Lorentz force* acting on an electric charge q:

$$\mathbf{F} = q\mathbf{E} + q\mathbf{B}\lfloor \mathbf{v}. \tag{3.10}$$

An analogous formula can be written for the "Lorentz-like" force acting on a magnetic charge g:

$$\mathbf{F}' = g\mathbf{H} + g\mathbf{D}\lfloor \mathbf{v}.$$

Both terms on the right-hand side are nontwisted one-forms.

A *contraction of a three-form* $\mathbf{T} = \mathbf{g} \wedge \mathbf{B}$ *with a vector* $\mathbf{a}$ is

$$\mathbf{a}\rfloor \mathbf{T} = \mathbf{a}\rfloor(\mathbf{g} \wedge \mathbf{B}) = \mathbf{g}[[\mathbf{a}]] \mathbf{B} - \mathbf{g} \wedge (\mathbf{a}\rfloor \mathbf{B}). \tag{3.11}$$

We can equally well write $\mathbf{T} = \mathbf{B} \wedge \mathbf{g}$ We define a contraction from the other side too:

$$\mathbf{T} \lfloor \mathbf{a} = (\mathbf{B} \wedge \mathbf{g}) \lfloor \mathbf{a} = \mathbf{B}\, \mathbf{g}[[\mathbf{a}]] - (\mathbf{B} \lfloor \mathbf{a}) \wedge \mathbf{g}. \tag{3.12}$$

The last term is the exterior product of two one-forms. It is anticommutative, hence we obtain

$$\mathbf{T} \lfloor \mathbf{a} = \mathbf{g}[[\mathbf{a}]]\, \mathbf{B} + \mathbf{g} \wedge (\mathbf{B} \lfloor \mathbf{a}).$$

Now we use (3.6):

$$\mathbf{T} \lfloor \mathbf{a} = \mathbf{g}[[\mathbf{a}]]\, \mathbf{B} - \mathbf{g} \wedge (\mathbf{a} \rfloor \mathbf{B}).$$

After comparing this with (3.11), we recognize the commutativity:

$$\mathbf{T} \lfloor \mathbf{a} = \mathbf{a} \rfloor \mathbf{T}. \tag{3.13}$$

The result of this contraction is obviously a two-form.

The factorization $\mathbf{T} = \mathbf{g} \wedge \mathbf{B}$ is arbitrary; we choose $\mathbf{B}$ parallel to $\mathbf{a}$. Then $\mathbf{a} \rfloor \mathbf{B} = 0$ and (3.11) reduces to

$$\mathbf{a} \rfloor \mathbf{T} = \mathbf{g}[[\mathbf{a}]]\, \mathbf{B},$$

showing that the contraction of the three-form with a vector is parallel to the vector. A three-form has a point as its attitude, and each plane can be considered parallel to a point. Hence the considered contraction is parallel also to the three-form. As a result, the *contraction of a three-form with a vector is parallel to both factors*.

Similarly, we can define the *contraction of a two- or three-form* $\mathbf{N}$ *with a bivector* $\mathbf{S} = \mathbf{a} \wedge \mathbf{b}$,

$$\mathbf{S} \rfloor \mathbf{N} = (\mathbf{a} \wedge \mathbf{b}) \rfloor \mathbf{N} = \mathbf{a} \rfloor (\mathbf{b} \rfloor \mathbf{N}). \tag{3.14}$$

Similarly, from the other side we define

$$\mathbf{N} \lfloor \mathbf{S} = \mathbf{N} \lfloor (\mathbf{a} \wedge \mathbf{b}) = (\mathbf{N} \lfloor \mathbf{a}) \lfloor \mathbf{b}. \tag{3.15}$$

Let us compare (3.14) and (3.15). If $\mathbf{N}$ is a two-form $\mathbf{N} = \mathbf{f} \wedge \mathbf{g}$:

$$\mathbf{S} \rfloor \mathbf{N} = \mathbf{a} \rfloor (\mathbf{b} \rfloor \mathbf{N}) = \mathbf{a} \rfloor \{\mathbf{b} \rfloor (\mathbf{f} \wedge \mathbf{g})\} = \mathbf{a} \rfloor \{\mathbf{f}[[\mathbf{b}]]\mathbf{g} - \mathbf{g}[[\mathbf{b}]]\mathbf{f}\}$$

$$= \mathbf{f}[[\mathbf{b}]]\, \mathbf{a} \rfloor \mathbf{g} - \mathbf{g}[[\mathbf{b}]]\, \mathbf{a} \rfloor \mathbf{f} = \mathbf{f}[[\mathbf{b}]]\, \mathbf{g}[[\mathbf{a}]] - \mathbf{g}[[\mathbf{b}]]\, \mathbf{f}[[\mathbf{a}]].$$

Apply $\mathbf{S}$ to the other side

$$\mathbf{N}\lfloor\mathbf{S} = (\mathbf{N}\lfloor\mathbf{a})\lfloor\mathbf{b} = \{(\mathbf{f} \wedge \mathbf{g})\lfloor\mathbf{a}\}\lfloor\mathbf{b} = \{\mathbf{f}\,\mathbf{g}[[\mathbf{a}]] - \mathbf{g}\,\mathbf{f}[[\mathbf{a}]]\}\lfloor\mathbf{b}.$$

Thus,

$$\mathbf{N}\lfloor\mathbf{S} = \mathbf{f}[[\mathbf{b}]]\,\mathbf{g}[[\mathbf{a}]] - \mathbf{g}[[\mathbf{b}]]\,\mathbf{f}[[\mathbf{a}]]. \tag{3.16}$$

In this manner we establish commutativity:

$$\mathbf{S}\rfloor\mathbf{N} = \mathbf{N}\lfloor\mathbf{S}. \tag{3.17}$$

Exercise 3.7 Show that for $\mathbf{N} = \mathbf{f} \wedge \mathbf{g} \wedge \mathbf{h}$ and $\mathbf{S} = \mathbf{a} \wedge \mathbf{b}$ the following equalities hold:

$$\mathbf{S}\rfloor\mathbf{N} = \mathbf{N}\lfloor\mathbf{S} = (\mathbf{g}[[\mathbf{b}]]\,\mathbf{h}[[\mathbf{a}]] - \mathbf{g}[[\mathbf{a}]]\,\mathbf{h}[[\mathbf{b}]])\,\mathbf{f}$$

$$+(\mathbf{f}[[\mathbf{a}]]\,\mathbf{h}[[\mathbf{b}]] - \mathbf{f}[[\mathbf{b}]]\,\mathbf{h}[[\mathbf{a}]])\,\mathbf{g} + (\mathbf{f}[[\mathbf{b}]]\,\mathbf{g}[[\mathbf{a}]] - \mathbf{f}[[\mathbf{a}]]\,\mathbf{g}[[\mathbf{b}]])\,\mathbf{h}.$$

We apply this result to the basis nontwisted three-form $\mathbf{N} = \mathbf{f}^1 \wedge \mathbf{f}^2 \wedge \mathbf{f}^3$:

$$\mathbf{f}^{123}\lfloor(\mathbf{a} \wedge \mathbf{b}) = (a^3b^2 - a^2b^3)\,\mathbf{f}^1 + (a^1b^3 - a^3b^1)\,\mathbf{f}^2 + (a^2b^1 - a^1b^2)\,\mathbf{f}^3.$$

This can be expressed by the coordinates of nontwisted bivector $\mathbf{S} = \mathbf{a} \wedge \mathbf{b}$:

$$\mathbf{f}^{123}\lfloor\mathbf{S} = -S^{23}\mathbf{f}^1 - S^{31}\mathbf{f}^2 - S^{12}\mathbf{f}^3. \tag{3.18}$$

Similar calculations performed for twisted bivector $\mathbf{S}^* = \mathbf{a}^* \wedge \mathbf{b}$ with twisted factor $\mathbf{a}^*$ lead to a similar formula with the twisted basis three-form:

$$\mathbf{f}_*^{123}\lfloor\mathbf{S}^* = -S^{23}\mathbf{f}^1 - S^{31}\mathbf{f}^2 - S^{12}\mathbf{f}^3. \tag{3.19}$$

Exercise 3.8 Let $\mathbf{D} = \frac{1}{2}D_{kl}\mathbf{f}^k \wedge \mathbf{f}^l$ and $\mathbf{S} = \frac{1}{2}S^{ij}\mathbf{e}_i \wedge \mathbf{e}_j$. Prove the identity $\mathbf{S}\rfloor\mathbf{D} = \frac{1}{2}S^{ij}D_{ji}$.

After comparing (3.16) with (2.7) we find the equality

$$\mathbf{D}\lfloor\mathbf{S} = -\mathbf{D}[[\mathbf{S}]], \tag{3.20}$$

unfortunately with the minus sign.[1]

---

[1] We remind the reader that $\mathbf{N}[[\mathbf{S}]]$ is the value of two-form $\mathbf{N}$ on bivector $\mathbf{S}$.

We now pass to contractions of multivectors with forms. The procedure is again inductive. A *contraction of a bivector* $\mathbf{S} = \mathbf{a} \wedge \mathbf{b}$ *with a one-form* $\mathbf{f}$ is

$$\mathbf{S}\lfloor \mathbf{f} = (\mathbf{a} \wedge \mathbf{b})\lfloor \mathbf{f} = \mathbf{a}\, \mathbf{f}[[\mathbf{b}]] - \mathbf{b}\, \mathbf{f}[[\mathbf{a}]], \quad \mathbf{f}\rfloor \mathbf{S} = \mathbf{f}\rfloor(\mathbf{a} \wedge \mathbf{b}) = \mathbf{f}[[\mathbf{a}]]\, \mathbf{b} - \mathbf{f}[[\mathbf{b}]]\, \mathbf{a},$$

and *of a bivector* $\mathbf{S}$ *with a two-form* $\mathbf{f} \wedge \mathbf{g}$:

$$\mathbf{S}\lfloor(\mathbf{f} \wedge \mathbf{g}) = (\mathbf{S}\lfloor \mathbf{f})\lfloor \mathbf{g}, \quad (\mathbf{f} \wedge \mathbf{g})\rfloor \mathbf{S} = \mathbf{f}\rfloor(\mathbf{g}\rfloor \mathbf{S}),$$

Exercise 3.9 For a bivector $\mathbf{S}$ and a two-form $\mathbf{D}$ show the commutativity: $\mathbf{S}\lfloor \mathbf{D} = \mathbf{D}\rfloor \mathbf{S}$.

Exercise 3.10 Let $\mathbf{D} = \frac{1}{2}D_{kl}\mathbf{f}^k \wedge \mathbf{f}^l$ and $\mathbf{S} = \frac{1}{2}S^{ij}\mathbf{e}_i \wedge \mathbf{e}_j$. Prove the identity $\mathbf{S}\lfloor \mathbf{D} = \frac{1}{2}S^{ij}D_{ji}$.

After comparing Exercises 3.7 and 3.9 we are convinced that the non-symmetric multiplication sign is not necessary. Hence we may denote the two contractions by a vertical bar:

$$\mathbf{S}|\mathbf{D} = \mathbf{D}|\mathbf{S} = \mathbf{S}\lfloor \mathbf{D} = \mathbf{S}\rfloor \mathbf{D}.$$

Exercise 3.11 Let $\mathbf{D}$ be a two-form, $\mathbf{E}$ a one-form and $\mathbf{S}$ a bivector. Prove the identity

$$(\mathbf{E} \wedge \mathbf{D})\lfloor \mathbf{S} = -\mathbf{E}\, \mathbf{D}[[\mathbf{S}]] - \mathbf{D}\lfloor(\mathbf{S}\lfloor \mathbf{E}). \tag{3.21}$$

Exercise 3.12 Let $\mathbf{D} = \mathbf{g} \wedge \mathbf{h}$ be a two-form, $\mathbf{E}$ a one-form and $\mathbf{S}$ a bivector. Prove the identity

$$(\mathbf{E} \wedge \mathbf{D})\lfloor \mathbf{S} = -\mathbf{E}\, \mathbf{D}[[\mathbf{S}]] + \mathbf{g}\,(\mathbf{E} \wedge \mathbf{h})[[\mathbf{S}]] - \mathbf{h}\,(\mathbf{E} \wedge \mathbf{g})[[\mathbf{S}]]. \tag{3.22}$$

A *contraction of a trivector* $V = \mathbf{S} \wedge \mathbf{c} = \mathbf{c} \wedge \mathbf{S}$ *with a one-form* $\mathbf{f}$ is:

$$V\lfloor \mathbf{f} = (\mathbf{S} \wedge \mathbf{c})\lfloor \mathbf{f} = \mathbf{S}\, \mathbf{f}[[\mathbf{c}]] + \mathbf{c} \wedge (\mathbf{S}\lfloor \mathbf{f}), \quad \mathbf{f}\rfloor V = \mathbf{f}\rfloor(\mathbf{c} \wedge \mathbf{S}) = \mathbf{f}[[\mathbf{c}]]\, \mathbf{S} + (\mathbf{f}\rfloor \mathbf{S}) \wedge \mathbf{c}. \tag{3.23}$$

The contraction of $V = \mathbf{a} \wedge \mathbf{b} \wedge \mathbf{c}$ with $\mathbf{f}$ from the right yields

$$V\lfloor \mathbf{f} = (\mathbf{a} \wedge \mathbf{b} \wedge \mathbf{c})\lfloor \mathbf{f} = (\mathbf{a} \wedge \mathbf{b})\mathbf{f}[[\mathbf{c}]] + (\mathbf{c} \wedge \mathbf{a})\mathbf{f}[[\mathbf{b}]] + (\mathbf{b} \wedge \mathbf{c})\mathbf{f}[[\mathbf{a}]] \tag{3.24}$$

whereas from the left:

$$\mathbf{f}\rfloor V = \mathbf{f}\rfloor(\mathbf{a} \wedge \mathbf{b} \wedge \mathbf{c}) = \mathbf{f}[[\mathbf{a}]](\mathbf{b} \wedge \mathbf{c}) + \mathbf{f}[[\mathbf{b}]](\mathbf{c} \wedge \mathbf{a}) + \mathbf{f}[[\mathbf{c}]](\mathbf{a} \wedge \mathbf{c}). \tag{3.25}$$

We see that both give the same result:

$$\mathbf{f} \lrcorner V = V \llcorner \mathbf{f}.$$

A *contraction of a trivector V with a two-form* $\mathbf{B} = \mathbf{f} \wedge \mathbf{h}$ is defined as

$$V \llcorner \mathbf{B} = V \llcorner (\mathbf{f} \wedge \mathbf{h}) = (V \llcorner \mathbf{f}) \llcorner \mathbf{h}, \qquad \mathbf{B} \lrcorner V = (\mathbf{f} \wedge \mathbf{h}) \lrcorner V = \mathbf{f} \lrcorner (\mathbf{h} \lrcorner V). \qquad (3.26)$$

Exercise 3.13 For $V = \mathbf{a} \wedge \mathbf{b} \wedge \mathbf{c}$, show the identity

$$V \llcorner \mathbf{B} = \mathbf{B} \lrcorner V = (\mathbf{f}[[\mathbf{c}]]\mathbf{h}[[\mathbf{b}]] - \mathbf{f}[[\mathbf{b}]]\mathbf{h}[[\mathbf{c}]])\mathbf{a} + (\mathbf{f}[[\mathbf{a}]]\mathbf{h}[[\mathbf{c}]] - \mathbf{f}[[\mathbf{c}]]\mathbf{h}[[\mathbf{a}]])\mathbf{b}$$

$$+ (\mathbf{f}[[\mathbf{b}]]\mathbf{h}[[\mathbf{a}]] - \mathbf{f}[[\mathbf{a}]]\mathbf{h}[[\mathbf{b}]])\mathbf{c}. \qquad (3.27)$$

Hint: Use (3.24) and (3.25).

The formula (3.27) shows that the contraction of a trivector with a two-form yields a vector. In particular for the nontwisted basis trivector $\mathbf{e}_{123} = \mathbf{e}_1 \wedge \mathbf{e}_2 \wedge \mathbf{e}_3$ we obtain

$$\mathbf{e}_{123} \llcorner (\mathbf{f} \wedge \mathbf{h}) = (\mathbf{e}_1 \wedge \mathbf{e}_2 \wedge \mathbf{e}_3) \llcorner (\mathbf{f} \wedge \mathbf{h}) = (\mathbf{f}[[\mathbf{e}_3]]\mathbf{h}[[\mathbf{e}_2]] - \mathbf{f}[[\mathbf{e}_2]]\mathbf{h}[[\mathbf{e}_3]])\mathbf{e}_1$$

$$+ (\mathbf{f}[[\mathbf{e}_1]]\mathbf{h}[[\mathbf{e}_3]] - \mathbf{f}[[\mathbf{e}_3]]\mathbf{h}[[\mathbf{e}_1]])\mathbf{e}_2 + (\mathbf{f}[[\mathbf{e}_2]]\mathbf{h}[[\mathbf{e}_1]] - \mathbf{f}[[\mathbf{e}_1]]\mathbf{h}[[\mathbf{e}_2]])\mathbf{e}_3,$$

$$\mathbf{e}_{123} \llcorner (\mathbf{f} \wedge \mathbf{h}) = (f_3 h_2 - f_2 h_3)\mathbf{e}_1 + (f_1 h_3 - f_3 h_1)\mathbf{e}_2 + (f_2 h_1 - f_1 h_2)\mathbf{e}_3.$$

This can be expressed by the coordinates of $\mathbf{B} = \mathbf{f} \wedge \mathbf{h}$:

$$\mathbf{e}_{123} \llcorner \mathbf{B} = B_{32}\mathbf{e}_1 + B_{13}\mathbf{e}_2 + B_{21}\mathbf{e}_3. \qquad (3.28)$$

Analogously, for the basic twisted trivector we have

$$\mathbf{e}^*_{123} \llcorner (\mathbf{f} \wedge \mathbf{h}) = (f_3 h_2 - f_2 h_3)\mathbf{e}^*_1 + (f_1 h_3 - f_3 h_1)\mathbf{e}^*_2 + (f_2 h_1 - f_1 h_2)\mathbf{e}^*_3. \qquad (3.29)$$

This can be expressed by the coordinates of $\mathbf{B}$:

$$\mathbf{e}^*_{123} \llcorner \mathbf{B} = B_{32}\mathbf{e}^*_1 + B_{13}\mathbf{e}^*_2 + B_{21}\mathbf{e}^*_3.$$

The transposition of the indices at B_{ij} implies change of the sign:

$$\mathbf{e}^*_{123} \llcorner \mathbf{B} = -B_{23}\mathbf{e}^*_1 - B_{31}\mathbf{e}^*_2 - B_{12}\mathbf{e}^*_3.$$

Compare this with (2.73). We see that the right-hand sides differ only by a sign. Accordingly, we may claim that the contraction

$$- \mathbf{e}^*_{123} \lfloor \mathbf{B} = B_{23}\mathbf{e}^*_1 + B_{31}\mathbf{e}^*_2 + B_{12}\mathbf{e}^*_3 = \mathbf{B} \tag{3.30}$$

performs the mapping (2.73). In this manner we have discovered how to write the change of two-forms into twisted vectors—we need to have a basis (the same that determines a scalar product), take the basic twisted trivector and contract it with the two-forms. The fact that we choose the twisted trivector, not the nontwisted one, expresses an independence of the mapping of any three-dimensional orientation.

Exercise 3.14 Let $V = \frac{1}{6} V^{ijk}\mathbf{e}_i \wedge \mathbf{e}_j \wedge \mathbf{e}_k$, $\mathbf{B} = \frac{1}{2} B_{lm}\mathbf{f}^l \wedge \mathbf{f}^m$. Prove the identity

$$V \lfloor \mathbf{B} = -\frac{1}{2} V^{jkl} B_{kl}\mathbf{e}_j. \tag{3.31}$$

Exercise 3.15 Let $V = \mathbf{S} \wedge \mathbf{l}$ where $\mathbf{S}$ is a bivector, $\mathbf{l}$ a vector and $\mathbf{j}$ a two-form parallel to $\mathbf{l}$. Prove the equality

$$V \lfloor \mathbf{j} = (\mathbf{S} \wedge \mathbf{l}) \lfloor \mathbf{j} = -\mathbf{j}[[\mathbf{S}]] \mathbf{l}.$$

Exercise 3.16 Prove that an inverse mapping to (3.30) is performed as the following contraction:

$$\mathbf{f}^{123}_* \lfloor \mathbf{B} = \mathbf{B}. \tag{3.32}$$

Exercise 3.17 Show that the mapping (2.74) of a twisted two-form $\mathbf{D}$ into vector $\mathbf{D}$ can be written as

$$- \mathbf{e}^*_{123} \lfloor \mathbf{D} = D_{23}\mathbf{e}_1 + D_{31}\mathbf{e}_2 + D_{12}\mathbf{e}_3 = \mathbf{D}. \tag{3.33}$$

Exercise 3.18 Prove that an inverse mapping to (3.33) is performed as the following contraction

$$\mathbf{f}^{123}_* \lfloor \mathbf{D} = \mathbf{D}. \tag{3.34}$$

We can check one after the other that in all introduced contractions the grades add up. Recall that the grades of multivectors are nonnegative, and those of forms are nonpositive.

Exercise 3.19 Let $\mathbf{D}$ be a two-form, $\mathbf{E}$—a one-form, $\mathbf{S}$—a bivector, $\mathbf{l}$—a vector. Prove the identity

$$(\mathbf{D} \wedge \mathbf{E})[[\mathbf{S} \wedge \mathbf{l}]] = \mathbf{D}[[\mathbf{S}]] \mathbf{E}[[\mathbf{l}]] + \mathbf{D}[[(\mathbf{E}\lfloor\mathbf{S}) \wedge \mathbf{l}]]. \tag{3.35}$$

Let $\mathbf{B}$ be a two-form, and $\mathbf{H}$ a one-form. *The contraction of a trivector V with the three-form* $\mathbf{B} \wedge \mathbf{H}$ is

$$V \lfloor (\mathbf{B} \wedge \mathbf{H}) = (V \lfloor \mathbf{B}) \lfloor \mathbf{H} = \mathbf{H}[[V \lfloor \mathbf{B}]]. \tag{3.36}$$

Exercise 3.20 Define the contraction of a three-form with a trivector. Show that for an arbitrary trivector V and an arbitrary three-form J the identity

$$V \lfloor J = J \lfloor V = -J[[V]] \tag{3.37}$$

is fulfilled.

This result justifies denoting of the two contractions by the vertical bar:

$$V | J = J | V = V \lfloor J = J \lfloor V$$

Exercise 3.21 Let $\mathbf{B}$ be a nontwisted two-form, $\mathbf{D}$ a twisted two-form, and V a twisted trivector. Show that the expression $\mathbf{B} \lfloor (V \lfloor \mathbf{D})$ is a nontwisted one-form. Check the following anticommutativity:

$$\mathbf{D} \lfloor (V \lfloor \mathbf{B}) = -\mathbf{B} \lfloor (V \lfloor \mathbf{D}).$$

Express this quantity in terms of coordinates B^i and D^j of vectors $\mathbf{B}$ and $\mathbf{D}$ replacing forms according to (2.73) and (2.74).
Answer $\{(D^2 B^3 - D^3 B^2)\mathbf{f}^1 + (D^3 B^1 - D^1 B^3)\mathbf{f}^2 + (D^1 B^2 - D^2 B^1)\mathbf{f}^3\} | V|$.
We see from this that such an operation between two-forms corresponds to the vector product $\mathbf{D} \times \mathbf{B}$ in traditional language.

Summing up: we have become acquainted with four kinds of products in the premetric formalism: the exterior product of multivectors, the exterior product of forms, the contractions of forms with multivectors and vice versa. In all of them the grades sum up and the result is parallel to the factors. The mixed product—one factor nontwisted the other twisted—yields the twisted result; the same parity factors—nontwisted-nontwisted and twisted-twisted—yield the nontwisted result.

A physical application of these considerations is the relation between force and pressure. We know that the force F exerted on the surface of a gas or liquid is proportional to its area S: $F = p'S$, where the proportionality coefficient is the scalar of pressure p'. As a directed quantity for the area we should choose the twisted bivector $\mathbf{S}$, since the natural orientation here is an arrow piercing the surface from the container to the outside. If we treat the force as a nontwisted one-form $\mathbf{F}$, the pressure p ought to be a twisted three-form $p = p'\mathbf{f}_*^{123}$. In such a case an operation between p and $\mathbf{S}$ could be the contraction:

$$\mathbf{F} = p \lfloor \mathbf{S} = p'\mathbf{f}_*^{123} \lfloor \mathbf{S}.$$

After comparing this with (3.19), we change the sign in order to obtain nontwisted one-form **F** with the same orientation as the twisted bivector **S**:

$$\mathbf{F} = -p\lfloor\mathbf{S}. \tag{3.38}$$

The parallelity of the contraction to its factors mentioned above yields the attitude of **F** (the plane) parallel to the surface on which the force is exerted. Thus, pressure as a twisted three-form should be added to our list of directed physical quantities. It is natural to assume that it is a positive twisted three-form: then **F** and **S** have the same direction for arbitrary **S**.

Another physical example is the contraction of a twisted trivector of the volume V with the twisted two-form of the current density **j**. Let us consider a cylindrical conductor with a uniform current. We represent the volume V as an exterior product of a twisted bivector **S** of the cross-section with a nontwisted vector **l** of the directed length: $V = \mathbf{S} \wedge \mathbf{l}$. We choose compatible orientations of **S** and **l** in order to obtain a positive product. In such a situation, the current I flowing through the conductor is given by $I = \mathbf{j}[[\mathbf{S}]]$. Moreover, the quantities **j** and **l** are parallel to the cylinder axis and parallel to each other. Thus, by virtue of Exercise 3.15 we may write

$$-V\lfloor\mathbf{j} = \mathbf{j}[[\mathbf{S}]]\mathbf{l} = I\mathbf{l}. \tag{3.39}$$

The quantity $I\mathbf{l}$ is called the *moment of electric current*. Thus, the contraction (with the minus sign) of the volume of a rectilinear segment of a conductor with the current density yields the ordinary vector of the moment of current of the segment. This quantity can be used to find an expression for the magnetic part of the Lorentz force acting on a current segment. We rewrite $\mathbf{F} = q\,\mathbf{B}\lfloor\mathbf{v}$ as $\mathbf{F} = \mathbf{B}\lfloor(q\mathbf{v})$ and transform the second factor:

$$q\mathbf{v} = q\,\frac{d\mathbf{l}}{dt} = \frac{d}{dt}(q\mathbf{l}).$$

We assume now that in the product $q\mathbf{l}$, only first factor changes in time:

$$\frac{d}{dt}(q\mathbf{l}) = \frac{dq}{dt}\mathbf{l} = I\mathbf{l}.$$

Accordingly, we replace $q\mathbf{v}$ by $I\mathbf{l}$ and obtain for the magnetic Lorentz force acting on a segment of an electric current

$$\mathbf{F} = \mathbf{B}\lfloor(I\mathbf{l}). \tag{3.40}$$

We now address a relation between the charge density ρ and the current density **j**. If there are charge carriers of the same kind moving with velocity **v**, in the traditional treatment (ρ is a scalar, **v** and **j** are vectors) the relation between them is $\mathbf{j} = \rho\mathbf{v}$. In the present approach, ρ is the twisted three-form and **j** the twisted two-form; hence a proper relation is the contraction:

$$\mathbf{j} = \rho \lfloor \mathbf{v}.$$

A similar relation holds between the twisted three-form w of the energy density, the twisted two-form $\mathbf{S}$ of the energy flux density (the Poynting form), and the velocity vector $\mathbf{v}$ of the energy transport:

$$\mathbf{S} = w \lfloor \mathbf{v}.$$

We may note four other physical quantities. The *electric polarization* $\mathbf{P}$ should be multiplied by volume V in order to yield the nontwisted vector $\mathbf{d}$ of the electric moment of electric dipoles contained in V. An appropriate operation is the contraction:

$$V \lfloor \mathbf{P} = \mathbf{d}, \quad (3^* - 2^* = 1).$$

Accordingly, $\mathbf{P}$ is a twisted two-form. (The relation between the grades is placed in parentheses; a star denotes a twisted quantity.)

The *magnetic polarization* or *magnetization* $\mathbf{M}$ is a twisted one-form, because it is contracted with volume V,

$$V \lfloor \mathbf{M} = \mathbf{m}, \quad (3^* - 1^* = 2),$$

to give the nontwisted bivector $\mathbf{m}$ of the magnetic moment of electric currents contained in V.

If we admit (at least theoretically!) the existence of magnetic charges, we may also consider the magnetic counterparts of the above polarizations. A new kind of polarization $\mathbf{P}_m$ is a nontwisted one-form, because the volume contracted with it,

$$V \lfloor \mathbf{P}_m = \mathbf{d}_m, \quad (3^* - 1 = 2^*),$$

gives the twisted bivector $\mathbf{d}_m$ of the electric moment of magnetic currents contained in V. Another new polarization $\mathbf{M}_m$ is a nontwisted two-form, because the volume contracted with it,

$$V \lfloor \mathbf{M}_m = \mathbf{m}_m, \quad (3^* - 2 = 1^*)$$

yields the twisted vector $\mathbf{m}_m$ of the magnetic moment of magnetic dipoles contained in V.

3.3 Linear Operators as Tensors

A linear operator is a mapping of one linear space into another linear space, which is additive and homogeneous. Generally, finite dimensional linear spaces over a field of reals are treated identically—their elements are called vectors and represented as rows or columns of reals. Then all linear operators are represented as matrices and the fact is lost that linear spaces can be of different nature. However, we are not going to forget this.

First we consider the mapping of the linear space of vectors into itself. Let $\mathbf{k}$ be a linear form and $\mathbf{v}$ a vector. We introduce an operator $\mathbf{v} \otimes \mathbf{k}$ by means of the rule

$$(\mathbf{v} \otimes \mathbf{k})[[\mathbf{r}]] = \mathbf{v}\, \mathbf{k}[[\mathbf{r}]]. \tag{3.41}$$

Accordingly, we first take the value of $\mathbf{k}$ on $\mathbf{r}$, and then multiply $\mathbf{v}$ by the obtained number $\mathbf{k}[[\mathbf{r}]]$.

Exercise 3.22 Check that the mapping (3.41) is linear.

The expression $\mathbf{v} \otimes \mathbf{k}$ is called the *tensor product* of $\mathbf{v}$ and $\mathbf{k}$ or *simple tensor*.[2] The operator (3.41) gives a zero vector where the linear form $\mathbf{k}$ gives the number zero. Therefore, its kernel is two-dimensional, a plane M. Hence one cannot obtain an arbitrary linear operator by means of (3.41).

The mapping (3.41) resembles a projection operator on the attitude of $\mathbf{v}$, because its image is parallel to $\mathbf{v}$. Since it is zero for vectors lying in M, we expect that this projection is parallel to M. The only property, demanded from a projection but not necessarily satisfied by (3.41), is that it should leave the vector $\mathbf{v}$ invariant. It yields $(\mathbf{v} \otimes \mathbf{k})[[\mathbf{v}]] = \mathbf{v}\, \mathbf{k}[[\mathbf{v}]]$, which in general is not $\mathbf{v}$. Thus we impose the condition $\mathbf{k}[[\mathbf{v}]] = 1$ and we obtain now the *projection operator*

$$P[[\mathbf{r}]] = (\mathbf{v} \otimes \mathbf{k})[[\mathbf{r}]] = \mathbf{v} \quad \text{if} \quad \mathbf{k}[[\mathbf{v}]] = 1.$$

We are also allowed to write the operator without the vector on which it acts:

$$P = \mathbf{v} \otimes \mathbf{k} \quad \text{with} \quad \mathbf{k}[[\mathbf{v}]] = 1. \tag{3.42}$$

Now P is exactly the projection parallel to $\mathbf{k}$ on the attitude of $\mathbf{v}$ (see Fig. 3.10).

Exercise 3.23 Demonstrate that P is *idempotent*, which means $P[[\, P[[\mathbf{r}]]\,]] = P[[\mathbf{r}]]$.

Lemma 3.1 *An arbitrary linear operator mapping n-dimensional vector space into itself can be represented as a sum of n operators of the form (3.41)*

---

[2] The adjective "simple" is used because other operators are possible that contain more tensor products.

Fig. 3.10 Action of
projection operator P

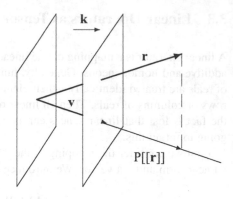

Proof Take a basis $\{\mathbf{e}_1, \ldots, \mathbf{e}_n\}$ in the vector space. Introduce the dual basis $\{\mathbf{f}^1, \ldots, \mathbf{f}^n\}$. For the arbitrary linear operator A find n vectors $\mathbf{v}_i = A\mathbf{e}_i$. We check how the sum $\mathbf{v}_1 \otimes \mathbf{f}^1 + \ldots + \mathbf{v}_n \otimes \mathbf{f}^n$ acts on a basis vector $\mathbf{e}_i$:

$$\sum_{j=1}^{n} (\mathbf{v}_j \otimes \mathbf{f}^j)[[\mathbf{e}_i]] = \sum_{j=1}^{n} \mathbf{v}_j \, \mathbf{f}^j [[\mathbf{e}_i]] = \sum_{j=1}^{n} \mathbf{v}_j \, \delta_i^j = \mathbf{v}_i.$$

We see that this sum acts on $\mathbf{e}_i$ in the same manner as A. Because of the linearity of A, we may write, for an arbitrary vector $\mathbf{r}$ that is a linear combination of basis vectors, the following:

$$A[[\mathbf{r}]] = \sum_{j=1}^{n} (\mathbf{v}_j \otimes \mathbf{f}^j)[[\mathbf{r}]].$$

Since $\mathbf{r}$ is arbitrary, we obtain the operator equality

$$A = \sum_{j=1}^{n} \mathbf{v}_j \otimes \mathbf{f}^j. \tag{3.43}$$

■

When written in such a way, the operator A is called a *tensor*. Subsequently we shall omit the summation sign, because the repeating index j stays on different levels, so the Einstein convention applies:

$$A = \mathbf{v}_j \otimes \mathbf{f}^j. \tag{3.44}$$

Exercise 3.24 Show that the set of all operators of the form (3.44) is a linear space. Find its dimension. *Answer*: Dimension is n^2.

Exercise 3.25 When is the sum of two projectors $P_1 = \mathbf{v}_1 \otimes \mathbf{f}_1$ and $P_2 = \mathbf{v}_2 \otimes \mathbf{f}_2$ idempotent? *Answer*: If $\mathbf{f}_1[[\mathbf{v}_2]] = \mathbf{f}_2[[\mathbf{v}_1]] = 0$.

In particular, the identity operator 1_V on the linear space of vectors has the property $1_V[[\mathbf{e}_i]] = \mathbf{e}_i$; thus we write it as

$$1_V = \sum_{j=1}^{n} \mathbf{e}_j \otimes \mathbf{f}^j. \qquad (3.45)$$

This is obviously the sum of n projection operators.

For the given vector basis $\mathbf{e}_1, \mathbf{e}_2, \ldots, \mathbf{e}_n$ and the dual basis $\mathbf{f}^1, \mathbf{f}^2, \ldots, \mathbf{f}^n$ natural projection operators exist:

$$P_1 = \mathbf{e}_1 \otimes \mathbf{f}^1, \quad P_2 = \mathbf{e}_2 \otimes \mathbf{f}^2, \quad \ldots \quad P_n = \mathbf{e}_n \otimes \mathbf{f}^n$$

which act as follows:

$$P_1[[\mathbf{r}]] = x^1 \mathbf{e}_1, \quad P_2[[\mathbf{r}]] = x^2 \mathbf{e}_2 \quad \ldots \quad P_n[[\mathbf{r}]] = x^n \mathbf{e}_n. \qquad (3.46)$$

By analogy with the sentence below Eq. (2.7) (*forms of the dual basis are devices generating the coordinates of vectors*) we can state: the projection operators (3.46) are devices generating the components of vectors in the given basis.

Similar results can be obtained for other linear spaces. For instance, an arbitrary linear operator B in the linear space of twisted vectors may be represented as

$$B = \mathbf{v}_j^* \otimes \mathbf{f}_*^j.$$

Here $\mathbf{v}_j^*$ are twisted vectors given by $B[[\mathbf{e}_j^*]] = \mathbf{v}_j^*$, and $\mathbf{f}_*^j$ are twisted forms dual to basis twisted vectors $\mathbf{e}_i^*$. In particular, the identity operator on this linear space has the form

$$1_P = \mathbf{e}_j^* \otimes \mathbf{f}_*^j.$$

Exercise 3.26 Show that the operator 1_P can also be written in the form (3.45).

An arbitrary linear operator C in the linear space of nontwisted one-forms can be represented as

$$C = \mathbf{h}^j \otimes \mathbf{e}_j,$$

where $\mathbf{h}^j$ are nontwisted one-forms given by $\mathbf{h}^j = C[[\mathbf{f}^j]]$. The identity operator on this space is

$$1_F = \mathbf{f}^j \otimes \mathbf{e}_j.$$

Similar formulas are valid for linear operators in the linear space of twisted one-forms.

One may write similar expressions for linear operators in the linear space of bivectors. Let **B** be a two-form and **s** a bivector. Then the operator $\mathbf{s} \otimes \mathbf{B}$ is introduced as follows:

$$(\mathbf{s} \otimes \mathbf{B})[[\mathbf{b}]] = \mathbf{s}\, \mathbf{B}[[\mathbf{b}]]. \tag{3.47}$$

One can prove a lemma similar to 3.1—an arbitrary linear operator in the linear space of bivectors is a sum of three operators of the form (3.47)—but we shall not dwell on it.

It is also possible to consider linear operators between linear spaces of different directed quantities, in particular those representing mappings (2.69)–(2.74). For example

$$U = \mathbf{e}_3^* \otimes \mathbf{f}^{12} + \mathbf{e}_1^* \otimes \mathbf{f}^{23} + \mathbf{e}_2^* \otimes \mathbf{f}^{31} = \mathbf{e}_3 \otimes \mathbf{f}_*^{12} + \mathbf{e}_1 \otimes \mathbf{f}_*^{23} + \mathbf{e}_2 \otimes \mathbf{f}_*^{31} \tag{3.48}$$

represents (2.69) and (2.70), whereas

$$S = \mathbf{e}_1 \otimes \mathbf{e}_1 + \mathbf{e}_2 \otimes \mathbf{e}_2 + \mathbf{e}_3 \otimes \mathbf{e}_3, \tag{3.49}$$

represents (2.71) and (2.72).

There are some mappings of two-forms into vectors for which the tensor form of the operators is not as useful as the contraction with the unit twisted trivector. For instance, formula (3.30)

$$\mathbf{B} = -\mathbf{e}_{123}^* \lfloor \mathbf{B} \tag{3.50}$$

yields the following relation between the coordinates B^i of the twisted vector **B** and coordinates B_{jk} of the two-form **B**:

$$B^1 = B_{23}, \quad B^2 = B_{31}, \quad B^3 = B_{12}. \tag{3.51}$$

These can be collected into a single formula with the aid of the *Levi-Civita symbol*:

$$B^k = \frac{1}{2} \epsilon^{kij} B_{ij} \quad \text{or} \quad B_{ij} = \epsilon_{ijk} B^k, \tag{3.52}$$

with the summation over i, j in the left-hand formula and over k in the right-hand one. In this manner (3.50) realizes mapping (2.73).

The mapping (3.33) or

$$\mathbf{D} = -\mathbf{e}_{123}^* \lfloor \mathbf{D} \tag{3.53}$$

yields similar relations between coordinates D^i of **D** and coordinates D_{jk} of **D**:

$$D^k = \frac{1}{2}\epsilon^{kij}D_{ij} \quad \text{or} \quad D_{ij} = \epsilon_{ijk}D^k \qquad (3.54)$$

and implements (2.74).

Bilinear functionals can also be represented as tensors. Let **f, h** be one-forms, and **u, v** vectors. We introduce a bilinear functional **f** ⊗ **h**:

$$(\mathbf{f} \otimes \mathbf{h})[[\mathbf{u}, \mathbf{v}]] = \mathbf{f}[[\mathbf{u}]]\,\mathbf{h}[[\mathbf{v}]], \qquad (3.55)$$

with the product of two values **f**[[**u**]] and **h**[[**v**]] of the two forms on the respective vectors on the right-hand side. It is obvious that this expression is bilinear. The functional **f** ⊗ **h** is called the *tensor product* of the forms **f** and **h** or a *simple tensor*. If **f** is not proportional to **h** the functional (3.55) is not symmetric. We can create the expression $\frac{1}{2}(\mathbf{f} \otimes \mathbf{h} + \mathbf{h} \otimes \mathbf{f})$, which defines a symmetric functional and is called the *symmetric part* of **f** ⊗ **h**.

We apply these consideration to formula (1.1) for three dimensions. That is the scalar product expressed by coordinates of vectors in a chosen basis $\{\mathbf{e}_i\}$:

$$G[[\mathbf{u}, \mathbf{v}]] = u^1 v^1 + u^2 v^2 + u^3 v^3.$$

Making use of (2.7), we express the coordinates through the dual basis one-forms: $u^i = \mathbf{f}^i(\mathbf{u})$, $v^j = \mathbf{f}^j(\mathbf{v})$. Hence

$$G[[\mathbf{u}, \mathbf{v}]] = \mathbf{f}^1[[\mathbf{u}]]\mathbf{f}^1[[\mathbf{v}]] + \mathbf{f}^2[[\mathbf{u}]]\mathbf{f}^2[[\mathbf{v}]] + \mathbf{f}^3[[\mathbf{u}]]\mathbf{f}^3[[\mathbf{v}]]$$

$$= (\mathbf{f}^1 \otimes \mathbf{f}^1 + \mathbf{f}^2 \otimes \mathbf{f}^2 + \mathbf{f}^3 \otimes \mathbf{f}^3)[[\mathbf{u}, \mathbf{v}]].$$

Thus, the scalar product as the bilinear functional can be written as the tensor

$$G = \mathbf{f}^1 \otimes \mathbf{f}^1 + \mathbf{f}^2 \otimes \mathbf{f}^2 + \mathbf{f}^3 \otimes \mathbf{f}^3. \qquad (3.56)$$

As a sum of symmetric tensors this is symmetric. It could be called the *scalar product tensor*, but the name *metric tensor* has been adopted.

Expression (3.56) can also be treated as a linear operator. Since the right-hand factors are one-forms, it may act on vectors and, since the left-hand factors are also one-forms, it gives one-forms as a result. Therefore, *the metric tensor serves as a change of the nontwisted vectors into nontwisted one-forms and, simultaneously, of the twisted vectors into twisted one-forms.* Is it an inverse operator to (3.49)?

To answer this question, we need to know a composition of operators $A = \mathbf{h} \otimes \mathbf{f}$ (where $\mathbf{h}$, $\mathbf{f}$ are one-forms) and $B = \mathbf{u} \otimes \mathbf{v}$ (where $\mathbf{u}$, $\mathbf{v}$ are vectors). Now, for an arbitrary form $\mathbf{k}$:

$$(A \circ B)[[\mathbf{k}]] = \{(\mathbf{h} \otimes \mathbf{f}) \circ (\mathbf{u} \otimes \mathbf{v})\}[[\mathbf{k}]] = (\mathbf{h} \otimes \mathbf{f})\{(\mathbf{u} \otimes \mathbf{v})[[\mathbf{k}]]\} = (\mathbf{h} \otimes \mathbf{f})\{\mathbf{u}\,\mathbf{k}[[\mathbf{v}]]\}$$

$$= \mathbf{k}[[\mathbf{v}]](\mathbf{h} \otimes \mathbf{f})[[\mathbf{u}]] = \mathbf{k}[[\mathbf{v}]]\,\mathbf{h}\,\mathbf{f}[[\mathbf{u}]] = \mathbf{f}[[\mathbf{u}]]\,\mathbf{h}\,\mathbf{k}[[\mathbf{v}]] = \mathbf{f}[[\mathbf{u}]](\mathbf{h} \otimes \mathbf{v})[[\mathbf{k}]].$$

Since $\mathbf{k}$ is arbitrary, the operator equality remains:

$$A \circ B = (\mathbf{h} \otimes \mathbf{f}) \circ (\mathbf{u} \otimes \mathbf{v}) = \mathbf{f}[[\mathbf{u}]]\,\mathbf{h} \otimes \mathbf{v}.$$

We see that to calculate the composition, the middle factors $\mathbf{f}$, $\mathbf{u}$, are to be contracted and the external factors are left for the tensor product. Similar calculations for the product in reverse order lead to the result

$$B \circ A = (\mathbf{u} \otimes \mathbf{v}) \circ (\mathbf{h} \otimes \mathbf{f}) = \mathbf{h}[[\mathbf{v}]]\,\mathbf{u} \otimes \mathbf{f}.$$

We can now calculate the composition of (3.49) and (3.56):

$$S \circ G = \sum_{i=1}^{3}(\mathbf{e}_i \otimes \mathbf{e}_i) \circ \sum_{j=1}^{3}(\mathbf{f}^j \otimes \mathbf{f}^j) = \sum_{i,j=1}^{3}(\mathbf{e}_i \otimes \mathbf{e}_i) \circ (\mathbf{f}^j \otimes \mathbf{f}^j)$$

$$= \sum_{i,j=1}^{3}\mathbf{f}^j[[\mathbf{e}_i]]\mathbf{e}_i \otimes \mathbf{f}^j = \sum_{i,j=1}^{3}\delta_i^j\,\mathbf{e}_i \otimes \mathbf{f}^j = \sum_{j=1}^{3}\mathbf{e}_j \otimes \mathbf{f}^j = 1_V.$$

The last equality is written by virtue of (3.45). The composition in reverse order is:

$$G \circ S = \sum_{i=1}^{3}(\mathbf{f}^i \otimes \mathbf{f}^i) \circ \sum_{j=1}^{3}(\mathbf{e}_j \otimes \mathbf{e}_j) = \sum_{i,j=1}^{3}(\mathbf{f}^i \otimes \mathbf{f}^i) \circ (\mathbf{e}_j \otimes \mathbf{e}_j)$$

$$= \sum_{i,j=1}^{3}\mathbf{f}^i[[\mathbf{e}_j]]\mathbf{f}^i \otimes \mathbf{e}_j = \sum_{i,j=1}^{3}\delta_j^i\mathbf{f}^i \otimes \mathbf{e}_j = \sum_{j=1}^{3}\mathbf{f}^j \otimes \mathbf{e}_j = 1_F.$$

We have checked that $S \circ G = 1_V$ and $G \circ S = 1_F$; hence we write the operator equality $S = G^{-1}$ and (3.49) gives:

$$G^{-1} = \mathbf{e}_1 \otimes \mathbf{e}_1 + \mathbf{e}_2 \otimes \mathbf{e}_2 + \mathbf{e}_3 \otimes \mathbf{e}_3, \tag{3.57}$$

We can express our observation as follows: *the inverse of the metric tensor serves to change of the nontwisted one-forms into nontwisted vectors or the twisted one-*

forms into twisted vectors. In this manner the operator G^{-1} fulfills the role of mappings (2.71) and (2.72). Hence we may write

$$\mathbf{E} = G^{-1}[[\mathbf{E}]] = \tilde{G}[[\mathbf{E}]], \qquad \mathbf{H} = G^{-1}[[\mathbf{H}]] = \tilde{G}[[\mathbf{H}]]. \tag{3.58}$$

The inverse of the metric tensor serves also as the scalar product of one-forms, that is, the bilinear mapping of one-forms $\mathbf{k}, \mathbf{l}$ into scalars:

$$G^{-1}[[\mathbf{k}, \mathbf{l}]] = k_1 l_1 + k_2 l_2 + k_3 l_3. \tag{3.59}$$

Let us consider another scalar product G', for which $\{\mathbf{m}_i\}$ is an orthonormal basis, connected with $\{\mathbf{e}_k\}$ by relation (2.18):

$$\mathbf{m}_i = (A)_i{}^j \mathbf{e}_j.$$

In this case, instead of (3.56) we should assume

$$G' = \mathbf{h}^1 \otimes \mathbf{h}^1 + \mathbf{h}^2 \otimes \mathbf{h}^2 + \mathbf{h}^3 \otimes \mathbf{h}^3,$$

where $\{\mathbf{h}^j\}$ is the basis dual to $\{\mathbf{m}_i\}$. Calculate how it acts on vector $\mathbf{e}_i$:

$$G'[[\mathbf{e}_i]] = \sum_j (\mathbf{h}^j \otimes \mathbf{h}^j)[[\mathbf{e}_i]] = \sum_j \mathbf{h}^j[[\mathbf{e}_i]] \, \mathbf{h}^j.$$

(We cannot apply the summation convention here, because it demands the summation indices to be on different levels.) We express $\mathbf{e}_i$ by $\mathbf{m}_j$:

$$G'[[\mathbf{e}_i]] = \sum_j \mathbf{h}^j[[(A^{-1})_i{}^k \mathbf{m}_k]] \, \mathbf{h}^j = \sum_j (A^{-1})_i{}^k \delta_k^j \mathbf{h}^j = \sum_j (A^{-1})_i{}^j \mathbf{h}^j.$$

We use now (2.23), i.e., $\mathbf{h}^j = ((A^{-1})^T)^j{}_k \mathbf{f}^k$:

$$G'[[\mathbf{e}_i]] = \sum_j (A^{-1})_i{}^j ((A^{-1})^T)^j{}_k \mathbf{f}^k = (A^{-1}(A^{-1})^T)_{ik} \mathbf{f}^k.$$

By virtue of (2.45) we obtain the elements of the scalar product matrix $(A^{-1}(A^{-1})^T)_{ik} = g_{ik}$, hence

$$G'[[\mathbf{e}_i]] = g_{ik} \mathbf{f}^k. \tag{3.60}$$

From this the inverse relation follows:

$$\tilde{G}'[[\mathbf{f}^k]] = \tilde{g}^{kj} \mathbf{e}_j \tag{3.61}$$

where $\tilde{G}' = G'^{-1}$.

It is in order now to devote some space to an operator transforming nontwisted one-forms $\mathbf{E}$ into twisted two-forms $\mathbf{D}$, since this will be needed in considering the constitutive relations of the type $\mathbf{D} = \varepsilon(\mathbf{E})$. We can perform this transformation in two steps: change the one-form $\mathbf{E}$ into vector $\mathbf{E}$ by means of (3.58) and then contract the twisted three-form $\mathbf{h}_*^{123}$ with the result:

$$\mathbf{D} = \mathbf{h}_*^{123} \lfloor \mathbf{E} = \mathbf{h}_*^{123} \lfloor \tilde{G}[[\mathbf{E}]]. \tag{3.62}$$

The same formula can be applied to linear transformation of $\mathbf{H}$ into $\mathbf{B}$ with a possible other choice of $\tilde{G}$. The character of directed quantities is preserved in this too: the first step changes a twisted one-form into a twisted vector, and the second step changes it into a nontwisted two-form.

Let us denote the mapping (3.62) by one symbol M:

$$M[[\mathbf{E}]] = \mathbf{h}_*^{123} \lfloor \tilde{G}[[\mathbf{E}]]. \tag{3.63}$$

This plays the role of the *Hodge operator*, denoted usually by a star, changing p-forms into $(3 - p)$-forms,[3] but is different because it changes a nontwisted form into a twisted one (in the case of $\mathbf{E}, \mathbf{D}$) or a twisted one into a nontwisted one (in the case of $\mathbf{H}, \mathbf{B}$). It contains the twisted basic trivector. Therefore it does not refer to any handedness. Thus, it is *parity invariant*, which means that: it does not depend on the handedness of the vector basis. This property was advocated by Burke [6], stressing that electromagnetism is a parity invariant theory.

Lemma 3.2 *If the matrix $\tilde{G}$ is symmetric, the operator (3.63) fulfills the identity*

$$\mathbf{E} \wedge M[[\mathbf{E}']] = M[[\mathbf{E}]] \wedge \mathbf{E}'.$$

Proof Calculate the left-hand side:

$$L = \mathbf{E} \wedge M[[\mathbf{E}']] = \mathbf{E} \wedge \{\mathbf{h}_*^{123} \lfloor (\mathbf{e}_i \, \tilde{g}^{ij} E'_j)\}.$$

Use the proportionality of all twisted trivectors, so $\mathbf{h}_*^{123} = \beta \mathbf{f}_*^{123}$ for some scalar β, and raise the indices of E:

$$L = \beta \, \mathbf{E} \wedge \{\mathbf{f}_*^{123} \lfloor (E'^i \mathbf{e}_i\} = \beta \, (E_1 \mathbf{f}^1 + E_2 \mathbf{f}^2 + E_3 \mathbf{f}^3) \wedge (E'^1 \mathbf{f}_* 23 + E'^2 \mathbf{f}_*^{31} + E'^3 \mathbf{f}_*^{12})$$

$$= \beta \, (E_1 E'^1 + E_2 E'2 + E_3 E'^3)_*^{123} = \beta \, E_i \tilde{g}^{ij} E'_j \mathbf{f}_*^{123}.$$

---

[3] In n-dimensional vector space the Hodge operator changes p-forms into $(n - p)$-forms.

Calculate the right-hand side:

$$R = M[[\mathbf{E}]] \wedge \mathbf{E}' = \beta \{\mathbf{f}_*^{123} \lfloor (E^i \mathbf{e}_i)\} \wedge \mathbf{E}'$$

$$= \beta (E^1 \mathbf{f}_*^{23} + E^2 \mathbf{f}_*^{31} + E^3 \mathbf{f}_*^{12}) \wedge (E_1' \mathbf{f}^1 + E_2' \mathbf{f}^2 + E_3' \mathbf{f}^3) = \beta (E^1 E_1' + E^2 E_2' + E^3 E_3') \mathbf{f}_*^{123}$$

$$= \beta (\tilde{g}^{ji} E_i E_j') \mathbf{f}_*^{123} = \beta E_i \tilde{g}^{ji} E_j' \mathbf{f}_*^{123}.$$

If the matrix $\tilde{G}$ is symmetric, then $L = R$. ∎

Since the exterior product of two-forms with one-forms is commutative, we may write this result as

$$\mathbf{E} \wedge M[[\mathbf{E}']] = \mathbf{E}' \wedge M[[\mathbf{E}]]. \tag{3.64}$$

When discussing examples of physical quantities in Sect. 1.4, we had a dilemma regarding the character of force. In one situation it could be considered as a nontwisted one-form in the relation of potential energy with displacement: $dU = -\mathbf{F}[[d\mathbf{r}]]$, while in another it is seen as a nontwisted vector in the relation of force and acceleration: $\mathbf{F} = m\mathbf{a}$. A possible way out of the dilemma is an assumption that *mass* is not a scalar, but a linear operator $\tilde{m}$ turning nontwisted vectors into nontwisted one-forms: $\mathbf{F} = \tilde{m}[[\mathbf{a}]]$. Mass is also present as a quadratic form in the expression for the kinetic energy of a particle: $T = \frac{1}{2}\tilde{m}[[\mathbf{v}, \mathbf{v}]]$. There are in physics examples when the effective inertia of particles depends on interactions with an environment. For instance, electrons in a solid body behave sometimes as though their inertia depends on direction. Therefore, in the theory of crystals it is assumed that electrons possess so-called *effective mass*, which is a second rank tensor.

What kind of directed quantity is momentum? Its role as a one-form mapping vectors into scalars is visible in the formula expressing increase of kinetic energy $dT = \mathbf{p}[[d\mathbf{v}]]$ by the increase of velocity $d\mathbf{v}$. The mass also occurs in the definition of momentum, $\mathbf{p} = m\mathbf{v}$, where the nontwisted vector $\mathbf{v}$ is turned into the one-form $\mathbf{p}$. We should write it rather as $\mathbf{p} = \tilde{m}[[\mathbf{v}]]$: then momentum as a nontwisted one-form fits with the nontwisted one-form of force in the relation $\mathbf{F} = \frac{d\mathbf{p}}{dt}$. Summarizing, force and momentum should be accepted as nontwisted one-forms and the two formulas

$$\mathbf{F} = \tilde{m}[[\mathbf{a}]], \quad \mathbf{p} = \tilde{m}[[\mathbf{v}]] \tag{3.65}$$

should be treated as linear mappings of vectors into one-forms, where mass is an operator.

If we wish to write the mass operator as a tensor, we can do it as follows:

$$\tilde{m} = \mathbf{h}^1 \otimes \mathbf{f}^1 + \mathbf{h}^2 \otimes \mathbf{f}^2 + \mathbf{h}^3 \otimes \mathbf{f}^3, \tag{3.66}$$

where $\mathbf{f}^i$ are three linearly independent nontwisted one-forms and $\mathbf{h}^j$ are three other ones. This formula is similar to (3.56), so it is not excluded that in empty space the mass tensor is proportional to the scalar product tensor. Then we would write $\tilde{m} = mG$ and the coefficient m could here be an ordinary scalar.

3.4 Unit System

E. J. Post [44], p.3 wrote: "(...) dimensional argument suggests the use of action and charge as the two invariant physical units, whereas length and time are solely associated with the transformation behavior of the fields." We are going to present arguments that in the three-dimensional approach, action, charge and time are scalar quantities, while the appropriate power of length is connected with the geometric character of a physical quantity. In the four-dimensional approach (see Chap. 7), time is not a scalar quantity and belongs to the other category. Nice accounts of this subject are contained in an articles by F. W. Hehl and Y. N. Obukhov [21, 25].

In Table 3.2 we have collected the physical directed quantities that do not need any scalar product for their definition.

Notice that the magnetic moment of an electric circuit as a directed quantity is different from the magnetic dipole moment that could be introduced if magnetic charges were to exist.[4] In the traditional language both would be pseudoscalars.

Other physical quantities exists that are not of the named types—they are linear operators, that is linear mappings between the directed quantities. For instance, the permittivity ε is a linear operator between one-forms $\mathbf{E}$ and two-forms $\mathbf{D}$: $\mathbf{D} = \varepsilon[[\mathbf{E}]]$. The same is true of permeability μ: $\mathbf{B} = \mu[[\mathbf{H}]]$. No specific direction can be ascribed to such quantities. A similar situation occurs with symmetric tensors, such as scalar product or mass.[5]

We divide the units into two classes: *scalar units*—not containing the metre in any power—and *extensional units*—containing various powers of metre. The latter can be divided further: if the metre has power one the unit can be called *vectorial*, if it has power two, *bivectorial* and so on. Then the directed type of a physical quantity will be recognized by its unit. For instance, the Cm (coulomb-metre) is the product of a scalar and a vectorial unit, so it corresponds to a vector; indeed, it is the unit of electric dipole moment known as a vectorial quantity. The magnitude of a one-form is linear density, so in its unit the metre should have power minus one, and the other

---

[4] See remark under Fig. 1.65.

[5] Twenty types of directed quantities exist in four-dimensional space because there are quadrivectors and four-forms nontwisted and twisted. Moreover, time, energy, and currents cease to be scalars or zero-forms.

Table 3.2 Premetric directed quantities

	Nontwisted quantities	Twisted quantities
Scalars	Time, action, electric charge	Axion, magnetic charge*
Vectors	Position, velocity, acceleration, electric dipole moment	Directed length of solenoid, magnetic dipole moment*
Bivectors	Directed area of electr. circuit. magnetic moment of electr. circuit	Section area of a flow, electr. moment of magnetic circuit*
Trivectors	Spirality of a trajectory	Volume
Zero-forms	Energy, electric potential, electric current	Magnetic current*
One-forms	Wave covector, momentum, force, electric field, magnetic potential	Magnetic field strength, magnetization
Two-forms	Magnetic induction, magn. current density*	Electr. induction, electr. polarization, electr. current density, Poynting form
Three-forms	Magnetic charge density*	Electric charge density, energy density

* Quantities occurring in the presence of magnetic charges

units should only be scalars. Here are examples from the SI system:

$$[k] = \frac{1}{m}, \quad E = \frac{V}{m}, \quad [H] = \frac{A}{m}.$$

Similarly, two-forms are area density and this fits with the units of appropriate physical quantities:

$$[j] = \frac{A}{m^2}, \quad [D] = \frac{C}{m^2}, \quad [B] = \frac{Vs}{m^2} = \frac{Wb}{m^2}. \tag{3.67}$$

We have used Vs = Wb, where the weber is the unit of magnetic flux. The electric charge density is a three-form, as reflected in the units:

$$[\rho] = \frac{C}{m^3}.$$

In the case of momentum and force the units

$$[p] = \frac{kg\,m}{s}, \quad [F] = \frac{kg\,m}{s^2}$$

suggest that they are vectors, but under the condition that mass is a scalar. The unit expressing energy:

$$[\mathcal{E}] = J = \frac{kg\,m^2}{s^2} \tag{3.68}$$

does not suggest that energy is a scalar; if mass is a scalar then energy should be rather a bivector.

If one doubts that mass is a scalar, one can cease treating mass as a fundamental unit and choose for this role instead the unit of action, since action is a scalar not only spatially but also spatio-temporally. Already Post in his book [44] suggested the use of action as one of the two invariant physical units. "Invariant" means for us "scalar" unit. A natural unit of action is the Planck constant h, but in quantum physics another constant is present, namely 2π times smaller: $\hbar = h/2\pi$. We do not decide now which of them is better, but let us follow the authors of [25] to use the notation Pl and the term *Planck* for a fundamental unit of action Js available in the SI system of units.

If we want to have as many scalar fundamental units as possible, we omit the ampere because the magnitude of electric current is not a spatio-temporal scalar. Therefore we take the coulomb as a fundamental unit. In this way, in place of the four electromechanical fundamental units of SI:

$$m, \quad s, \quad kg, \quad A$$

another four fundamental units are more appropriate:

$$m, \quad s, \quad Pl, \quad C \tag{3.69}$$

In the three-dimensional formulation, the first unit in (3.69) is extensional and the three others are scalars. In the four-dimensional formulation in space-time, the first two units are extensional because they correspond to spatio-temporal features of physical quantities, and the last two are scalar units.

Now we can identify by units, what kinds of directed quantities are energy, momentum and force. In the proposed unit system, energy is expressed in the units:

$$[\mathcal{E}] = \frac{Pl}{s},$$

from which is visible that it is a spatial scalar. The units of momentum and force are:

$$[p] = \frac{Pl}{m}, \qquad [F] = \frac{Pl}{ms},$$

which confirms that they are one-forms. In this unit system, the de Broglie relations are easy to recognize:

$$\mathcal{E} = \hbar\omega, \qquad \mathbf{p} = \hbar\mathbf{k}.$$

According to Dirac [12, 13], *magnetic charge g* must be quantized and, together with the quantum of electric charge e, satisfies the relation: $eg = \frac{1}{2}n\hbar$, where n is a natural number. It is visible from here that the magnetic charge has the unit:

$$[g] = \frac{\text{Pl}}{\text{C}}.$$

Hence, it is also a spatio-temporal scalar (ordinary or twisted). If we write this traditionally: $[g] = \frac{\text{Js}}{\text{C}} = \text{Vs} = \text{Wb}$, we may state that the *weber* is a scalar unit of magnetic charge. One should express it as $\text{Wb} = \text{Pl}/\text{C}$. Since magnetic charges have not been discovered, we are allowed to claim only that the weber is the scalar unit of magnetic flux.

The coulomb C is a scalar unit, hence the ampere $A = \text{Cs}^{-1}$ is of the same character. Now the electromagnetic field quantities, in the proposed unit system, have the following units:

$$[E] = \frac{\text{Pl}\,\text{C}^{-1}\text{s}^{-1}}{\text{m}}, \qquad [H] = \frac{\text{Cs}^{-1}}{\text{m}}.$$

$$[D] = \frac{\text{C}}{\text{m}^2}, \qquad [B] = \frac{\text{Pl}\,\text{C}^{-1}}{\text{m}^2}.$$

We consider also the character of an axion. According to left-hand equation (1.9):

$$[\alpha] = \left[\frac{D}{B}\right] = \frac{\text{C/m}^2}{\text{Pl}/\text{Cm}^2} = \frac{\text{C}^2}{\text{Pl}},$$

and the right-hand equation (1.9) implies

$$[\alpha] = \left[\frac{H}{E}\right] = \frac{\text{C/sm}}{\text{Pl}/\text{Csm}} = \frac{\text{C}^2}{\text{Pl}}.$$

This proves the consistency of the two relations (1.9).

It is worth to notice that the unit of electric resistance, ohm, should be expressed as:

$$\Omega = \left[\frac{U}{I}\right] = \frac{\text{V}}{\text{A}} == \frac{\text{Pl}/\text{Cs}}{\text{C/s}} = \frac{\text{Pl}}{\text{C}^2}.$$

We see that it is a scalar unit. Its inverse, the unit of electric conductance

$$[\sigma] = \frac{\text{C}^2}{\text{Pl}}$$

is the same as the unit of axion. Is some meaning hidden behind this coincidence?
The importance of this physical quantity stems from the discovery of Klaus von
Klitzing with coworkers that the co called Hall conductivity is quantized according
to the formula:

$$\sigma_{xy} = (e^2/h)\,n,$$

where e is the elementary charge, h—the Planck constant and n—a natural number,
see [33].

We now repeat Table 3.2, adding an appropriate unit for each quantity. The reader
could ask: why are some quantities known traditionally as scalars or pseudoscalars
put in the table as (nontwisted od twisted) scalars, while others are given as
(nontwisted or twisted) zero-forms? The answer is: the latter become the ingredients
of one-forms in four-dimensional space-time (see Sect. 7.2).

Mass as the ratio of momentum to velocity has the unit:

$$[m] = \frac{\text{Pl}\,\text{m}^{-1}}{\text{ms}^{-1}} = \frac{\text{Pl}\,\text{s}}{\text{m}^2}.$$

(The symbol m on the left-hand side, printed in italic, denotes mass and m denotes
metre.) Mass, as the bilinear form (3.66) ought to have the extensional unit, metre,
with power minus two, the same as the scalar product tensor, which has $[G] = \text{m}^{-2}$
as the unit.[6]

Another operator quantity, namely the permittivity, ε is the ratio of electric induc-
tion to electric field, so it has the unit $\text{Cm}^{-2}/\text{Pl}\,\text{C}^{-1}\,\text{s}^{-1}\,\text{m}^{-1} = \text{Pl}^{-1}\text{C}^2\,\text{sm}^{-1}$,
whereas the permeability μ is expressed in $\text{Pl}\,\text{C}^{-1}\,\text{m}^2/\text{Cs}^{-1}\,\text{m}^{-1} = \text{Pl}\,\text{C}^{-2}\,\text{sm}^{-1}$.
Their product has the unit

$$[\varepsilon\mu] = \frac{\text{s}^2}{\text{m}^2}$$

which fits the formula $\varepsilon\mu = c^{-2}$ relating them to the speed of light.

Since the system (3.69) of electromechanical fundamental units reflects well the
geometric character of directed physical quantities, it highlights the close relation
of a physical unit with the geometric character of a physical quantity. The presented
unit system deserves a name. Most proper one is *unit system compatible with
geometry*. I propose also the shorter name: *geometric unit system*.

We should, however, mention quantities that have units not fitting to their directed
type. The first is planar angle, which in a natural way can be seen as a bivector
because one can point its direction as a plane together with a curved arrow with

---

[6] Is it possible that the scalar product is determined by the mass? Then the mass may be treated
as a product of some scalar μ and the scalar product tensor G: $m = \mu G$. In this case the scalar
product tensor absorbs the extensional unit m^{-2}, and the scalar μ plays the role of the scalar mass
with unit $[\mu] = \text{Ps}$.

Table 3.3 Directed quantities with their units

	Nontwisted quantities	Twisted quantities
Scalars	Time s, action Pl, electric charge C	Axion $C^2 Pl^{-1}$, magnetic charge $Pl\,C^{-1}$ *
Vectors	Position m, velocity ms^{-1}, acceleration ms^{-2}, electric dipole moment mC	Directed length of solenoid m, magnetic dipole moment $mPl\,C^{-1}$ *
Bivectors	Directed area of electr. circuit m^2 magn. moment of electr. circuit $m^2 Cs^{-1}$	Section area of a flow m^2, electric moment of magn. circuit $m^2 Pl\,C^{-1} s^{-1}$ *
Trivectors	Spirality of a trajectory $m^3 s^{-6}$	Volume m^3
Zero-forms	Energy $Pl\,s^{-1}$, electr. potential $Pl\,C^{-1}s^{-1}$, electr. current Cs^{-1}, circular frequency s^{-1}	Magnetic current $Pl\,C^{-1}s^{-1}$ *
One-forms	Wave covector $\frac{1}{m}$, momentum $\frac{Pl}{m}$, force $\frac{Pl\,s^{-1}}{m}$, electric field $\frac{Pl\,C^{-1}s^{-1}}{m}$, magnetic potential $\frac{Pl\,C^{-1}}{m}$	Magnetic field $\frac{Cs^{-1}}{m}$, magnetization $\frac{Cs^{-1}}{m}$
Two-forms	Magnetic induction $\frac{Pl\,C^{-1}}{m^2}$, magnetic current density $\frac{Pl\,C^{-1}s^{-1}}{m^2}$ *	Electric induction $\frac{C}{m^2}$, electric polarization $\frac{C}{m^2}$, electr. current density, $\frac{Cs^{-1}}{m^2}$, Poynting form $\frac{Cs^{-2}}{m^2}$
Three-forms	Magnetic charge density $\frac{Pl\,C^{-1}}{m^3}$ *	Electric charge density $\frac{C}{m^3}$, energy density $\frac{Pl\,s^{-1}}{m^3}$

* Quantities occurring in the presence of magnetic charges

the orientation of rotation. Unfortunately, its magnitude is not measured in square metres but in radians treated as dimensionless units. This occurs in SI as well as in the geometric unit system.

Another such quantity is angular momentum, which also should be a bivector. Its unit in the SI system is $kg\,m^2\,s^{-1}$, which could be good if the kilogram were a scalar, but in the considered system its unit is Planck, which suggests rather a scalar. Probably the disparate character of the two named quantities stems from that in their definition a scalar product is necessary, which is not the case for the quantities named in Table 3.3. Therefore it seems justified to call the directed properties of those quantities *premetric properties*.

3.5 Differential Forms

We proceed to consider form fields by analogy with vector fields, that is, to ascribe an l-form to each point of the space. Such objects are known in the mathematical and physical literature by the name of *differential forms*. The symbol $\mathbf{E}(\mathbf{r})$ will denote a one-form $\mathbf{E}$ at the point with radius vector $\mathbf{r}$. The usual bracket helps to distinguish this from the value of this form on a vector. The symbol $\mathbf{E}(\mathbf{r})[[\mathbf{v}]]$ will denote the value of $\mathbf{E}$ at point $\mathbf{r}$ if contracted with $\mathbf{v}$. Each differential form can be written in terms of basic forms; for instance, a nontwisted one-form:

$$\mathbf{E}(\mathbf{r}) = E_i(\mathbf{r})\mathbf{f}^i, \tag{3.70}$$

a twisted one-form:

$$\mathbf{H}(\mathbf{r}) = H_i(\mathbf{r})\mathbf{f}^i_*, \tag{3.71}$$

and similarly for forms of other grades. A differential l-form is called *continuous* if its coordinates are continuous as the scalar functions. From now on we consider only such forms.

When a one-form depends on position, it is not appropriate to take its value on a radius vector of arbitrary length. A form field is better suited for infinitesimal vectors; therefore, for visualization of such an object, it is not proper to draw parallel planes separated by value one, but rather by a smaller number, so small that in a spatial separation corresponding to this number, the neighbouring forms are almost parallel. In this manner, instead of $\mathbf{E}[[\Delta\mathbf{l}]]$ we should consider rather $\mathbf{E}[[d\mathbf{l}]]$. We see that the differential one-forms are well suited for integration over directed curves; in fact, they were invented for this purpose. Similarly, differential two-forms are introduced for integration over directed surfaces and three-forms over directed volumes.

In Sect. 1.3 we became acquainted with simplified geometric images of forms; they are shown in Figs. 1.40 and 1.43. We repeat here a cone as the simplified image of a nontwisted one-form in Fig. 3.11, just as a reminder. We may now draw images of a differential one-form putting such cones at various points in space. I shall illustrate the differential one-form $\mathbf{E}(\mathbf{r}) = x^2\mathbf{f}^1$ for the radius vector $\mathbf{r} = x^i\mathbf{e}_i$ in some nontwisted vector basis $\{\mathbf{e}_i\}$; this form increases in the direction of $\mathbf{e}_2$. Since it is difficult to make a three-dimensional drawing, I show this on Fig. 3.12 as a section in the $(\mathbf{e}_1, \mathbf{e}_2)$-plane. In this case the planes of $\mathbf{E}$ reduce to straight lines, and the cones to triangles with one side highlighted, treated as the base of the cone. Also two basis vectors and two basis one-forms are shown on the figure. If you join the vertices of the cones (of the triangles in this drawing) with the bases, being a continuation of one another on the line AB, you obtain a curve CD not coinciding with the line DE linking bases passing through D, the vertex of one of the cones. In such a situation we say that the differential form is *nonfitted*.

Fig. 3.11 Simplified image
of a nontwisted one-form

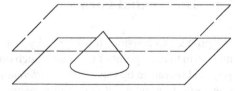

Fig. 3.12 The nonfitted
differential form

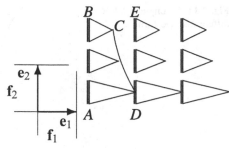

Fig. 3.13 The fitted
differential form

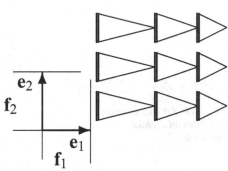

If—conversely—the lines obtained by joining the vertices of cones from one family (with bases being a continuation of each other) coincide with the attitudes of forms defined in the vertices, the differential form will be called *fitted*. An example of such a one-form is given by formula $\mathbf{E}(\mathbf{r}) = x^1 \mathbf{f}^1$ and shown in Fig. 3.13. It increases in the direction of $\mathbf{e}_1$.

In this manner we see that the fitted differential one-form can be depicted as a family of non intersecting surfaces with the property that the attitude of the form at each point is tangent to the surface passing through this point. If the surfaces intersect somewhere, the field there is singular. The inverse of the distance between the surfaces is a measure of the magnitude of the form. Arrows are also needed between the surfaces to define the orientation. We show in Figs. 3.14 and 3.15 two examples of differential nontwisted one-forms. Figure 3.15 can illustrate the electric field intensity for a charge distribution with spherical symmetry. The ellipses visible on it are artificial openings made to show the next surfaces inside. For the differential twisted one-forms the picture is similar; the only difference is in marking the orientations—these should be curved arrows on the surfaces.

Fig. 3.14 Example of a fitted
differential one-form

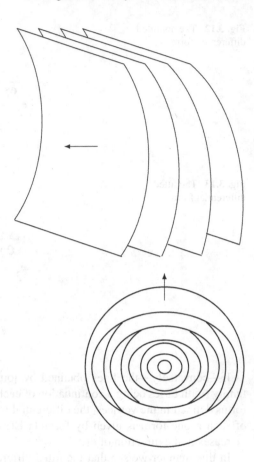

Fig. 3.15 Another example
of a fitted differential
one-form

A differential two-form can also be fitted or non-fitted. The former can be
visualized by two families of surfaces, where the surfaces of one family do not
intersect each other but intersect surfaces of the other family (see Fig. 3.16). The
differential two-form has its attitudes tangent to the lines of intersection of the
surfaces.

If the differential two-form (two-form field) $\mathbf{B}(\mathbf{r})$ is fitted, it is visualized by tubes
of variable cross-section, to which the field $\mathbf{B}$ is tangent. The natural convention for
tubes is that a single cross-section Δ has an area such that $\int_\Delta \mathbf{B}(\mathbf{r}) \, [[d\mathbf{s}]] = 1$ unit
of magnetic flux when the orientations of $\mathbf{B}$ and Δ are concordant. Then the integral
of this field over a surface S, i.e., $\int_S \mathbf{B}(\mathbf{r}) \, [[d\mathbf{s}]]$, is equal (with an appropriate sign)
to the number of tubes intersected by S.

The differential two-form $\mathbf{D}$ is also depicted by the family of tubes. Such a picture
of electric induction is in accordance with Faraday's notion of "tubes of inductive
force".

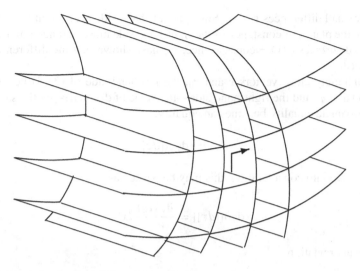

Fig. 3.16 Example of a fitted differential two-form

3.6 Exterior Derivative

The differential l-form is called *differentiable* if its coordinates are differentiable as scalar functions, that is, all partial derivatives of their coordinates in an arbitrary basis exist.

Consider a differentiable function $\varphi : \mathbf{R}^3 \to \mathbf{R}$, which maps radius vectors into scalars (or, alternatively into zero-forms) $\varphi : \mathbf{r} \to \varphi(\mathbf{r})$. Its derivative is a directed quantity, but of what nature? To answer this question, we expand it in a Taylor series up to the first-order terms:

$$\varphi(\mathbf{r}) = \varphi(\mathbf{r}_0) + \frac{\partial \varphi(\mathbf{r}_0)}{\partial x^i}(x^i - x_0^i) + \dots . \tag{3.72}$$

The first derivative term is a linear functional of the difference of the radius vectors $\mathbf{r} - \mathbf{r}_0$. We now see that the object traditionally called the *gradient*, $\nabla \varphi = (\frac{\partial \varphi}{\partial x^1}, \frac{\partial \varphi}{\partial x^2}, \frac{\partial \varphi}{\partial x^3})$ is, in fact *a one-form*, which we shall denote subsequently by $\mathbf{d}\varphi$, with boldface $\mathbf{d}$. This directed differentiation in three-dimensional space becomes a one-form operator $\mathbf{d}$ changing a scalar field φ into the differential one-form $\mathbf{d}\varphi$. We call it a *one-form derivative*. If φ is treated as a zero-form field or differential zero-form, we are allowed to claim that the exterior product is between $\mathbf{d}$ and φ, namely $\mathbf{d}\varphi = \mathbf{d} \wedge \varphi$. Therefore $\mathbf{d}\varphi$ is also called an *exterior derivative* of φ. We should stress that the exterior derivative makes sense only for differentiable functions φ.

A geometric image of the differential form $\mathbf{d}\varphi$ is the family of surfaces described by the equation $\varphi(\mathbf{r}) =$const. Justification for this claim rests on the equality

$$\varphi(\mathbf{r}) - \varphi(\mathbf{r}_0) \cong (\mathbf{d}\varphi)[[\mathbf{r} - \mathbf{r}_0]], \tag{3.73}$$

valid for small differences $\mathbf{r} - \mathbf{r}_0$. Since the left-hand side is constant where $\varphi(\mathbf{r})$ is constant, the planes of constancy of the right-hand side must be tangent to the above mentioned surfaces $\varphi(\mathbf{r}) =$ const. Hence, we have shown that the differential form $\mathbf{d}\varphi$ is fitted.

For infinitely close vectors $\mathbf{r}$ and $\mathbf{r}_0$, the left-hand side of (3.73) becomes the differential of φ, and the right-hand side—the value of $\mathbf{d}\varphi$ on $d\mathbf{r}$—at the same time, the approximate equality becomes an equality:

$$d\varphi = (\mathbf{d}\varphi)[[d\mathbf{r}]]. \tag{3.74}$$

Taking (3.72) into consideration, this may be written as:

$$(\mathbf{d}\varphi)[[d\mathbf{r}]] = \frac{\partial \varphi(\mathbf{r}_0)}{\partial x^i} dx^i. \tag{3.75}$$

With the abbreviation

$$\frac{\partial \varphi}{\partial x^i} = \partial_i \varphi$$

we write it more simply:

$$(\mathbf{d}\varphi)[[d\mathbf{r}]] = \partial_i \varphi(\mathbf{r}_0)\, dx^i. \tag{3.76}$$

From (2.7), i.e., from $\mathbf{f}^i[[\mathbf{r}]] = x^i$, we have $\mathbf{f}^i[[\mathbf{r} - \mathbf{r}_0]] = x^i - x_0^i$, or on infinitely small increments (differentials), $\mathbf{f}^i[[d\mathbf{r}]] = dx^i$. Insert this in (3.76):

$$(\mathbf{d}\varphi)[[d\mathbf{r}]] = \partial_i \varphi(\mathbf{r}_0)\, \mathbf{f}^i[[d\mathbf{r}]].$$

Since the differential $d\mathbf{r}$ is arbitrary, we may omit it on both sides:

$$\mathbf{d}\varphi = \partial_1 \varphi\, \mathbf{f}^1 + \partial_2 \varphi\, \mathbf{f}^2 + \partial_3 \varphi\, \mathbf{f}^3 = \partial_i \varphi\, \mathbf{f}^i. \tag{3.77}$$

Thus, we have succeeded in expressing the exterior derivative of φ by basis one-forms. If, in the presence of a scalar product, we change the one-forms into vectors according to (2.71), we obtain the gradient of φ.

In particular, the constant differential one-form $\mathbf{f}^i$, is an exterior derivative of some function, namely of the *coordinate* x^i; hence we obtain

$$\mathbf{f}^i = \mathbf{d}x^i. \tag{3.78}$$

We can now rewrite (3.77) as

$$\mathbf{d}\varphi = \partial_i \varphi\, \mathbf{d}x^i.$$

This is the typical formula in mathematical textbooks (about differential forms) for the exterior derivative of scalar function. If it is not typed in boldface, it can be confused with the ordinary differential of a function. What makes understanding more difficult is that it is also called a differential. That is why I prefer to call (3.77) the *exterior derivative*. Only its value on a differential of **r** gives the differential of the function

$$d\varphi = \mathbf{d}\varphi[[d\mathbf{r}]] = \partial_i \varphi f^i [[d\mathbf{r}]] = \partial_i \varphi \, dx^i.$$

Equation (2.7) can be now displayed as

$$\mathbf{d}x^i[[d\mathbf{r}]] = dx^i.$$

Exercise 3.27 Let **k** be a nontwisted one-form, g a scalar function of one scalar variable x, $\varphi_1(\mathbf{r}) = \mathbf{k}[[\mathbf{r}]] + a$, $\varphi_2(\mathbf{r}) = g(\mathbf{k}[[\mathbf{r}]])$. Show that $\mathbf{d}\varphi_1 = \mathbf{k}$, $\mathbf{d}\varphi_2 = \frac{dg}{dx}\mathbf{k}$.

When acting on a twisted scalar field $\psi = \lambda\varphi$ (here φ is an ordinary scalar field, λ is the unit twisted scalar), the one-form derivative yields a differential twisted one-form $\mathbf{d}\psi$ according to the following prescription: $\mathbf{d}(\lambda\varphi) = \lambda\mathbf{d}\varphi$.

For a differentiable differential (nontwisted or twisted) one-form **E**, we define its *exterior derivative* as

$$\mathbf{d} \wedge \mathbf{E} = \mathbf{d}E_i \wedge \mathbf{f}^i. \tag{3.79}$$

The one-form operator $\mathbf{d}\wedge$ now changes a differential nontwisted one-form into a differential nontwisted two-form or a differential twisted one-form into a differential twisted two-form. We call it *one-form differentiation*. After decomposing (3.79) onto basis two-forms we obtain

$$\mathbf{d} \wedge \mathbf{E} = \frac{\partial E_i}{\partial x^j}\mathbf{f}^{ji} = (\partial_1 E_2 - \partial_2 E_1)\,\mathbf{f}^{12} + (\partial_2 E_3 - \partial_3 E_2),\mathbf{f}^{23} + (\partial_3 E_1 - \partial_1 E_3)\,\mathbf{f}^{31}. \tag{3.80}$$

The appearance of the coordinates allows to notice that if, in the presence of a metric, two-forms are replaced by twisted vectors according to (2.73), the curl of the vector field **E** is obtained.

Exercise 3.28 Let g and **k** be as in Exercise 3.27, let $\mathbf{E}_0$ be a constant one-form. Introduce the differential one-form $\mathbf{E}(\mathbf{r}) = g(\mathbf{k}[[\mathbf{r}]])\mathbf{E}_0$. Show that $\mathbf{d} \wedge \mathbf{E} = \frac{dg}{dx}\mathbf{k} \wedge \mathbf{E}_0$.

In the particular case of $\mathbf{E} = \mathbf{d}\varphi$, where φ is a doubly differentiable function, we have according to (3.77): $E_i = \partial_i\psi$ and again due to (3.77) we obtain

$$\mathbf{d}E_i = \partial_{1i}^2\varphi\,\mathbf{f}^1 + \partial_{2i}^2\varphi\,\mathbf{f}^2 + \partial_{3i}^2\varphi\,\mathbf{f}^3 = \partial_{ji}^2\,\mathbf{f}^j,$$

where we have used the notation $\partial_j \partial_i \varphi = \partial_{ji}^2 \varphi$. This, substituted into (3.79), yields

$$\mathbf{d} \wedge \mathbf{E} = \partial_{ji}^2 \varphi \, \mathbf{f}^j \wedge \mathbf{f}^i.$$

In this double sum, the expression $\partial_{ji}^2 \varphi$ is symmetric under the interchange of indices, and $\mathbf{f}^j \wedge \mathbf{f}^i$ is antisymmetric; therefore the sum is zero. In this manner we have shown that

$$\mathbf{d} \wedge \mathbf{d}\varphi = 0. \qquad (3.81)$$

For a differentiable differential (nontwisted or twisted) two-form

$$\mathbf{B}(\mathbf{r}) = B_{12}(\mathbf{r})\mathbf{f}^{12} + B_{23}(\mathbf{r})\mathbf{f}^{23} + B_{31}(\mathbf{r})\mathbf{f}^{31} = \frac{1}{2} B_{ij}(\mathbf{r})\mathbf{f}^{ij}$$

we define its *exterior derivative* as

$$\mathbf{d} \wedge \mathbf{B} = \frac{1}{2}\mathbf{d}B_{ij} \wedge \mathbf{f}^{ij}. \qquad (3.82)$$

The one-form operator $\mathbf{d}\wedge$ now changes a nontwisted two-form into a nontwisted three-form or a twisted two-form into a twisted three-form.

Calculate the only coordinate of (3.82):

$$\mathbf{d} \wedge \mathbf{B} = \partial_i B_{12} \, \mathbf{f}^i \wedge \mathbf{f}^{12} + \partial_i B_{23} \mathbf{f}^i \wedge \mathbf{f}^{23} + \partial_i B_{31} \, \mathbf{f}^i \wedge \mathbf{f}^{31}.$$

The summation over i leaves only one non-zero term:

$$\mathbf{d} \wedge \mathbf{B} = \partial_3 B_{12} \, \mathbf{f}^{312} + \partial_1 B_{23} \, \mathbf{f}^{123} + \partial_2 B_{31} \, \mathbf{f}^{231} = (\partial_3 B_{12} + \partial_1 B_{23} + \partial_2 B_{31}) \, \mathbf{f}^{123}. \qquad (3.83)$$

The expression in the bracket is the coordinate sought. When the two-form $\mathbf{B}$ is replaced by twisted vector $\mathbf{B}$ with coordinates $B_1 = B_{23}$, $B_2 = B_{31}$, $B_3 = B_{12}$, we recognize in this the divergence of $\mathbf{B}$.

Exercise 3.29 Let g and $\mathbf{k}$ be as in Exercise 3.27, and let $\mathbf{B}_0$ be a constant two-form. Consider the differential two-form $\mathbf{B}(\mathbf{r}) = g(\mathbf{k}[[\mathbf{r}]])\mathbf{B}_0$. Show that $\mathbf{d} \wedge \mathbf{B} = \frac{dg}{dx}\mathbf{k} \wedge \mathbf{B}_0$.

Take $\mathbf{B} = \mathbf{d} \wedge \mathbf{E}$ where $\mathbf{E}$ is a twice differentiable differential one-form. Then according to (3.80) $B_{ij} = \partial_i E_j - \partial_j E_i$ and due to (3.77)

$$\mathbf{d}B_{ij} = \partial_k \left(\partial_i E_j - \partial_j E_i \right) \mathbf{f}^k.$$

This substituted in (3.82) yields

$$\mathbf{d} \wedge \mathbf{B} = \frac{1}{2} \left(\partial_{ki}^2 E_j - \partial_{kj}^2 E_i \right) \mathbf{f}^k \wedge \mathbf{f}^{ij} = \frac{1}{2} \left(\partial_{ki}^2 E_j \, \mathbf{f}^{kij} - \partial_{kj}^2 E_i \, \mathbf{f}^{kij} \right).$$

The first term in the bracket contains the pair of indices ki at the derivatives, where they are symmetric, and at the exterior three-form, where they are antisymmetric; hence the sum is zero. The same situation occurs with the second term. In this manner we have shown that

$$\mathbf{d} \wedge \mathbf{d} \wedge \mathbf{E} = 0. \tag{3.84}$$

An exterior derivative of an arbitrary differential three-form $\mathbf{T(r)}$ in the three-dimensional space is zero: $\mathbf{d} \wedge \mathbf{T} = 0$ by virtue of the algebraic property (lack of four-forms in such a space). In particular we may write

$$\mathbf{d} \wedge \mathbf{d} \wedge \mathbf{B} = 0 \tag{3.85}$$

for an arbitrary differential two-form $\mathbf{B}$. Joining Eqs. (3.81), (3.84) and (3.85), we are allowed to claim that the double action of the one-form derivative gives zero. In this context it is the famous theorem known as the *Poincaré lemma* [41]:

Theorem 3.1 *If a differential l-form* $\mathbf{B}$ *is differentiable in an open star-shaped region and* $\mathbf{d} \wedge \mathbf{B} = 0$, *then a differential* $(l-1)$-*form* A *exists, such that* $\mathbf{B} = \mathbf{d} \wedge \mathbf{A}$.

The theorem is known for nontwisted forms, but it can be applied also to twisted forms due to the property $\mathbf{d} \wedge (\lambda \mathbf{A}) = \lambda \mathbf{d} \wedge \mathbf{A}$ for multiplication by the unit twisted scalar λ and for an appropriate nontwisted k-form $\mathbf{A}$.

Lemma 3.3 *Let* E *be a nontwisted differential one-form and* H *an twisted differential one-form. If both* E *and* H *are differentiable, then*

$$\mathbf{d} \wedge (\mathbf{E} \wedge \mathbf{H}) = (\mathbf{d} \wedge \mathbf{E}) \wedge \mathbf{H} - \mathbf{E} \wedge (\mathbf{d} \wedge \mathbf{H}).$$

Proof Expand $\mathbf{E}$ and $\mathbf{H}$ into basis forms

$$\mathbf{E} \wedge \mathbf{H} = E_i \mathbf{f}^i \wedge H_j \mathbf{f}_*^j = E_i H_j \mathbf{f}_*^{ij},$$

and calculate

$$\mathbf{d} \wedge (\mathbf{E} \wedge \mathbf{H}) = \mathbf{f}^k \partial_k \wedge (E_i H_j) \mathbf{f}_*^{ij} = \mathbf{f}^k \left[(\partial_k E_i) H_j + E_i (\partial_k H_j) \right] \wedge \mathbf{f}_*^{ij}$$

$$= \mathbf{f}^k \partial_k E_i \wedge \mathbf{f}^i \wedge H_j \mathbf{f}_*^j + \mathbf{f}^k \wedge E_i \mathbf{f}^i \wedge \partial_k H_j \, \mathbf{f}_*^j$$

$$= \mathbf{f}^k \partial_k E_i \wedge \mathbf{f}^i \wedge H_j \mathbf{f}^j_* - E_i \mathbf{f}^i \wedge \mathbf{f}^k \partial_k H_j \wedge \mathbf{f}^j_*$$

$$= (\mathbf{d} \wedge \mathbf{E}) \wedge \mathbf{H} - \mathbf{E} \wedge (\mathbf{d} \wedge \mathbf{H}).$$

∎

We now proceed to consider specific examples of differential forms. If we have a basis $\{\mathbf{m}_1, \ \mathbf{m}_2, \ \mathbf{m}_3\}$ and determined by it coordinates ξ^i occurring in the radius-vector:

$$\mathbf{r} = \xi^i \mathbf{m}_i,$$

we introduce *quasi-cylindrical* coordinates through the well known formulas:

$$\xi^1 = \rho \cos\alpha, \qquad \xi^2 = \rho \sin\alpha, \qquad \xi^3 = z;$$

$$\rho = \sqrt{(\xi^1)^2 + (\xi^2)^2}, \qquad z = \xi^3, \qquad \alpha = \arctan\frac{\xi^2}{\xi^1}.$$

The last function is singular on the Z-axis, i.e., for $\xi^1 = \xi^2 = 0$; moreover, it is non-unique, because somewhere the value 2π must touch 0. Its exterior derivative, however, is unique, though still singular on the Z-axis.

Let $\{\mathbf{h}^1, \ \mathbf{h}^2, \ \mathbf{h}^3\}$ be dual forms, i.e., $\mathbf{h}^i[[\mathbf{m}_j]] = \delta^i_j$. Calculate the exterior derivative of α:

$$\mathbf{d}\alpha = \frac{\partial\alpha}{\partial\xi^1}\mathbf{h}^1 + \frac{\partial\alpha}{\partial\xi^2}\mathbf{h}^2 = -\frac{\xi^2}{(\xi^1)^2}\frac{1}{1+\left(\frac{\xi^2}{\xi^1}\right)^2}\mathbf{h}^1 + \frac{1}{\xi^1}\frac{1}{1+\left(\frac{\xi^2}{\xi^1}\right)^2}\mathbf{h}^2$$

$$\mathbf{d}\alpha = \frac{-\xi^2\mathbf{h}^1 + \xi^1\mathbf{h}^2}{(\xi^1)^2 + (\xi^2)^2}.$$

After expressing this in quasi-cylindrical coordinates we obtain

$$\mathbf{d}\alpha = \frac{-\rho\sin\alpha\,\mathbf{h}^1 + \rho\cos\alpha\,\mathbf{h}^2}{\rho^2} = \frac{1}{\rho}(-\sin\alpha\,\mathbf{h}^1 + \cos\alpha\,\mathbf{h}^2). \tag{3.86}$$

We cannot read off the geometric image of $\mathbf{d}\alpha$ directly from (3.86). But from the considerations after (3.73) we know that the surfaces $\alpha = \text{const}$ form the image, that is half-planes passing through Z-axis as determined by the basis vector $\mathbf{m}_3$. We illustrate this in Fig. 3.17 as parallelograms with an oblique attitude of $\mathbf{m}_3$ and as half-circles with the vertical attitude of $\mathbf{m}_3$.

Exercise 3.30 Show that the exterior derivative of ρ is

$$\mathbf{d}\rho = \cos\alpha\,\mathbf{h}^1 + \sin\alpha\,\mathbf{h}^2. \tag{3.87}$$

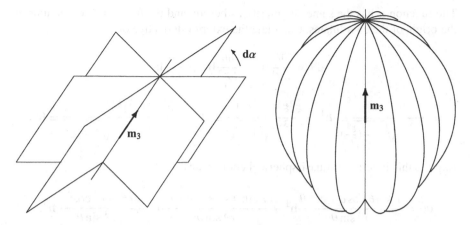

Fig. 3.17 Two ways of depicting the geometric image of the one-form $d\alpha$

Calculate the magnitude $|d\alpha|$ of (3.86) in the metric given by $\mathbf{m}_i$:

$$|d\alpha|^2 = \frac{\sin^2 \alpha + \cos^2 \alpha}{\rho^2} = \frac{1}{\rho^2}.$$

Therefrom we find $|d\alpha| = 1/\rho$. We see that it is constant on quasi-circles $\rho = \text{const}$. In such a case we see that the differential one-form (3.86) has a quasi-rotational symmetry around the Z-axis and translational symmetry along the same axis.

Consider now *quasi-spherical* coordinates chosen from the basis $\{\mathbf{m}_i\}$:

$$r = \sqrt{(\xi^1)^2 + (\xi^2)^2 + (\xi^3)^2}, \quad \theta = \arccos \frac{\xi^3}{r}, \quad \alpha = \arctan \frac{\xi^2}{\xi^1}.$$

The inverse relations are:

$$\xi^1 = r \sin \theta \cos \alpha, \quad \xi^2 = r \sin \theta \sin \alpha, \quad \xi^3 = r \cos \theta.$$

The exterior derivative of r is

$$\mathbf{d}r = \frac{\partial r}{\partial \xi^1}\mathbf{h}^1 + \frac{\partial r}{\partial \xi^2}\mathbf{h}^2 + \frac{\partial r}{\partial \xi^3}\mathbf{h}^3 = \frac{\xi^1}{r}\mathbf{h}^1 + \frac{\xi^2}{r}\mathbf{h}^2 + \frac{\xi^3}{r}\mathbf{h}^3,$$

$$\mathbf{d}r = \sin \theta \cos \alpha \, \mathbf{h}^1 + \sin \theta \sin \alpha \, \mathbf{h}^2 + \cos \theta \, \mathbf{h}^3. \tag{3.88}$$

The function α has the same singularity as before, and the function θ is singular in the origin of the coordinates. Calculate the exterior derivative of θ:

$$d\theta = \frac{\partial \theta}{\partial \xi^1}\mathbf{h}^1 + \frac{\partial \theta}{\partial \xi^2}\mathbf{h}^2 + \frac{\partial \theta}{\partial \xi^3}\mathbf{h}^3$$

$$= \frac{\xi^1 \xi^3}{r^3}\frac{1}{\sqrt{1-(\frac{\xi^3}{r})^2}}\mathbf{h}^1 + \frac{\xi^2 \xi^3}{r^3}\frac{1}{\sqrt{1-(\frac{\xi^3}{r})^2}}\mathbf{h}^2 - \frac{r^2 - (\xi^3)^2}{r^3}\frac{1}{\sqrt{1-(\frac{\xi^3}{r})^2}}\mathbf{h}^3.$$

Express this in terms of quasi-spherical coordinates:

$$d\theta = \frac{r\sin\theta\cos\alpha\, r\cos\theta}{r^3\sin\theta}\mathbf{h}^1 + \frac{r\sin\theta\sin\alpha\, r\cos\theta}{r^3\sin\theta}\mathbf{h}^2 - \frac{r^2 - r^2\cos^2\theta}{r^3\sin\theta}\mathbf{h}^3,$$

$$d\theta = \frac{1}{r}(\cos\theta\cos\alpha\,\mathbf{h}^1 + \cos\theta\sin\alpha\,\mathbf{h}^2 - \sin\theta\,\mathbf{h}^3). \tag{3.89}$$

A geometric image of this one-form is a family of cones $\theta = $ const with vertices in the coordinate origin (see Fig. 3.18). This form is not defined on the Z-axis, because, for example for $\theta = 0$ only $d\theta = r^{-1}(\cos\alpha\,\mathbf{h}^1 + \sin\alpha\,\mathbf{h}^2)$ remains from (3.89), and the angle α is not defined on the Z-axis.

The differential two-form $d\theta \wedge d\alpha$ is determined as the exterior product of (3.89) and (3.86):

$$d\theta \wedge d\alpha = \frac{1}{r}(\cos\theta\cos\alpha\,\mathbf{h}^1 + \cos\theta\sin\alpha\,\mathbf{h}^2 - \sin\theta\,\mathbf{h}^3) \wedge \frac{1}{\rho}(-\sin\alpha\,\mathbf{h}^1 + \cos\alpha\,\mathbf{h}^2)$$

$$= \frac{1}{r\rho}(-\cos\theta\sin^2\alpha\,\mathbf{h}^2\wedge\mathbf{h}^1 + \sin\theta\sin\alpha\,\mathbf{h}^3\wedge\mathbf{h}^1 + \cos\theta\cos^2\alpha\,\mathbf{h}^1\wedge\mathbf{h}^2 - \sin\theta\cos\alpha\,\mathbf{h}^3\wedge\mathbf{h}^2).$$

Fig. 3.18 Geometric image of the one-form $d\theta$

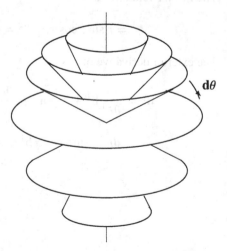

Fig. 3.19 Geometric image of the two-form $\mathbf{d}\theta \wedge \mathbf{d}\alpha$.

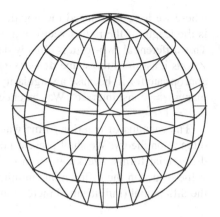

We substitute $\rho = r\sin\theta$:

$$\mathbf{d}\theta \wedge \mathbf{d}\alpha = \frac{1}{r^2\sin\theta}(\cos\theta\,\mathbf{h}^{12} + \sin\theta\sin\alpha\,\mathbf{h}^{31} + \sin\theta\cos\alpha\,\mathbf{h}^{23}). \qquad (3.90)$$

A geometric image of this two-form is obtained by superimposing the cones of Fig. 3.18 on the half-planes of Fig. 3.17. In this manner one obtains expanding tubes going radially outwards from the origin (see Fig. 3.19). This two-form is not defined on the Z-axis, because for $\theta = 0$ zero appears in the denominator. Calculate the magnitude of (3.90):

$$|\mathbf{d}\theta \wedge \mathbf{d}\alpha|^2 = \frac{\cos^2\theta + \sin^2\theta\sin^2\alpha + \sin^2\theta\cos^2\alpha}{r^4\sin^2\theta} = \frac{\cos^2\theta + \sin^2\theta}{r^4\sin^2\theta} = \frac{1}{r^4\sin^2\theta},$$

which gives $|\mathbf{d}\theta \wedge \mathbf{d}\alpha| = 1/r^2\sin\theta$. We see that it is not constant on quasi-spheres $r = \text{const}$. We can, however, multiply it by $\sin\theta$ and then we obtain a two-form, the magnitude of which, $|\sin\theta\,\mathbf{d}\theta \wedge \mathbf{d}\alpha| = 1/r^2$, is constant on quasi-spheres. Thus, it has quasi-spherical symmetry.

Using the equality $\frac{d(-\cos\theta)}{d\theta} = \sin\theta$, we write the new two-form as

$$\mathbf{d}(-\cos\theta) \wedge \mathbf{d}\alpha = \frac{1}{r^2}(\cos\theta\,\mathbf{h}^{12} + \sin\theta\sin\alpha\,\mathbf{h}^{31} + \sin\theta\cos\alpha\,\mathbf{h}^{23}). \qquad (3.91)$$

Making use of the formulas

$$\xi^1 = r\sin\theta\cos\alpha, \quad \xi^2 = r\sin\theta\sin\alpha, \quad \xi^3 = r\cos\theta,$$

we may write this two-form differently

$$\mathbf{d}(-\cos\theta) \wedge \mathbf{d}\alpha = \frac{\xi^3\mathbf{h}^{12} + \xi^2\mathbf{h}^{31} + \xi^1\mathbf{h}^{23}}{|\mathbf{r}|_m^3}, \qquad (3.92)$$

where we have introduced the notation $r = |\mathbf{r}|_m$, marking by the index m that it is the length of the radius-vector $\mathbf{r}$ in terms of the metric given by the basis $\{\mathbf{m}_i\}$. This differential two-form is already defined everywhere outside the origin. On the Z-axis it assumes the shape $\mathbf{d}(-\cos\theta) \wedge \mathbf{d}\alpha|_{\theta=0} = \mathbf{h}^1 \wedge \mathbf{h}^2/r^2$. Figure 3.19 is not appropriate for it, because a singularity occurs there on the Z-axis, where the half-planes $\alpha = \text{const}$ intersect. We can check from (3.92) more easily that $|\mathbf{d}(-\cos\theta) \wedge \mathbf{d}\alpha| = 1/r^2$.

The calculus of exterior forms was developed expressly to unify and simplify integration theory. We shall present now briefly the linear and surface integrals of differential forms. We know from (3.74) that for a differentiable function φ the value of $\mathbf{d}\varphi$ on the differential (infinitesimal increment) of the radius-vector $d\mathbf{r}$ is the differential (infinitesimal increment) of the function φ itself:

$$d\varphi = (\mathbf{d}\varphi)[[d\mathbf{r}]].$$

(Notice the difference between boldface $\mathbf{d}$ which denotes the exterior derivative, and italic d which is for traditional differential used at introductory calculus.) Therefore, such an expression can be easily integrated over curves, because a curve is an infinite sum of infinitesimal increments of the radius-vector. Hence the *integral of the one-form* $\mathbf{d}\varphi$ *over curve* C from $\mathbf{r}_A$ to $\mathbf{r}_B$ will be the expression

$$\int_C \mathbf{d}\varphi[[d\mathbf{r}]] = \varphi(\mathbf{r}_B) - \varphi(\mathbf{r}_A).$$

Such an integral does not depend on the shape of C, but only on its end points if the whole curve is contained in the star-shaped domain of φ.

As an example we may consider the one-form $\mathbf{d}\alpha$, given in (3.86), where α is the angular coordinate, one of the quasi-cylindrical or quasi-spherical coordinates. Note that the function $\alpha(\mathbf{r})$ is not defined on the Z-axis, so its domain is not star-shaped. Since the value of $\mathbf{d}\alpha$ on $d\mathbf{r}$ is the differential $d\alpha$, we conclude that the integral $\int_C \mathbf{d}\alpha[[d\mathbf{r}]]$ over curve C with ends $\mathbf{r}_A$ and $\mathbf{r}_B$ is the *dihedral visual angle* of C from the Z-axis. For closed curves not surrounding the Z-axis, this integral is zero, since the initial point coincides with the end point, and the whole curve is contained in the star-shaped domain of α. If, however, the closed curve surrounds the Z-axis (for instance once), the integral over such a curve must be represented as a sum of two integrals, each contained in a different star-shaped domain. If one of them gives the value $\pm\beta$ (the sign depending on the sense of circulation around Z), then the other gives the complementary value $\pm(2\pi - \beta)$; hence their sum is $\pm 2\pi$. Summarizing, the integral of the one-form $\mathbf{d}\alpha$ over a closed curve C surrounding the Z-axis once is $\pm 2\pi$, depending on the sense of circulation. In the case of multiple circulation around the Z-axis we obtain 2π times the number of windings of C on the Z-axis, and that winding number can be positive or negative.

Similarly, as the value of a one-form $\mathbf{d}\alpha$ on the element of the radius-vector $d\mathbf{r}$ is the differential $d\alpha$ of the dihedral angle, the value of a two-form $\mathbf{d}(-\cos\theta) \wedge \mathbf{d}\alpha$ on the bivector element of a directed surface $d\mathbf{s}$ is the differential

$d(-\cos\theta)d\alpha = \sin\theta d\theta d\alpha$ of the *solid visual angle* of the surface element. The integral $\int_S \mathbf{d}(-\cos\theta) \wedge \mathbf{d}\alpha[[ds]]$ over a surface S is the solid visual angle of this surface from the origin of coordinates. By reasoning similar to that of the previous paragraph, we may conclude that the integral of (3.92) over a closed surface not surrounding the origin is zero, and over a closed surface winding once around the origin is $\pm 4\pi$.

I have forgotten to mention earlier that a differential form N is called *closed* if $\mathbf{d} \wedge N = 0$. My conjecture is that a differential form is fitted if and only if it is closed.

It is worth finishing this section by formulating Stokes' theorem. But before some definitions are needed. A *manifold* V is the set of points described by radius-vectors depending on several real parameters $\lambda_1, \ldots \lambda_k$, i.e.,

$$\mathbf{r} = \mathbf{r}(\lambda_1, \ldots \lambda_k). \tag{3.93}$$

The number k of parameters determines the *dimension* of the manifold. An example of a two-dimensional manifold is a surface of the ellipsoid described by Eq. (2.38). The manifold is *oriented* if we have a way of continuously prescribing a nontwisted or twisted k-vector $\mathbf{s}_k$ tangent to the manifold at each point. The manifold is *of class* C^l if functions (3.93) have continuous lth derivatives. The boundary ∂V of V can also be a manifold, but of dimension $k - 1$. We know already that the orientation of a manifold determines the orientation of its boundary. Now we enunciate the *generalized Stokes' theorem*:

Theorem 3.2 *Let V be k-dimensional oriented manifold of class C^2 with a boundary of class C^1. Let ω be a differentiable nontwisted or twisted differential $(k-1)$-form defined on a neighbourhood of V. Then*

$$\int_V (\mathbf{d} \wedge \omega)[[d\mathbf{s}_k]] = \int_{\partial V} \omega[[d\mathbf{s}_{k-1}]].$$

In standard textbooks, see e.g., [50], the generalized Stokes' theorem is formulated and proved only for ordinary manifolds and nontwisted forms, without twisted orientations. Generalization for twisted quantities is not difficult—it suffices to multiply in an appropriate place the integral or form by the unit twisted scalar r or l. This theorem is a generalization of the primary Stokes' theorem linking the curvilinear integral of a vector field with the surface integral of the curl of this field. One particular case of the generalized Stokes' theorem is also Gauss's theorem.

Exercise 3.30 Show that for $k = 2$ Theorem 3.2 can be reduced to Stokes' theorem and for $k = 3$ to Gauss's theorem.

Chapter 4
Selected Problems of Electrodynamics

Section 4.1 is devoted to describing electromagnetism as far as possible in a metric-free manner. This includes definitions of relevant physical quantities and principal relations: field quantities, potentials, densities of charge, current, energy and energy flux, Maxwell's equations, Lorentz force, continuity equation, and gauge transformation.

To get solutions to Maxwell's equations, constitutive relations are needed $\mathbf{D} = \varepsilon(\mathbf{E})$ and $\mathbf{B} = \mu(\mathbf{H})$ for dielectric and magnetic materials, respectively. Each of them is built as a Hodge map containing a metric. A special metric is proposed in Sect. 4.2 for an anisotropic dielectric in which the solutions are formally the same as for an isotropic material in a natural metric. Static solutions of electric field are found for the simplest configurations. Similar considerations are performed in Sect. 4.3 for an anisotropic magnetic material.

4.1 Premetric Electrodynamics

In the last three decades a presentation of electrodynamics has become popular, based on a wide use of differential forms [1–9, 11–13, 52]. Such a formulation allows for deep synthesis of formulae (Maxwell's equations obtain a unified form) and simplification of deductions (e.g., derivation of differential Maxwell equations from integral ones, conservation of charge etc.).

In the vast majority of presentations, however, the distinction between twisted and nontwisted quantities is missing. For instance, the electromagnetic fields are chosen as the following quantities: $\mathbf{E}$ and $\mathbf{H}$—one-forms, $\mathbf{D}$ and $\mathbf{B}$—two-forms, in supposition—nontwisted forms. This leads sometime to misunderstandings. As we have shown in Sect. 3.3, when a scalar product is present, the nontwisted one-forms can be changed into ordinary vectors (commonly called polar vectors) and nontwisted two-forms into twisted vectors (called axial vectors or pseudovectors).

© Springer Nature Switzerland AG 2021
B. Jancewicz, *Directed Quantities in Electrodynamics*,
https://doi.org/10.1007/978-3-030-90471-5_4

Therefore the authors of work [11] concluded that **H** is a polar vector whereas **D** is an axial vector, contrary to the general belief of physicists.

We shall use the results of the named references with only two changes: **D** and **H** are twisted forms. Thus, according to the rules of Sect. 3.3, **D** is changed into a polar vector, and **H** into an axial vector in the presence of the scalar product. I am going to address the question: how far can one go in electrodynamics without invoking any scalar product and, therefore, a metric. This part of the theory will be called *premetric electrodynamics*.

Since we have already discussed in Sect. 1.4 how the electromagnetic field quantities can be defined, we proceed immediately to the integral Maxwell equations. When **E**, **D**, **H** and **B** are differential forms, the integrals can be written without any scalar product.

Gauss's law assumes the form

$$\oint_S \mathbf{D}[[d\mathbf{s}^*]] = Q, \tag{4.1}$$

where S is a closed surface, $d\mathbf{s}^*$ is its twisted bivector element of the surface area with outer orientation and Q is the electric charge encompassed by S. It is worthwhile to look at this law from the perspective of the operational definition of electric induction **D**. Imagine a closed surface ∂V covered by double discs (see Fig. 4.1). The integral over this surface is summing up the charges induced on the external discs. Then Eq. (4.1) says that the sum of charges collected on the discs is equal to the charge contained in the volume V.

Fig. 4.1 Double plates deployed on a spherical surface

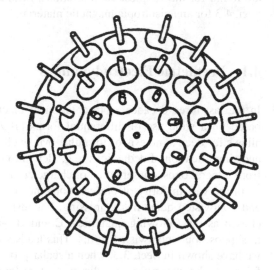

Magnetic Gauss Law

$$\oint_S \mathbf{B}[[d\mathbf{s}]] = 0. \tag{4.2}$$

Here $d\mathbf{s}$ is a bivector element of the surface area, orientation not important, since the right-hand side is zero anyway. If it is non zero, i.e., when magnetic charges exist, this integration element should be a twisted bivector (with outer orientation) in order to give a twisted scalar as the result of integration.

Faraday's Law

$$\frac{d}{dt}\int_S \mathbf{B}[[d\mathbf{s}]] = -\oint_{\partial S} \mathbf{E}[[d\mathbf{l}]], \tag{4.3}$$

where S is an arbitrary surface, $d\mathbf{s}$ is its bivector integration element, and $d\mathbf{l}$ is the vector element of the boundary ∂S.

Ampère-Oersted Law (Supplemented by Maxwell with the Last Term)

$$\oint_{\partial S} \mathbf{H}[[d\mathbf{l}^*]] = I + \frac{d}{dt}\int_S \mathbf{D}[[d\mathbf{s}^*]], \tag{4.4}$$

where I is the electric current flowing through a surface S, $d\mathbf{s}^*$ is a twisted bivector element of the surface area and $d\mathbf{l}^*$ a twisted vector element of its boundary. The operational definition of the magnetic field intensity $\mathbf{H}$ allows us to consider superconducting solenoids "strung" on the curve ∂S (see Fig. 4.2). The curvilinear

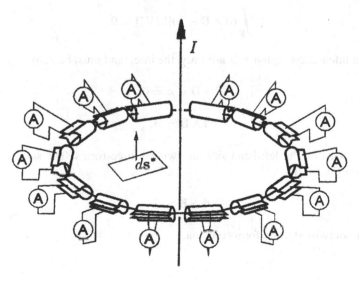

Fig. 4.2 Superconducting solenoids "string" onto a curve

integral denotes summing up the currents. Equation (4.4) says that the sum is equal to the current and displacement current flowing through the surface S.

Equations (4.1)–(4.4) have the common name *integral Maxwell equations*. We shall derive differential equations corresponding to them.

Knowing that each closed surface S is a boundary ∂V of some spatial region V, we rewrite *Gauss's law* in the form

$$\int_{\partial V} \mathbf{D}[[d\mathbf{s}^*]] = Q.$$

We apply the generalized Stokes' theorem

$$\int_V (\mathbf{d} \wedge \mathbf{D})[[dV]] = Q.$$

We represent the electric charge Q as the integral over the same region of the charge density ρ (which is a twisted three-form):

$$\int_V (\mathbf{d} \wedge \mathbf{D})[[dV]] = \int_V \rho[[dV]].$$

$$\int_V (\mathbf{d} \wedge \mathbf{D})[[dV]] - \int_V \rho[[dV]] = 0.$$

The integration region is the same, so we can write both expressions under a single integration sign

$$\int_V (\mathbf{d} \wedge \mathbf{D} - \rho)[[dV]] = 0.$$

Since the integration region V is arbitrary, the integrand must be zero:

$$\mathbf{d} \wedge \mathbf{D} - \rho = 0,$$

$$\mathbf{d} \wedge \mathbf{D} = \rho. \tag{4.5}$$

The two terms on the left-hand side are twisted three-forms. The same reasoning leads from Eq. (4.2) to

$$\mathbf{d} \wedge \mathbf{B} = 0. \tag{4.6}$$

This is a nontwisted three-form equation.

We assume in Faraday's law that the integration surface S is stationary; hence the time derivation can be done under the integration sign:

$$\int_S \dot{\mathbf{B}}[[ds]] = -\oint_{\partial S} \mathbf{E}[[dl]].$$

We apply the generalized Stokes' theorem to the right-hand side

$$\int_S \dot{\mathbf{B}}[[ds]] = -\int_S (\mathbf{d} \wedge \mathbf{E})[[ds]],$$

$$\int_S \dot{\mathbf{B}}[[ds]] + \int_S (\mathbf{d} \wedge \mathbf{E})[[ds]] = 0,$$

$$\int_S (\dot{\mathbf{B}} + \mathbf{d} \wedge \mathbf{E})[[ds]] = 0.$$

The integration surface S is arbitrary, so the integrand must be zero:

$$\dot{\mathbf{B}} + \mathbf{d} \wedge \mathbf{E} = 0.$$

The same equation can be written out as:

$$\mathbf{d} \wedge \mathbf{E} + \frac{\partial \mathbf{B}}{\partial t} = 0. \tag{4.7}$$

Its two terms are nontwisted two-forms.

We transform similarly (4.4). The electric current I is expressed as the surface integral of the current density twisted two-form $\mathbf{j}$: $I = \int_S \mathbf{j}[[ds^*]]$, so Eq. (4.4) takes the form

$$\oint_{\partial S} \mathbf{H}[[dl^*]] - \int_S \dot{\mathbf{D}}[[ds^*]] - \int_S \mathbf{j}[[ds^*]] = 0,$$

and by virtue of the generalized Stokes' theorem,

$$\int_S (\mathbf{d} \wedge \mathbf{H})[[ds^*]] - \int_S \dot{\mathbf{D}}[[ds^*]] - \int_S \mathbf{j}[[ds^*]] = 0,$$

$$\int_S (\mathbf{d} \wedge \mathbf{H} - \dot{\mathbf{D}} - \mathbf{j})[[ds^*]] = 0.$$

Again, the arbitrariness of S implies vanishing of the integrand:

$$\mathbf{d} \wedge \mathbf{H} - \dot{\mathbf{D}} - \mathbf{j} = 0.$$

This twisted two-form equation can be rewritten as:

$$\mathbf{d} \wedge \mathbf{H} - \frac{\partial \mathbf{D}}{\partial t} = \mathbf{j}. \tag{4.8}$$

Equations (4.5)–(4.8) are called *differential Maxwell equations*. Each of them is of a distinct kind with respect to the contained directed quantities. Equations (4.5) and (4.7) bear the name *homogeneous Maxwell equations* because they contain only the field quantities, whereas Eqs. (4.5) and (4.8) are called *inhomogeneous Maxwell equations*, since they contain also charge and current densities treated as sources of the electromagnetic field.

Calculate the exterior derivative of (4.8):

$$\mathbf{d} \wedge \mathbf{d} \wedge \mathbf{H} - \mathbf{d} \wedge \frac{\partial \mathbf{D}}{\partial t} = \mathbf{d} \wedge \mathbf{j}.$$

The first term vanishes and the time derivative commutes with the spatial differentiation, so

$$-\frac{\partial}{\partial t} \mathbf{d} \wedge \mathbf{D} = \mathbf{d} \wedge \mathbf{j}.$$

Take the time derivative of (4.5):

$$\frac{\partial}{\partial t} \mathbf{d} \wedge \mathbf{D} = \frac{\partial \rho}{\partial t}.$$

By adding the two last formulas we arrive at the equality

$$\mathbf{d} \wedge \mathbf{j} + \frac{\partial \rho}{\partial t} = 0, \tag{4.9}$$

that is, to the *continuity equation* of the electric charge.

Equation (4.6) is satisfied in the whole space, and this is a star-shaped region. Hence by virtue of the Poincaré lemma we obtain from it the existence of a one-form potential $\mathbf{A}$ such that

$$\mathbf{B} = \mathbf{d} \wedge \mathbf{A}. \tag{4.10}$$

We call the differential form $\mathbf{A}$ the *magnetic potential*. Substituting (4.10) into (4.7) yields $\mathbf{d} \wedge \mathbf{E} + \frac{\partial}{\partial t} \mathbf{d} \wedge \mathbf{A} = 0$ or $\mathbf{d} \wedge (\mathbf{E} + \frac{\partial \mathbf{A}}{\partial t}) = 0$. This equality is fulfilled in the whole space, hence again Poincaré lemma ensures existence of a scalar (or rather zero-form, for coherence with other quantities) potential Φ such that $E + \frac{\partial A}{\partial t} = -d\Phi$ and we obtain

$$E = -d\Phi - \frac{\partial A}{\partial t}. \tag{4.11}$$

The zero-form Φ is called *electric potential*.

The magnetic potential is not determined uniquely: if the exterior derivative of a zero-form χ is added to A, then B given by (4.10) remains unchanged, since $d \wedge (A + d \wedge \chi) = d \wedge A = B$ by dint of (3.81). Therefore, the magnetic potential is defined up to the exterior derivative of a zero-form. Of course, an appropriate change of the electric potential must accompany the change $A \rightarrow A + d\chi$ in order that relation (4.11) is preserved: $\Phi \rightarrow \Phi - \frac{\partial \chi}{\partial t}$. In this manner, the following simultaneous change of the two potentials

$$A \rightarrow A + d\chi, \qquad \Phi \rightarrow \Phi - \frac{\partial \chi}{\partial t}, \tag{4.12}$$

is possible without affecting the forms E and B. This change is called *gauge transformation* and the fact that E and B remain unchanged is called *gauge invariance*.

As we mentioned in Sect. 1.4, the energy density of the electromagnetic field, written in traditional language as the scalar $w' = \frac{1}{2}(E \cdot D + B \cdot H)$, in the form language is the twisted three-form

$$w = \frac{1}{2}(E \wedge D + B \wedge H). \tag{4.13}$$

The energy flux density of the electromagnetic field—as all flux densities—should be a twisted two-form. Only the exterior product $E \wedge H$ (or $H \wedge E$) gives such a quantity, which replaces the traditional Poynting vector:

$$S = E \wedge H. \tag{4.14}$$

It could be called a *Poynting twisted two-form*. The proper order of factors can be established by considering the twisted vector H and nontwisted vector E on the background of Fig. 1.51. In traditional vector language, the expression $S = E \times H$ corresponds to (4.14).

The Lagrangian density of the electromagnetic field is taken from the Landau and Lifshitz book [27]: $L_{em} = \frac{1}{8\pi}(E^2 - H^2)$, written for the vacuum in CGS units, should be rewritten in the SI system for an arbitrary medium as $L_{em} = \frac{1}{2}(E \cdot D - B \cdot H)$. This is an expression similar to the energy density of the electromagnetic field, but the difference is in place of the sum. Hence we are able to write the *Lagrangian density* in the premetric electrodynamics as

$$L_{em} = \frac{1}{2}(E \wedge D - B \wedge H). \tag{4.15}$$

This is a twisted three-form, corresponding to one of the Lorentz transformation invariants of the electromagnetic field quantities. Another term in the Lagrangian density exists which describes the interaction of electric charges with the electromagnetic field, which in traditional language contains the vector quantities $\mathbf{A}$, $\mathbf{j}$ and the scalar charge density ρ':

$$L_{int} = \mathbf{j} \cdot \mathbf{A} - \rho' \Phi.$$

In the language of premetric electrodynamics, this should be rewritten with the use of a nontwisted one-form $\mathbf{A}$, a twisted two-form $\mathbf{j}$ and a twisted three-form $\rho = \rho' \mathbf{f}_*^{123}$:

$$L_{int} = \mathbf{j} \wedge \mathbf{A} - \rho \, \Phi. \tag{4.16}$$

This is also a twisted three-form. In this manner, the total Lagrangian density, in which the electromagnetic field quantities are present, is the sum

$$L = L_{em} + L_{int} = \frac{1}{2}(\mathbf{E} \wedge \mathbf{D} - \mathbf{B} \wedge \mathbf{H}) + \mathbf{j} \wedge \mathbf{A} - \rho \, \Phi. \tag{4.17}$$

Still another quantity, namely

$$\mathbf{q} = \mathbf{D} \times \mathbf{B} \tag{4.18}$$

exists in the vector language. This is the *momentum density* of the electromagnetic field. It is difficult to find its counterpart in the exterior form approach, because $\mathbf{D}$ and $\mathbf{B}$ are two-forms and their outer product is zero. But there is a possibility to compose a bilinear quantity of $\mathbf{D}$ and $\mathbf{B}$, however, with the explicit presence of a volume element dV, namely $\mathbf{D} \lfloor (dV \lfloor \mathbf{B})$, which is a nontwisted one-form. I claim that this is just the momentum element of the electromagnetic field contained in the volume dV:

$$d\mathbf{p} = \mathbf{D} \lfloor (dV \lfloor \mathbf{B}). \tag{4.19}$$

As checked in Exercise 3.16, this expression is antisymmetric under interchange of $\mathbf{D}$ and $\mathbf{B}$, which corresponds to the same property of (4.18). The total momentum of the electromagnetic field contained in region V is the integral

$$\mathbf{p} = \int_V \mathbf{D} \lfloor (dV \lfloor \mathbf{B}).$$

Exercise 4.1 Prove the identity

$$\mathbf{D} \lfloor (V \lfloor \mathbf{B}) = (\mathbf{D} \rfloor V) \rfloor \mathbf{B}.$$

If we consider all the possible outer products of forms describing the electromagnetic field, the products $\mathbf{E} \wedge \mathbf{B}$ and $\mathbf{D} \wedge \mathbf{H}$ are left, which give three-forms. Unfortunately they have distinct physical dimensions—that of the former in SI system is V^2s/m^3, that of the latter is A^2s/m^3. Hence one cannot form a sum of them similar to (4.15). We shall show in Sect. 7.2 that they are two Lorentz transformation invariants. In the vacuum they are proportional to each other: $\mu_0^{-1}\mathbf{E} \cdot \mathbf{B} = \mathbf{E} \cdot \mathbf{H} = \varepsilon_0^{-1}\mathbf{D} \cdot \mathbf{H}$.

We have already discussed the *Lorentz force* $\mathbf{F}$ acting on the electric charge q— see Eq. (3.10):

$$\mathbf{F} = q\mathbf{E} + q\mathbf{B}\lfloor\mathbf{v}. \tag{4.20}$$

Formulas (4.1)–(4.15), (4.19) exhaust the list of equations of premetric electrodynamics. With the exception of (4.13)–(4.19), they are equations to which solutions are to be found. It is at this stage that a scalar product or metric is needed.

Once we have admitted the possible existence of magnetic charges, let us see how differential Maxwell equations would be altered to accommodate them. We can no longer sustain $\mathbf{d} \wedge \mathbf{B} = 0$, but have instead

$$\mathbf{d} \wedge \mathbf{B} = \rho_m, \tag{4.21}$$

where ρ_m is a nontwisted three-form of magnetic charge density. Also, the other homogeneous equation $\mathbf{d} \wedge \mathbf{E} + \frac{\partial \mathbf{B}}{\partial t} = 0$ would obtain nonzero term at the right-hand side:

$$\mathbf{d} \wedge \mathbf{E} + \frac{\partial \mathbf{B}}{\partial t} = -\mathbf{j}_m, \tag{4.22}$$

which is a nontwisted two-form of magnetic current density. It is not a twisted form, as for other flux densities, because its value on a twisted bivector should give a twisted scalar for the magnetic current. The minus sign is necessary to ensure a continuity equation for magnetic charge. Indeed, from (4.22)

$$0 = \mathbf{d} \wedge (\mathbf{d} \wedge \mathbf{E}) = -\mathbf{d} \wedge \frac{\partial \mathbf{B}}{\partial t} - \mathbf{d} \wedge \mathbf{j}_m = -\frac{\partial}{\partial t}(\mathbf{d} \wedge \mathbf{B}) - \mathbf{d} \wedge \mathbf{j}_m,$$

and (4.21) yields

$$\frac{\partial \rho_m}{\partial t} + \mathbf{d} \wedge \mathbf{j}_m = 0.$$

Thus only two Maxwell equations, the homogeneous ones, had to be altered. Now, the generalized Maxwell equations become

$$\mathbf{d} \wedge \mathbf{D} = \rho, \qquad \mathbf{d} \wedge \mathbf{B} = \rho_m, \tag{4.23}$$

$$\mathbf{d} \wedge \mathbf{H} = \frac{\partial \mathbf{D}}{\partial t} + \mathbf{j}, \quad \mathbf{d} \wedge \mathbf{E} = -\frac{\partial \mathbf{B}}{\partial t} - \mathbf{j}_m. \tag{4.24}$$

By analogy with $\frac{\partial \mathbf{D}}{\partial t}$, the term $\frac{\partial \mathbf{B}}{\partial t}$ can be called the *magnetic displacement current density*.

Also, a counterpart of the Lorentz force acting on a particle bearing a magnetic charge q_m could be written:

$$\mathbf{F}_m = q_m \mathbf{H} + q_m \mathbf{D} \lfloor \mathbf{v}. \tag{4.25}$$

The magnetic charge q_m, as a twisted scalar, multiplied by twisted one-forms $\mathbf{H}$ and $\mathbf{D} \lfloor \mathbf{v}$, yields a nontwisted one-form, which is appropriate for the force.

4.2 Static Solutions in an Anisotropic Dielectric Medium

Solutions to the Maxwell equations, that is, specific electromagnetic fields, can be written down due to the scalar product, which we know in our physical space. At first sight it seems that the static uniform fields can be given without any metric at all. For instance, an electrostatic uniform field may be written $\mathbf{E} = \text{const}$, $\mathbf{D} = \text{const}$. But this is not enough, since $\mathbf{E}$, $\mathbf{D}$ are related to each other by the *constitutive relation*, which in the vector language was expressed as $\mathbf{D} = \varepsilon(\mathbf{E})$, and the passage from nontwisted and twisted forms to vectors demands a scalar product.

In an isotropic medium, the electric field vector $\mathbf{E}$ is parallel to the electric induction vector $\mathbf{D}$. This sentence must be reformulated in terms of the forms as follows: in an isotropic medium, the planes of the electric field one-form $\mathbf{E}$ are perpendicular to the lines of the electric induction two-form $\mathbf{D}$. Since we know that the fields $\mathbf{E}$ and $\mathbf{D}$ in an anisotropic medium are no longer parallel, we should express this so: the forms $\mathbf{E}$ and $\mathbf{D}$ are no longer perpendicular. But perpendicularity depends on a scalar product. In this manner a question arises: can we find another scalar product, appropriate for a given medium, such that the same forms become perpendicular to each other?

We tackle this question now. We all live in the three-dimensional space in which the *natural scalar product* exists, because we know which vectors are orthogonal and we ascribe a length to each of them independently of attitude. Knowing this natural scalar product, we may, in agreement with Sect. 2.3, change nontwisted and twisted one- and two-forms into nontwisted vectors $\mathbf{E}$, $\mathbf{D}$ or twisted vectors $\mathbf{B}$, $\mathbf{H}$, which are present in the traditional physics textbooks. With the known orthonormal basis, by dint of (2.71) and (2.72) we obtain vector coordinates (with the upper indices) from one-form coordinates (with lower indices) according to the formulas:

$$E^i = E_i, \qquad H^i = H_i. \tag{4.26}$$

Moreover, by virtue of (3.52) and (3.54), relations concerning the electric and magnetic induction can be written with the aid of Levi-Civita symbol:

$$D_{ij} = \epsilon_{ijk} D^k, \qquad B_{ij} = \epsilon_{ijk} B^k. \tag{4.27}$$

This is the vector notation in which the *constitutive relations* are traditionally written down: $\mathbf{D} = \varepsilon(\mathbf{E})$, $\mathbf{B} = \mu(\mathbf{H})$, in which, for so called *linear media*, the symbols ε and μ denote linear mappings. For the electric quantities this mapping has the form:

$$D^1 = \varepsilon_0(\varepsilon^{11} E^1 + \varepsilon^{12} E^2 + \varepsilon^{13} E^3),$$

$$D^2 = \varepsilon_0(\varepsilon^{21} E^1 + \varepsilon^{22} E^2 + \varepsilon^{23} E^3),$$

$$D^3 = \varepsilon_0(\varepsilon^{31} E^1 + \varepsilon^{32} E^2 + \varepsilon^{33} E^3),$$

where ε^{ij} are (dimensionless) elements of the *relative permittivity* matrix $\boldsymbol{\varepsilon}$ of the medium, whereas ε_0 is the *permittivity of the vacuum*. After using (4.26) we write this with lower indices:

$$D^1 = \varepsilon_0(\varepsilon^{11} E_1 + \varepsilon^{12} E_2 + \varepsilon^{13} E_3),$$

$$D^2 = \varepsilon_0(\varepsilon^{21} E_1 + \varepsilon^{22} E_2 + \varepsilon^{23} E_3),$$

$$D^3 = \varepsilon_0(\varepsilon^{31} E_1 + \varepsilon^{32} E_2 + \varepsilon^{33} E_3),$$

or, according to the summation convention

$$D^k = \varepsilon_0 \varepsilon^{kl} E_l. \tag{4.28}$$

By virtue of (4.27) we write this in terms of form coordinates:

$$D_{ij} = \varepsilon_0 \epsilon_{ijk} \varepsilon^{kl} E_l. \tag{4.29}$$

Calculate now the outer product $\mathbf{E} \wedge \mathbf{D}$:

$$\mathbf{E} \wedge \mathbf{D} = (E_1 \mathbf{f}^1 + E_2 \mathbf{f}^2 + E_3 \mathbf{f}^3) \wedge (D_{12} \mathbf{f}_*^{12} + D_{23} \mathbf{f}_*^{23} + D_{31} \mathbf{f}_*^{31})$$

$$= (E_1 D_{23} + E_2 D_{31} + E_3 D_{12}) \, \mathbf{f}_*^{123} = (E_1 D^1 + E_2 D^2 + E_3 D^3) \, \mathbf{f}_*^{123} = E_i D^i \, \mathbf{f}_*^{123}$$

Use (4.28):

$$\mathbf{E} \wedge \mathbf{D} = (\varepsilon_0 \, E_i \varepsilon^{ij} E_j) \, \mathbf{f}_*^{123}. \tag{4.30}$$

Hence the energy density of the electric field is

$$w_e = \frac{1}{2} \mathbf{E} \wedge \mathbf{D} = \frac{1}{2} \varepsilon_0 (E_i \varepsilon^{ij} E_j) \, \mathbf{f}_*^{123}. \tag{4.31}$$

This is a twisted three-form, but its single scalar coordinate,

$$|w_e| = \frac{1}{2} \varepsilon_0 E_i \varepsilon^{ij} E_j, \tag{4.32}$$

is an image of the bilinear mapping of one-forms into scalars. We see that a scalar product of one-forms should be introduced for which the permittivity matrix plays the role of the metric matrix. Comparing (4.32) with (2.51) allows us to write $\tilde{G} = \boldsymbol{\varepsilon}$, and due to (2.50),

$$G = \boldsymbol{\varepsilon}^{-1}. \tag{4.33}$$

We formulate our observation as follows: the *matrix G inverse to the matrix of relative permittivity determines a scalar product appropriate for a given anisotropic dielectric.*

It has been stated in Sect. 3.3 that the inverse $G^{-1} = \tilde{G}$ of the metric tensor serves to change one-forms into vectors. Thus, in accordance with (3.58) we write $\mathbf{E} = G^{-1}[\mathbf{E}]$. This is not the normal vector of the electric field intensity, which we use in the traditional approach, hence it is safer to denote it differently, for instance by prime:

$$\mathbf{E}' = \tilde{G}[[\mathbf{E}]]. \tag{4.34}$$

This mapping assumes the following form on the coordinates

$$E'^i = \tilde{g}^{ij} E_j = \varepsilon^{ij} E_j.$$

After comparing with (4.28) we notice that only the coefficient ε_0 is lacking to write down

$$D^i = \varepsilon_0 \varepsilon^{ij} E_j.$$

This relation can be expressed physically (permittivity changes the electric field intensity into the electric induction) and mathematically: as the inverse of the metric tensor (with the additional coefficient ε_0) it changes the one-form $\mathbf{E}$ into the vector $\mathbf{D}$:

$$\mathbf{D} = \varepsilon_0 \tilde{G}[[\mathbf{E}]], \tag{4.35}$$

or

$$\mathbf{D} = \varepsilon_0 \mathbf{E}'. \tag{4.36}$$

Notice that the vectors $\mathbf{D}$ and $\mathbf{E}'$ are parallel, and the relation between them looks like it does in the vacuum.

We may combine the mapping (4.35) with (3.34) and obtain the equality

$$\mathbf{D} = \mathbf{f}_*^{123} \lfloor (\varepsilon_0 \tilde{G}[[\mathbf{E}]]) = \varepsilon_0 \mathbf{f}_*^{123} \lfloor \tilde{G}[[\mathbf{E}]] \tag{4.37}$$

as the *constitutive relation* between the nontwisted one-form $\mathbf{E}$ and the twisted two-form $\mathbf{D}$. This is a composition of two linear mappings: first the inverse of the metric tensor connected with a given dielectric (that is, such that $\tilde{G} = \boldsymbol{\varepsilon}$), and then the contraction with the twisted three-form f_*^{123} made from the one-form basis, orthonormal in ordinary metric, not connected with the dielectric. Finally, multiplication by the scalar permittivity ε_0 is performed.

Let us calculate the inner product of the forms $\mathbf{D}$ and $\mathbf{E}$ according to the scalar product given by the matrix $\boldsymbol{\varepsilon}$. We first rewrite formula (2.64):

$$(\mathbf{D} \cdot \mathbf{E})_k = D_{kj} \tilde{g}^{ji} E_i,$$

and substitute $\tilde{G} = \boldsymbol{\varepsilon}$ into it:

$$(\mathbf{D} \cdot \mathbf{E})_k = D_{kj} \varepsilon^{ji} E_i.$$

Use (4.27) and (4.28)

$$(\mathbf{D} \cdot \mathbf{E})_k = \varepsilon_0^{-1} \epsilon_{klj} D^l D^j.$$

The symbol ϵ_{klj} is antisymmetric and the product $D^l D^j$ is symmetric in the indices l, j; thus the sum is zero. Hence

$$\mathbf{D} \cdot \mathbf{E} = 0.$$

We have shown that *the twisted two-form $\mathbf{D}$ is perpendicular to the nontwisted one-form $\mathbf{E}$ in the metric given by the permittivity*. This is a formal proof of the expectation expressed in the second paragraph of this section.

Another question is connected with this observation. Let $\{\mathbf{m}_1,\ \mathbf{m}_2,\ \mathbf{m}_3\}$ be the orthonormal basis for the new scalar product. With the aid of this basis we change forms $\mathbf{E}$ and $\mathbf{D}$ into vectors $\mathbf{D}'$ and $\mathbf{E}'$. How are they related? We apply (4.34) to $\mathbf{E}$ and Exercise 3.11 to $\mathbf{D}$, that is, contraction with the basic twisted trivector $-\mathbf{m}_{123}^*$:

$$\mathbf{D}' = -\mathbf{m}_{123}^*\lfloor\mathbf{D}.$$

But by virtue of (2.52) we have $\mathbf{m}_{123}^* = (\det\tilde{\mathcal{G}})^{1/2}\mathbf{e}_{123}^*$, so

$$\mathbf{D}' = -(\det\tilde{\mathcal{G}})^{1/2}\mathbf{e}_{123}^*\lfloor\mathbf{D} = -(\det\varepsilon)^{1/2}\mathbf{e}_{123}^*\lfloor\mathbf{D}.$$

We invoke (3.53)

$$\mathbf{D}' = (\det\varepsilon)^{1/2}\mathbf{D}.$$

whence, by dint of (4.36), we have

$$\mathbf{D}' = \varepsilon_0(\det\varepsilon)^{1/2}\mathbf{E}'. \tag{4.38}$$

We notice that vectors $\mathbf{D}'$ and $\mathbf{E}'$ are parallel—this can be expressed by saying that in the metric given by the basis $\{\mathbf{m}_i\}$ *the medium appears to be isotropic* with the scalar permittivity $\varepsilon_0\sqrt{\det\varepsilon}$.

When comparing formulas (4.36) and (4.38) one may ask: which is better? Equation (4.36) has the advantage of having an ordinary vector $\mathbf{D}$, used in the traditional approach, whereas (4.38) has the merit of being formally consistent, because both vectors $\mathbf{D}'$ and $\mathbf{E}'$ are obtained from the corresponding forms by prescriptions of Sect. 3.3 using the same basis $\{\mathbf{m}_i\}$.

A theorem exists that relates the time derivative of the energy density of the electromagnetic field with the energy flux. We did not tackle it within the premetric electrodynamics, because the constitutive relations are needed in its proof. We can do this now. The electric energy density is $w_e = \frac{1}{2}\mathbf{E} \wedge \mathbf{D}$. Before differentiating, we rewrite (4.37), introducing the symbol N for shorter notation:

$$\mathbf{D} = \varepsilon_0 N[[\mathbf{E}]] = \varepsilon_0 \mathbf{f}_*^{123}\lfloor\tilde{G}[[\mathbf{E}]]. \tag{4.39}$$

The linear operator N should be compared with M defined in (3.63) They differ only by having $\mathbf{f}_*^{123}$ and $\mathbf{h}_*^{123}$, respectively. By virtue of (2.53), i.e., $\mathbf{f}_*^{123} = (\det\tilde{G})^{1/2}\mathbf{h}_*^{123}$, we see that $N = (\det\tilde{G})^{1/2}M$, hence Lemma 3.2 is true also for N replacing M. The operator N is linear and—because of this—continuous, hence it satisfies

$$\frac{\partial}{\partial t}N[[\mathbf{E}]] = N\left[\left[\frac{\partial\mathbf{E}}{\partial t}\right]\right].$$

We also assumed that the medium is independent of time—then the time derivation does not act on the permittivity contained in N.

We are now ready to calculate the derivative:

$$\frac{\partial}{\partial t}(\mathbf{E} \wedge \mathbf{D}) = \frac{\partial \mathbf{E}}{\partial t} \wedge \mathbf{D} + \mathbf{E} \wedge \frac{\partial \mathbf{D}}{\partial t} = \frac{\partial \mathbf{E}}{\partial t} \wedge \varepsilon_0 N[[\mathbf{E}]] + \mathbf{E} \wedge \varepsilon_0 N\left[\left[\frac{\partial \mathbf{E}}{\partial t}\right]\right].$$

We use (3.64) to change the first term

$$\frac{\partial}{\partial t}(\mathbf{E} \wedge \mathbf{D}) = \mathbf{E} \wedge \varepsilon_0 N\left[\left[\frac{\partial \mathbf{E}}{\partial t}\right]\right] + \mathbf{E} \wedge \varepsilon_0 N\left[\left[\frac{\partial \mathbf{E}}{\partial t}\right]\right] = 2\mathbf{E} \wedge \frac{\partial \mathbf{D}}{\partial t}.$$

A similar identity can be proved for the linear magnetic medium

$$\frac{\partial}{\partial t}(\mathbf{H} \wedge \mathbf{B}) = 2\mathbf{H} \wedge \frac{\partial \mathbf{B}}{\partial t}.$$

We obtain from this

$$\frac{\partial w}{\partial t} = \frac{1}{2}\frac{\partial}{\partial t}(\mathbf{H} \wedge \mathbf{B} + \mathbf{E} \wedge \mathbf{D}) = \mathbf{H} \wedge \frac{\partial \mathbf{B}}{\partial t} + \mathbf{E} \wedge \frac{\partial \mathbf{D}}{\partial t}. \tag{4.40}$$

We proceed now to prove the *Poynting theorem*.

Theorem 4.1 *If the medium is electrically and magnetically linear and stationary in time, then the following identity holds*

$$\mathbf{d} \wedge \mathbf{S} + \frac{\partial w}{\partial t} = -\mathbf{E} \wedge \mathbf{j}.$$

Proof The linearity of the medium implies the relations $\mathbf{D} = N[\mathbf{E}]$, $\mathbf{B} = N'[\mathbf{H}]$, which—along with the time independence of the medium—imply (4.40). We use Lemma 3.3 to calculate the exterior derivative of the Poynting two-form

$$\mathbf{d} \wedge \mathbf{S} = \mathbf{d} \wedge (\mathbf{E} \wedge \mathbf{H}) = (\mathbf{d} \wedge \mathbf{E}) \wedge \mathbf{H} - \mathbf{E} \wedge (\mathbf{d} \wedge \mathbf{H}).$$

We make use of the Maxwell equations (4.7) and (4.8)

$$\mathbf{d} \wedge \mathbf{S} = -\frac{\partial \mathbf{B}}{\partial t} \wedge \mathbf{H} - \mathbf{E} \wedge \left(\frac{\partial \mathbf{D}}{\partial t} + \mathbf{j}\right),$$

$$\mathbf{d} \wedge \mathbf{S} + \mathbf{H} \wedge \frac{\partial \mathbf{B}}{\partial t} + \mathbf{E} \wedge \frac{\partial \mathbf{D}}{\partial t} = -\mathbf{E} \wedge \mathbf{j}.$$

By virtue of (4.40) we obtain

$$\mathbf{d} \wedge \mathbf{S} + \frac{\partial w}{\partial t} = -\mathbf{E} \wedge \mathbf{j},$$

∎

The integral version of this theorem takes the form

$$\frac{dW}{dt} = -\int_{\partial V} \mathbf{S}[[ds]] - \int (\mathbf{E} \wedge \mathbf{j})[[dV]].$$

This states physically that the energy of the electromagnetic field contained in a region V increases by the energy transported through its boundary ∂V and decreases by the energy converted into heat.

In electrostatic problems, the Maxwell equation (4.7) assumes the form $\mathbf{d} \wedge \mathbf{E} = 0$ and, due to this, relation (4.11) becomes $\mathbf{E} = -\mathbf{d}\Phi$. Inserting this into (4.39) yields the following relation between the electric induction and the electric potential

$$\mathbf{D} = -\varepsilon_0 N[[\mathbf{d}\Phi]].$$

We plug this expression into the Maxwell equation (4.5):

$$\mathbf{d} \wedge N[[\mathbf{d}\Phi]] = -\varepsilon_0^{-1} \rho. \tag{4.41}$$

The expression $\mathbf{d} \wedge N[[\mathbf{d} \cdot]]$ is a second-order differential operator changing a scalar (or zero-form) field into a differential three-form. This could be called the *Laplace operator for anisotropic dielectric*. Its dependence on the dielectric is contained in the tensor $\tilde{G}$ of the matrix $\tilde{G} = \boldsymbol{\varepsilon}$. Equation (4.41) is the counterpart of the *Poisson equation* in an anisotropic medium.

Exercise 4.2 Let $\{\mathbf{e}_i\}$ be basis orthonormal in the ordinary metric. Show that in the coordinates x^i, occurring in $\mathbf{r} = x^i \mathbf{e}_i$, Eq. (4.41) can be written as

$$\varepsilon^{ij} \frac{\partial^2 \Phi}{\partial x^i \partial x^j} = -\varepsilon_0^{-1} \rho'. \tag{4.42}$$

where ρ' is the scalar present in the equality $\rho = \rho' \, \mathbf{f}_*^{123}$.

This equation (written in another system of units) can be found in the book of Landau and Lifshitz [43].

Exercise 4.3 Let $\{\mathbf{m}_i\}$ be the orthonormal basis in the dielectric metric. Show that in coordinates ξ^i present in $\mathbf{r} = \xi^i \mathbf{m}_i$ Eq. (4.41) can be written as

$$\delta^{ij} \frac{\partial^2 \Phi}{\partial \xi^i \partial \xi^j} = -\varepsilon_0^{-1} \rho'. \tag{4.43}$$

This result shows that the scalar product determined by the basis $\{\mathbf{m}_i\}$ implies that the Laplace operator has, in its coordinates, the same form as that in the isotropic medium.

We pass now to a particular physical situation. Consider the electrostatic field point electric charge Q placed in a homogeneous anisotropic dielectric in the origin of coordinates. We wish to solve Eq. (4.42) after inserting in it the charge density corresponding to the point charge:

$$\rho'(x^1, x^2, x^3) = Q\delta(x^1)\delta(x^2)\delta(x^3) = Q\delta^3(x^1, x^2, x^3),$$

where δ is the one-dimensional, and δ^3 the three-dimensional *Dirac distribution* or *Dirac delta*. Equation (4.42), however, is closely related to (4.43) which has the form well known from electrostatics in an isotropic medium. One must only write the charge density in terms of coordinates ξ^i.

We recall considerations from Sect. 2.2 where we assumed the following connection between the bases: $\mathbf{m}_i = (A)_i{}^j \mathbf{e}_j$ which implies the following relation between the coordinates:

$$x^j = (A^T)^j{}_k \xi^k.$$

It follows from the integration properties of the Dirac delta that then

$$\delta^3(x^1, x^2, x^3) = |\det A|^{-1}\delta^3(\xi^1, \xi^2, \xi^3).$$

By virtue of (2.47), i.e., $|\det A| = (\det G)^{-1/2}$, we have

$$\delta^3(x^1, x^2, x^3) = (\det G)^{1/2}\delta^3(\xi^1, \xi^2, \xi^3).$$

At last equality (4.33) gives

$$\delta^3(x^1, x^2, x^3) = (\det \boldsymbol{\varepsilon})^{-1/2}\delta^3(\xi^1, \xi^2, \xi^3).$$

We are able now to write the right-hand side of (4.43) in the ξ^i coordinates:

$$\frac{\partial^2 \Phi}{(\partial \xi^1)^2} + \frac{\partial^2 \Phi}{(\partial \xi^2)^2} + \frac{\partial^2 \Phi}{(\partial \xi^3)^2} = -\frac{Q}{\varepsilon_0(\det \boldsymbol{\varepsilon})^{1/2}}\delta^3(\xi^1, \xi^2, \xi^3).$$

Recalling the well-known solution of this equation in the form of the Coulomb potential we may write down

$$\Phi(\mathbf{r}) = \frac{Q}{4\pi\varepsilon_0(\det \boldsymbol{\varepsilon})^{1/2}[(\xi^1)^2 + (\xi^2)^2 + (\xi^3)^2]^{1/2}} = \frac{Q}{4\pi\varepsilon_0(\det \boldsymbol{\varepsilon})^{1/2}|\mathbf{r}|_{\varepsilon}},$$
$$(4.44)$$

Fig. 4.3 Quasi-spheres in the
anisotropic dielectric's metric

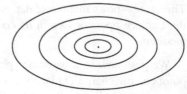

Fig. 4.4 Lines perpendicular
to quasi-spheres in a
dielectric's metric

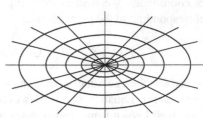

where $|\mathbf{r}|_\varepsilon$ is the length of $\mathbf{r}$ in the dielectric metric. If the matrix $\boldsymbol{\varepsilon}$ is diagonal with elements ε_i on the diagonal, then this length can be written in terms of x^i as

$$|\mathbf{r}|_\varepsilon^2 = \frac{(x^1)^2}{\varepsilon_1} + \frac{(x^2)^2}{\varepsilon_2} + \frac{(x^3)^2}{\varepsilon_3}.$$

The equipotential surfaces are quasi-spheres of constant "distance" from the origin of coordinates, that is ellipsoids in the natural metric—Fig. 4.3 shows their intersection with a plane passing through the charge.

From this we find the electric field intensity:

$$\mathbf{E}(\mathbf{r}) = -\mathbf{d}\Phi = -\mathbf{h}^i \frac{\partial \Phi}{\partial \xi^i} = \frac{Q}{4\pi\varepsilon_0 (\det\boldsymbol{\varepsilon})^{1/2}} \frac{\xi^1 \mathbf{h}^1 + \xi^2 \mathbf{h}^2 + \xi^3 \mathbf{h}^3}{[(\xi^1)^2 + (\xi^2)^2 + (\xi^3)^2]^{3/2}}. \qquad (4.45)$$

As a one-form field it is tangent to the above mentioned ellipsoids. Its coordinates are simultaneously coordinates of the primed vector field:

$$\mathbf{E}' = G^{-1}(\mathbf{E}) = \frac{Q\xi^i \mathbf{m}_i}{4\pi\varepsilon_0 (\det\boldsymbol{\varepsilon})^{1/2} |\mathbf{r}|_\varepsilon^3} = \frac{Q\mathbf{r}}{4\pi\varepsilon_0 (\det\boldsymbol{\varepsilon})^{1/2} |\mathbf{r}|_\varepsilon^3}.$$

This field is perpendicular in the dielectric metric to the equipotential surfaces; hence its lines disperse or converge radially from or to the origin, depending on the sign of the charge. They are shown in Fig. 4.4 on the background of the equipotential surfaces. Only the orientation is not marked. The lines of the non-primed vector field, that is, of the traditional $\mathbf{E}$, should be drawn as perpendicular to those surfaces in the natural metric; hence they are different—they are shown in Fig. 4.5. Due to relation (4.36) the lines of Fig. 4.4 are the lines of the $\mathbf{D}$ field. (Recall that two-forms have lines as their attitudes.) They are more natural for images of the electric field of a single charge.

Fig. 4.5 Lines perpendicular
to quasi-spheres in a natural
metric

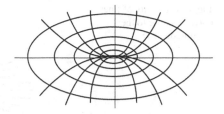

In order to find the electric induction we apply (4.37)–(4.45):

$$\mathbf{D} = \varepsilon_0 \mathbf{f}_*^{123} \lfloor \tilde{G}(\mathbf{E}) = \varepsilon_0 \mathbf{f}_*^{123} \lfloor \mathbf{E}'.$$

Due to (2.53) we have

$$\mathbf{f}_*^{123} = (\det \mathcal{G})^{-1/2} \mathbf{h}_*^{123} = (\det \boldsymbol{\varepsilon})^{1/2} \mathbf{h}_*^{123},$$

hence

$$\mathbf{D} = \frac{Q}{4\pi} \mathbf{h}_*^{123} \lfloor \frac{\xi^1 \mathbf{m}_1 + \xi^2 \mathbf{m}_2 + \xi^3 \mathbf{m}_3}{|\mathbf{r}|_\varepsilon^3} = \frac{Q}{4\pi} \frac{\xi^1 \mathbf{h}_*^{23} + \xi^2 \mathbf{h}_*^{31} + \xi^3 \mathbf{h}_*^{12}}{|\mathbf{r}|_\varepsilon^3}.$$

We may write this also as

$$\mathbf{D} = \frac{\lambda Q}{4\pi} \frac{\xi^1 \mathbf{h}^{23} + \xi^2 \mathbf{h}^{31} + \xi^3 \mathbf{h}^{12}}{|\mathbf{r}|_\varepsilon^3},$$

where λ is the unit nontwisted scalar with the handedness of the basis $\{\mathbf{m}_i\}$; we assume that it is the same as that of $\{\mathbf{e}_i\}$. We now invoke (3.92) and write down

$$\mathbf{D} = \frac{\lambda Q}{4\pi} \mathbf{d}(-\cos\theta) \wedge \mathbf{d}\alpha, \tag{4.46}$$

where θ and α are the angles of the quasi-spherical coordinates based on $\{\xi_i\}$. In this manner the electric induction of the Coulomb field is the solid visual angle twisted differential two-form from the position of the charge. This field is illustrated in Fig. 4.6 in some plane passing through the charge Q. The boundaries of its tubes are represented by straight lines outgoing radially from the charge. The rotationally non-symmetric density of the lines corresponds just to the "rotational" symmetry of (4.46) in the dielectric metric.

In this way we have arrived at a method of solving electrostatic problems in an anisotropic medium by applying a particular metric, in which the medium "appears" to be isotropic.

We may apply it to the next typical problem, namely the method of images. This is used to find the electrostatic field in the vicinity of conducting surfaces, and the simplest example is the field of a single charge q over an infinite conducting plane.

Fig. 4.6 Density of lines
corresponding to the
magnitude of **D**

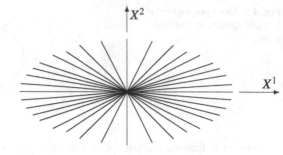

Fig. 4.7 Lines of the **D**-field
in the method of images for
an isotropic medium

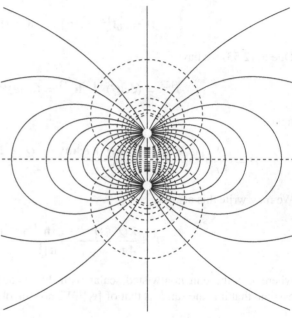

The essence of the method lies in the observation that for a pair of charges q and $-q$
placed in an isotropic medium, one of the equipotential surfaces is a plane of points
equally spaced from both charges (see Fig. 4.7), on which continuous lines are lines
of electric induction, and dotted lines are equipotential surfaces. The plane passes
through the centre of the segment joining the two charges and is perpendicular to the
segment. Since the conductor must have constant potential, it is sufficient to assume
that for a charge q placed at a distance a from the conductor, the electric field in
the half space above it is the same as the field produced by a pair of charges: one q
at distance a above the conductor, and a second, hypothetical $-q$, at distance a
inside the conductor. The latter is the mirror image of the first one, hence the name:
method of images.

To apply the method of images to a similar problem in an anisotropic dielectric,
one should find first the electrostatic field of two charges q and $-q$ in such a
medium. Now, whereas in Fig. 4.7 the equipotential surfaces in the vicinity of

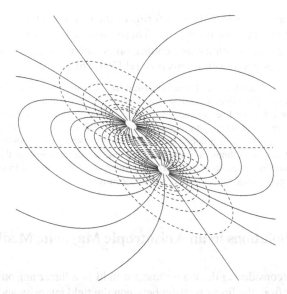

Fig. 4.8 Lines of the **D**-field in the method of images for an anisotropic medium

charges (visible as two empty circles) are approximately almost spheres, they are now almost ellipsoids, as shown in Fig. 4.3. Then, when approaching the middle point of the segment joining the charges, the surfaces become flatter, and exactly at half way one of the equipotentials becomes a plane perpendicular—in the sense of the dielectric metric—to the joining segment (see Fig. 4.8).

We are convinced that it is a plane because the field of a single charge is given in coordinates ξ_i by the same formula as the Coulomb field in the isotropic medium x_i coordinates. Therefore, if for two charges in one metric a particular equipotential is a plane, then in another metric it must also be a plane; only perpendicularity to the segment will be different. The continuous lines in this figure correspond to the attitudes of the electric induction, because they are perpendicular to the equipotentials in the dielectric metric. The lines of vectorial electric field intensity are different, because their perpendicularity in the natural metric is distinct. These lines are not shown in Fig. 4.8.

In this manner we arrive at the conclusion that the method of images applied to a charge q placed in the vicinity of a conducting plane in an anisotropic dielectric should be formulated as follows. The electric field of such a system is the same as that of a pair of charges q, $-q$, of which the second is the mirror reflection of the first one in the plane of the conductor, but the reflection is *perpendicular in the dielectric metric*.

Since the directions of vectors **D** do not depend on the metric (they are the same as the direction of the two-form **D**), they have an objective meaning and if we want to introduce electric field lines in anisotropic media, we should rather relate them to the electric induction. They can, for instance, be defined as follows: *the electric field* *lines* are such curves to which the two-forms of the electric induction are tangent at each point and have the same orientation. These lines have the property of beginning on positive charges or at infinity, and ending on negative charges or at infinity.

It is worth quoting Maxwell in this context [39], vol 1, p. 93:

> We have used the phrase Lines of Force because it has been used by Faraday and others. In strictness, however, these lines should be called Lines of Electric Induction.
>
> In the ordinary cases the lines of induction indicate the direction and magnitude of the resultant electromotive intensity at every point, because the intensity and the induction are in the same direction and in a constant ratio. There are other cases, however, in which it is important that these lines indicate primarily the induction, and that the intensity is directly indicated by the equipotential surfaces, [intensity] being normal to these surfaces and inversely proportional to the distances of consecutive surfaces.

4.3   Static Solutions in an Anisotropic Magnetic Medium

We move on to considering the magnetostatic field in a linear anisotropic medium. We should start from the linear relation between the field intensity and the magnetic induction, which in the traditional (twisted) vector coordinates is

$$B^1 = \mu_0(\mu^{11}H^1 + \mu^{12}H^2 + \mu^{13}H^3),$$

$$B^2 = \mu_0(\mu^{21}H^1 + \mu^{22}H^2 + \mu^{23}H^3),$$

$$B^3 = \mu_0(\mu^{31}H^1 + \mu^{32}H^2 + \mu^{33}H^3),$$

where μ^{ij} are (dimensionless) elements of the *relative permeability matrix* M of a given medium, whereas μ_0 is the *permeability of the vacuum*. Similarly as in the previous section we express the magnetic induction by the (twisted) vector coordinates:

$$B^k = \mu_0\mu^{kl}H_l, \tag{4.47}$$

or in two-form ones:

$$B_{ij} = \mu_0\epsilon_{ijk}\mu^{kl}H_l. \tag{4.48}$$

Likewise, we find the energy density of the magnetic field:

$$w_m = \frac{1}{2}\mathbf{H} \wedge \mathbf{B} = \frac{1}{2}\mu_0(H_i\mu^{ij}H_j)\,\mathbf{f}_*^{123}. \tag{4.49}$$

After introducing the matrix

$$G = M^{-1}$$

we arrive at the observation: the *inverse matrix to the relative permeability matrix determines a scalar product appropriate for a given anisotropic magnetic medium*. The basis $\{\mathbf{m}_i\}$ is now orthonormal for it.

This matrix determines the scalar product tensor G, which is present in the following relation between the twisted one-form $\mathbf{H}$ and the two-form $\mathbf{B}$:

$$\mathbf{H} = -\mu_0^{-1} G[[\mathbf{e}_{123}^* \lfloor \mathbf{B}]]. \tag{4.50}$$

Its inverse $\tilde{G} = G^{-1}$ stands in the inverse relation:

$$\mathbf{B} = \mu_0 \mathbf{f}_*^{123} \lfloor \tilde{G}[[\mathbf{H}]]. \tag{4.51}$$

One can check that, according to the scalar product $\tilde{G}$, the two-form $\mathbf{B}$ is perpendicular to $\mathbf{H}$, and the twisted vectors $\mathbf{B}'$ and $\mathbf{H}'$ satisfy the equality

$$\mathbf{B}' = \mu_0 (\det \mathcal{M})^{1/2} \mathbf{H}'. \tag{4.52}$$

which means that in this metric *the medium appears to be isotropic* with the scalar permeability $\mu_0 \sqrt{\det \mathcal{M}}$.

We proceed to consider the magnetostatic field produced in the anisotropic magnetic medium by an element $I\Delta\mathbf{l}$ of the electric current placed in the origin of coordinates. In other words, we seek a generalization of the *Biot-Savart* law to the anisotropic case. We assume that the magnetic field intensity has the form

$$\Delta\mathbf{H}(\mathbf{r}) = 1^* \eta \Delta\mathbf{l} \lfloor \{\mathbf{d}(-\cos\theta) \wedge \mathbf{d}\alpha\}, \tag{4.53}$$

where 1^* is the twisted unit scalar with the handedness of the basis $\{\mathbf{m}_i\}$ and η is a constant to be determined. We use here the result obtained for the electric induction. We cannot rewrite formula (4.46) (with a possible change of the coefficient), since point currents do not exist. Therefore we have added a contraction of the twisted two-form of the solid visual angle with the directed length of the current element. We know from the properties of contraction given in Sect. 3.2 that the twisted one-form (4.53) is parallel to $\Delta\mathbf{l}$.

To determine the coefficient η, we apply our formula to an infinite straight line conductor, which we choose parallel to $\mathbf{m}_3$ and, by this, to the one-form $\mathbf{d}\alpha$ (see Fig. 3.16). We have from the definition of the contraction

$$\Delta\mathbf{l} \lfloor \{\mathbf{d}(-\cos\theta) \wedge \mathbf{d}\alpha\} = \mathbf{d}(-\cos\theta)[[\Delta\mathbf{l}]] \, \mathbf{d}\alpha - \mathbf{d}\alpha[[\Delta\mathbf{l}]] \, \mathbf{d}(-\cos\theta).$$

(Recall: placing a vector in square brackets denotes the value of a form on that vector.) The second term vanishes, because now $\Delta\mathbf{l} \parallel \mathbf{d}\alpha$ and for the infinitesimal $d\mathbf{l}$ there remains

$$d\mathbf{H}(\mathbf{r}) = 1^* \eta \, \mathbf{d}(-\cos\theta)[[d\mathbf{l}]] \, \mathbf{d}\alpha.$$

Fig. 4.9 Twisted one-form **H** around a straight-line conductor

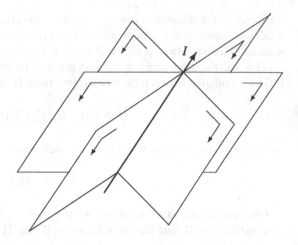

We already know that for the function $-\cos\theta$ the level surfaces are cones with common vertices in the origin and common axes coinciding with the attitude of $\mathbf{m}_3$. The one-form $\mathbf{d}(-\cos\theta)$ has the orientation from the positive part of ξ^3-axis to the negative one, hence for a vector $d\mathbf{l}$ with direction of $\mathbf{m}_3$, the value of this form on $d\mathbf{l}$ gives the differential of $\cos\theta$ without the minus. Thus we have $d\mathbf{H}(\mathbf{r}) = 1^*\eta\, d(\cos\theta)\mathbf{d}\alpha$. Integration over an infinite straight conductor yields

$$\mathbf{H}(\mathbf{r}) = 1^*\eta\cos\theta|_\pi^0\,\mathbf{d}\alpha = 2\cdot 1^*\eta\,\mathbf{d}\alpha. \qquad (4.54)$$

We see that the twisted differential one-form of the magnetic field intensity produced by an infinite straight-line current is proportional to the differential form $\mathbf{d}\alpha$ considered in Sect. 3.5. We copy here Fig. 3.16 as Fig. 4.9 adapted to the considered magnetic field. It is singular on the ξ^3-axis, that is, on the current itself, which is rather perspicuous. The conductor with the current constitutes a natural common boundary of half planes determining the attitude of **H**. Its orientation is compatible with that of the current on the boundary of the half-planes.[1]

We still do not know η. Use now the Ampere-Oersted law by integrating (4.54) over a closed curve surrounding the conductor once:

$$\oint_C \mathbf{H}[[d\mathbf{l}^*]] = 2\cdot 1^*\eta\oint_C \mathbf{d}\alpha[[d\mathbf{l}^*]] = 2\eta\oint_C \mathbf{d}\alpha[[d\mathbf{l}]] = I.$$

---

[1] The fact the surfaces of **H** have current as their boundaries is a common feature of all magnetic intensity fields. Such surfaces were called *force surfaces of the magnetic field* in Sect. 1.1 of my book [17] and depicted for several typical and non-typical configurations of currents.

The unit twisted scalar 1^* has changed the twisted vector $d\mathbf{l}^*$ into the nontwisted vector $d\mathbf{l}$. We know that the integral of one-form $d\alpha$ over such a curve is 2π, so we have $4\pi\eta = I$, whence

$$\eta = \frac{I}{4\pi}.$$

By inserting this into (4.54) we obtain

$$\mathbf{H}(\mathbf{r}) = \frac{1^*I}{2\pi}\, d\alpha. \tag{4.55}$$

We may now write (4.53) as

$$\Delta\mathbf{H}(\mathbf{r}) = \frac{1^*I}{4\pi}\, \Delta\mathbf{l}\rfloor\{d(-\cos\theta) \wedge d\alpha\}. \tag{4.56}$$

We obtain from (3.92)

$$\Delta\mathbf{H}(\mathbf{r}) = \frac{1^*I}{4\pi}\, \frac{\Delta\mathbf{l}\rfloor(\xi^3\,\mathbf{h}^{12} + \xi^2\,\mathbf{h}^{31} + \xi^1\,\mathbf{h}^{23})}{|\mathbf{r}|_\mu^3},$$

where $|\mathbf{r}|_\mu$ is length of the radius vector in the metric determined by the basis $\{\mathbf{m}_i\}$. In a particular case of the vacuum one may transform this expression to the form

$$\Delta\mathbf{H}(\mathbf{r}) = \frac{I}{4\pi}\, \frac{\Delta\mathbf{l} \times \mathbf{r}}{|\mathbf{r}|^3}, \tag{4.57}$$

in which we recognize the *Biot-Savart law*. Therefore, we may call (4.56) a *generalization of the Biot-Savart law on a homogeneous anisotropic medium.*

To find the magnetic induction two-form, we apply the mapping (4.50) to the found magnetic field intensity. I omit tedious and uninteresting calculations leading to the result:

$$\Delta\mathbf{B}(\mathbf{r}) = -\frac{\mu_0 I\,(\det\mathcal{M})^{1/2}}{4\pi}\, \frac{(\xi^1\mathbf{h}^1 + \xi^2\mathbf{h}^2 + \xi^3\mathbf{h}^3) \wedge \underline{\Delta\mathbf{l}}}{[(\xi^1)^2 + (\xi^2)^2 + (\xi^3)^2]^{3/2}},$$

where $\underline{\Delta\mathbf{l}} = G[\Delta\mathbf{l}]$ is a one-form obtained from vector $\Delta\mathbf{l}$ by the mapping (2.71). This may also be written as

$$\Delta\mathbf{B} = \underline{\Delta\mathbf{l}} \wedge d\Psi(\mathbf{r}), \tag{4.58}$$

where

$$\Psi(\mathbf{r}) = \frac{\mu_0 I\,(\det\mathcal{M})^{1/2}}{4\pi[(\xi^1)^2 + (\xi^2)^2 + (\xi^3)^2]^{1/2}}$$

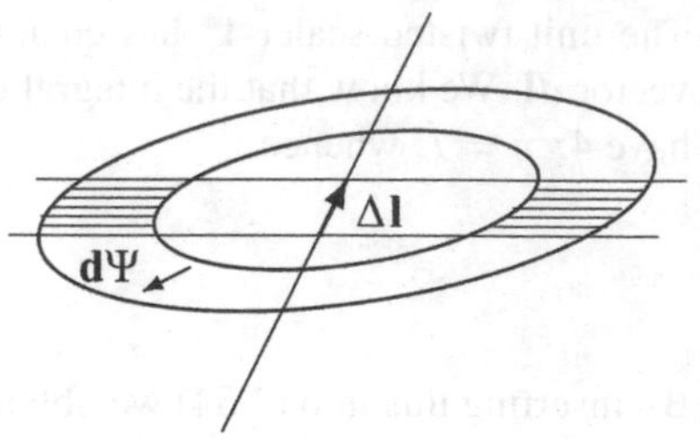

Fig. 4.10 Section of two
ellipsoids $|\mathbf{r}|_\mu = $ const by a
plane passing through $\Delta\mathbf{l}$

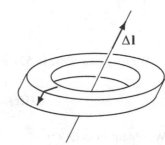

Fig. 4.11 One tube of the
two-form $\Delta\mathbf{B}$

is a scalar function proportional to (4.44); hence its level surfaces are ellipsoids.

The attitude of $\underline{\Delta\mathbf{l}}$ is formed by planes perpendicular (in $\{\mathbf{m}_i\}$ metric) to the vector $\Delta\mathbf{l}$. We show two of them in Fig. 4.10 in the intersection by a plane passing through the current element. The attitude of $\mathbf{d}\Psi(\mathbf{r})$ is determined by ellipsoids $|\mathbf{r}|_\mu = $ const, and we also show in Fig. 4.10 two such ellipsoids in a section by the named plane passing through the current element.

The intersection of layers between the ellipsoids with the layers between the planes is the two-part region filled in by lines. In three dimensions this intersection forms a tube of the outer product (4.58). One tube of the $\Delta\mathbf{B}$ two-form is shown in Fig. 4.11. An axis of this tube is the ellipse playing the role of the circle in the metric of the basis $\{\mathbf{m}_i\}$. The field (4.58) is perpendicular (in the medium metric) to the field (4.53) depicted in Fig. 4.9.

One may also ask how the twisted vector fields $\mathbf{H}$ and $\mathbf{B}$ would look, these fields being introduced by means of the natural metric. The lines of $\mathbf{H}$ are circles with centres on the line of $\Delta\mathbf{l}$ (see Fig. 4.12). This circles are orthogonal in the natural metric to the planes from Fig. 4.9. The lines of $\mathbf{B}$ are the same ellipses, discussed in Fig. 4.11, and depicted in Fig. 4.13. As we see on these two figures, there are two kinds of magnetic field lines in the natural metric. Which is better? In my opinion the lines of the pseudovector field $\mathbf{B}$ are better, because they are independent of a metric.

Observe that the choice of orientation for the pseudovectors (twisted vectors) discussed in Sect. 1.2 and illustrated in Figs. 1.7 and 1.8 gives a new orientation for the magnetic field lines: the oriented rings surrounding the lines have their arrows oriented compatibly with the currents generating the field. The word "compatible" means here that the arrow on the side of the ring that is closer to the current has the

Fig. 4.12 Field lines for the
twisted vector **H** in an
anisotropic medium

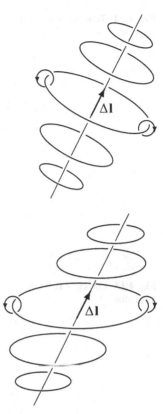

Fig. 4.13 Field lines for the
twisted vector **B** in an
anisotropic medium

same orientation as the current itself. We do not need any left or right hand for this
rule of determining the orientation of magnetic field lines.

The adoption of a particular metric in which an anisotropic magnetic medium
appears to be isotropic allows us to solve other problems in magnetostatics, for
instance the magnetic field of a solenoid filled by some anisotropic material. If
loops of the current are "magnetically" orthogonal to the solenoid axis, then we
are allowed to claim that the two-form **B** is parallel to the axis (see Fig. 4.14), and
the twisted one-form **H** is parallel to the current loops (see Fig. 4.15). Both fields are
homogenous inside the solenoid. After passing to pseudovector fields **H** and **B** we
may state that only the field **B** is parallel to the axis of the solenoid, but we cannot
say the same about **H**.

We see again in this example of a solenoid that the magnetic field surfaces
(embraced by ellipses of Fig. 4.15) have their boundaries on the currents.

Fig. 4.14 Tubes of **B** inside a
solenoid

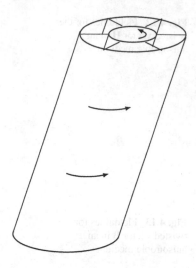

Fig. 4.15 Planes of **H** inside
a solenoid

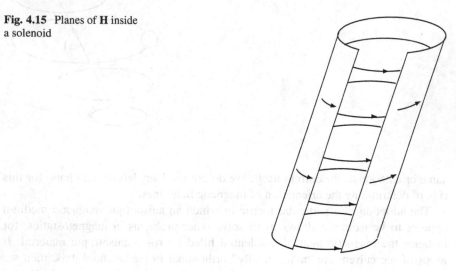

Chapter 5
Electromagnetic Waves

This chapter is devoted to electromagnetic waves, as far as possible treated without the use of metric. Essential notions, necessary to describe any wave are introduced in Sect. 5.1, namely: a phase, a wave covector and a one-from of slowness as a quantity dual to phase velocity (which can't be defined uniquely in a metric-free manner).

A general electromagnetic wave is considered in Sect. 5.2 with the use of three scalar functions η, ζ, ξ of position, which can be treated as curvilinear coordinates. The first, η, is present in the phase, whereas the exterior derivatives $\mathbf{d}\zeta$ and $\mathbf{d}\xi$ play the role of amplitudes for $\mathbf{E}$ and $\mathbf{H}$. After solving Maxwell's equations, the Poynting two-form, energy density and momentum density are found for the wave. It turns out that the energy ΔW and momentum $\Delta \mathbf{p}$ contained in a volume ΔV are proportional to ω and $\mathbf{k} = \mathbf{d}\eta$, respectively, with the same coefficient. This resembles the de Broglie relations.

Then the wave is considered in a doubly anisotropic medium, for which two metrics are needed: electric and magnetic. Some field forms are pairwise orthogonal with respect to one or other scalar product. Moreover the functions ζ and ξ must be harmonic (i.e., fulfilling the Laplace equation) in an appropriate metric. An example of cylindrical coordinates (modified to be harmonic) is considered. It is shown that a linearly polarized wave propagating radially from the cylinder axis Z does not exist, whereas a plane wave propagating along the Z axis is possible. Its surfaces of constant phase are planes $z-$ const, but the fields $\mathbf{E}, \mathbf{H}$ are not constant on those planes. Because of this the wave is called *semiplane*. The attempt to use harmonic spherical coordinates to find a wave propagating radially from the origin is not successful.

The truly plane electromagnetic wave (with constant wave covector and constant fields on planes of constant phase) is considered in Sect. 5.3 for a medium that is magnetically isotropic and electrically anisotropic. The need to introduce so-called *eigenwaves* appears, such that for a fixed wave covector the one-form of slowness has two magnitudes depending on the eigenvalues of a reduced permittivity matrix.

© Springer Nature Switzerland AG 2021

B. Jancewicz, *Directed Quantities in Electrodynamics*,
https://doi.org/10.1007/978-3-030-90471-5_5

The two eigenwaves have mutually orthogonal field quantities in electric metrics and their Poynting forms have diferent directions.

5.1 Phase of a Wave

Before we proceed to electromagnetic waves, let us consider a scalar-valued plane wave of the form

$$F(\mathbf{r}, t) = A \cos(\mathbf{k}[[\mathbf{r}]] - \omega t), \tag{5.1}$$

where A is amplitude, ω is circular frequency, $\mathbf{k}$ is the nontwisted one-form of phase density (see (1.6)), for which we shall use the traditional name *wave covector*, and $\mathbf{k}[[\mathbf{r}]]$ is its value of on the ordinary radius vector $\mathbf{r}$. This one-form is a premetric counterpart of the wave vector $\mathbf{k}$. Due to the Exercise 3.26 it satisfies equality $\mathbf{d} \wedge \mathbf{k}[[\mathbf{r}]] = \mathbf{k}$. By freezing time, that is by choosing a fixed moment $t = 0$, one obtains a new function

$$f(\mathbf{r}) = A \cos \mathbf{k}[[\mathbf{r}]], \tag{5.2}$$

which is shown in Fig. 5.1 for two spatial dimensions. It is visible that the form $\mathbf{k}$ is chosen parallel to the Y axis.

This diagram does not show either of the two possible orientations of $\mathbf{k}$. It could equally well be the same as or opposite to that of the X axis. This corresponds mathematically to the fact, that the cosine function is symmetric, hence instead of (5.2) we could consider the function

$$f'(\mathbf{r}) = A \cos \left(-\mathbf{k}[[\mathbf{r}]]\right), \tag{5.3}$$

Fig. 5.1 Scalar-valued wave in two dimensions for freezed time

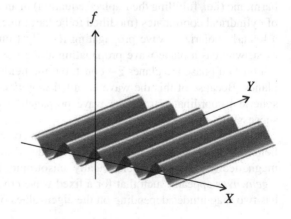

to which the same diagram corresponds. This observation is related to the fact that a snapshot photograph of the sinusoidal wave does not show the direction in which the wave is moving.

Since one may consider two functions (5.2) and (5.3) referring to the image of a given moment, one can also admit two functions with variable time. Therefore, besides (5.1) one should also consider the function

$$G(\mathbf{r}, t) = A \cos(-\mathbf{k}[[\mathbf{r}]] + \omega t), \tag{5.4}$$

which due to the symmetry of the cosine gives the same result as (5.1).

Expressions (5.1–5.4) are called *plane* waves because they are constant where the function $k(\mathbf{r}) = \mathbf{k}[[\mathbf{r}]]$ is constant, that is on planes. The use of the one-form $\mathbf{k}$ leaves open the question of the direction of propagation of the wave. A direction, understood as the one-dimensional direction of a vector, cannot be chosen perpendicular to the planes of constant field A, because a scalar product is not yet chosen.

The argument $\phi = \mathbf{k}[[\mathbf{r}]] - \omega t$ of the cosine in (5.1), called the *phase*, depends linearly on both on position and time. Another function $\phi' = \omega t - \mathbf{k}[[\mathbf{r}]]$ is the *phase* for (5.4). The loci of points of constant phase are still planes, but these planes move when time flows. Can a phase velocity be introduced? If position $\mathbf{r}$ becomes a function of time, this means that we introduce the motion $\mathbf{r}(t)$ of a fictitious particle. Introduce thus a motion (of course uniform) such that, the fictitious particle is always on a plane of fixed phase:

$$\phi = \mathbf{k}[[\mathbf{r}(t)]] - \omega t = \text{const} \quad \text{or} \quad \phi' = \omega t - \mathbf{k}[[\mathbf{r}(t)]] = \text{const}. \tag{5.5}$$

Differentiate this equality with respect to time:

$$\frac{d\mathbf{k}[[\mathbf{r}(t)]]}{dt} - \omega = 0 \quad \text{and} \quad \omega - \frac{d\mathbf{k}[[\mathbf{r}(t)]]}{dt} = 0.$$

Since the mapping $\mathbf{r} \to \mathbf{k}[\mathbf{r}]$ is linear and continuous, the derivative may be put under the argument of $\mathbf{k}$, which for both equalities gives the same relation:

$$\mathbf{k}\left[\left[\frac{d\mathbf{r}(t)}{dt}\right]\right] = \omega,$$

that is,

$$\mathbf{k}[[\mathbf{v}]] = \omega,$$

where $\mathbf{v} = d\mathbf{r}/dt$ denotes the velocity of the fictitious particle. We obtain:

$$\omega^{-1}\mathbf{k}[[\mathbf{v}]] = 1. \tag{5.6}$$

Fig. 5.2 The one-form $\omega^{-1}\mathbf{k}$
has the same value for many
arguments

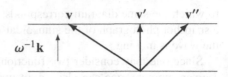

There are many velocities **v** for which the one-form $\omega^{-1}\mathbf{k}$ gives the value one
(see Fig. 5.2). Which one can we admit as the phase velocity? As long as the scalar
product is not present, none of them. Once a scalar product is introduced, we choose
the velocity perpendicular to the planes of constant phase. For such a velocity **v**,
relation (5.6) assumes the shape

$$\omega^{-1}\mathbf{k}\cdot\mathbf{v} = \omega^{-1}kv = 1,$$

where **k** is a vector also perpendicular to the planes of constant phase, $k = |\mathbf{k}|$, and
hence $v = \frac{\omega}{k}$, which is the well-known formula for the value of the *phase velocity*.
The equivalent formula $k/\omega = v^{-1}$ says that the quotient $\omega^{-1}k$ is the inverse of the
phase velocity. If we want to call $u = \omega^{-1}k$ by a separate word, *slowness* is most
suitable. (The lower the velocity, the greater the slowness—its physical dimension is
s/m.) It can be treated as a directed quantity of the same type as the wave covector **k**:

$$\mathbf{u} = \omega^{-1}\mathbf{k} \tag{5.7}$$

and may be called the linear form of *phase slowness*. This notion can be introduced
when no scalar product is present, which is not the case for the phase velocity. The
relation (5.6) yields

$$\mathbf{u}[[\mathbf{v}]] = 1.$$

It is possible to consider a more general wave from given (5.1) by writing down,
instead of cosine, an arbitrary scalar function ψ:

$$F(\mathbf{r}, t) = A\psi(\mathbf{k}[[\mathbf{r}]] - \omega t). \tag{5.8}$$

We admit in this manner an arbitrary time dependence determined by ψ.
An even more general wave is obtained when in place of $\mathbf{k}[[\mathbf{r}]]$ an arbitrary scalar
function η of position is taken:

$$F(\mathbf{r}, t) = A\psi(\eta(\mathbf{r}) - \omega t). \tag{5.9}$$

The argument $\phi = \eta(\mathbf{r}) - \omega t$ of ψ is still called the *phase*, depending linearly only
on time. If ψ is sine or cosine the wave is called *harmonic in time*. The loci of
constant phase, that is, level surfaces of η, are *wave fronts*; they move when time
flows. When η is not a linear function, the wave may not be called the plane wave.

Is it possible to introduce the phase velocity? Let us consider again the motion $\mathbf{r}(t)$ of a fictitious particle that moves (not necessarily uniformly) on a surface of constant phase:

$$\phi = \eta(\mathbf{r}(t)) - \omega t = \text{const.}$$

Differentiate this equality with respect to time:

$$\mathbf{d}\eta(\mathbf{r}(t))\lfloor\frac{d\mathbf{r}}{dt} - \omega = 0,$$

which gives

$$(\mathbf{d}\eta)[[\mathbf{v}]] = \omega,$$

where $\mathbf{v} = d\mathbf{r}/dt$ is the velocity of the particle. We want to accept this as the phase velocity of the wave. We obtain:

$$\omega^{-1}\mathbf{d}\eta[[\mathbf{v}]] = 1. \tag{5.10}$$

We know already from considerations about plane waves that there any many velocities $\mathbf{v}$ satisfying this equation, as long as a scalar product is not introduced.

It is worth denoting the exterior derivative by one symbol:

$$\mathbf{k} = \mathbf{d}\eta. \tag{5.11}$$

This can be nonconstant. Equation (5.10) is a counterpart of (5.6); hence the expression

$$\mathbf{u} = \omega^{-1}\mathbf{d}\eta = \omega^{-1}\mathbf{k} \tag{5.12}$$

can be accepted as the *phase slowness* of the wave (5.9). It is a premetric quantity.

5.2 General Waves

5.2.1 Premetric Description

When considering electromagnetic waves we have to admit the simultaneous presence of electric and magnetic fields. If the matrices of permittivity $\mathcal{E}$ and permeability $\mathcal{M}$ are not proportional, one cannot find a vector basis simultaneously orthonormal for the two scalar products. Therefore application of two scalar products is necessary: electric $g_\varepsilon(\cdot, \cdot)$ and magnetic $g_\mu(\cdot, \cdot)$.

Let $\psi(\cdot)$ be a scalar function of a scalar argument, ω a positive scalar constant, η, ζ and ξ given scalar functions of position $\mathbf{r}$ (they can be treated as curvilinear coordinates) and $\mathbf{B}_0$, $\mathbf{D}_0$ constant two-forms to be found. We are searching for solutions of the free Maxwell equations in the form:

$$\mathbf{E}(\mathbf{r}, t) = a\psi(\eta(\mathbf{r}) - \omega t)\,\mathbf{d}\zeta(\mathbf{r}), \qquad \mathbf{B}(\mathbf{r}, t) = \psi(\eta(\mathbf{r}) - \omega t)\,\mathbf{B}_0(\mathbf{r}), \qquad (5.13)$$

$$\mathbf{H}(\mathbf{r}, t) = b1^*\psi(\eta(\mathbf{r}) - \omega t)\,\mathbf{d}\xi(\mathbf{r}), \qquad \mathbf{D}(\mathbf{r}, t) = \psi(\eta(\mathbf{r}) - \omega t)\,\mathbf{D}_0(\mathbf{r}), \qquad (5.14)$$

where a, b are two scalar constants; they can be treated as amplitudes of oscillations. The symbol 1^* is the basic twisted scalar,[1] necessary to ensure that $\mathbf{H}$ is a twisted one-form. Note that time was "put" only into ψ. The fields (5.13, 5.14) can be interpreted as a *running electromagnetic wave* with an arbitrary time dependence determined by ψ. Let us ponder the physical dimensions. Since the second term in the combination $\eta(\mathbf{r}) - \omega t$ is dimensionless, the function η has the same character. If functions ψ, ξ, ζ are assumed to be dimensionless scalars, the constant a should be expressed in volts, and the constant b in amperes. Then the physical dimensions of fields are $[\mathbf{E}]$=V/m, $[\mathbf{H}]$=A/m, as they should be.

The one-forms $\mathbf{d}\zeta$ and $\mathbf{d}\xi$ are generally functions of position. The presence of the same scalar function ψ in all expressions (5.13, 5.14) is tantamount to the assumption that oscillations of the fields $\mathbf{E}$, $\mathbf{H}$, $\mathbf{B}$ and $\mathbf{D}$ are synchronous.

We proceed to consider Maxwell's equations. The exterior derivative of $\mathbf{E}$ is $\mathbf{d} \wedge \mathbf{E} = a\mathbf{d}\psi \wedge \mathbf{d}\zeta$, and due to Exercise 3.26 it can be written as:

$$\mathbf{d} \wedge \mathbf{E} = a\psi'\mathbf{d}\eta \wedge \mathbf{d}\zeta,$$

where ψ' is the derivative of ψ with respect to the whole scalar argument. We calculate also the time derivative of $\mathbf{B}$:

$$\frac{\partial \mathbf{B}}{\partial t} = \frac{\partial \psi}{\partial t}\mathbf{B}_0 = -\omega\psi'\mathbf{B}_0.$$

The Maxwell equation $\mathbf{d} \wedge \mathbf{E} + \partial \mathbf{B}/\partial t = 0$ yields the condition

$$a\psi'\,\mathbf{d}\eta \wedge \mathbf{d}\zeta - \omega\psi'\mathbf{B}_0 = 0$$

and division by ψ' allows us to write down

$$\mathbf{B}_0(\mathbf{r}) = a\omega^{-1}\mathbf{d}\eta \wedge \mathbf{d}\zeta = a\omega^{-1}\mathbf{k} \wedge \mathbf{d}\zeta \qquad (5.15)$$

---

[1] See the definition preceding Exercise 3.2.

Fig. 5.3 Forms **E** and **B** for given **k**

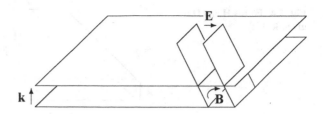

and

$$\mathbf{B} = \omega^{-1}\mathbf{d}\eta \wedge \mathbf{E} = \omega^{-1}\mathbf{k} \wedge \mathbf{E} = \mathbf{u} \wedge \mathbf{E}. \tag{5.16}$$

We see from (5.16) that two forms (5.13) are parallel: $\mathbf{B} \parallel \mathbf{E}$. Moreover, $\mathbf{B} \parallel \mathbf{u}$. This observations is illustrated in Fig. 5.3.

In order to check the second Maxwell equation $\mathbf{d} \wedge \mathbf{B} = 0$ we notice that

$$\mathbf{d} \wedge \mathbf{B} = \psi' \, \mathbf{d}\eta \wedge \mathbf{B}_0 + \psi \, \mathbf{d} \wedge \mathbf{B}_0.$$

By virtue of (5.15) the second term is zero; hence there remains

$$\mathbf{d} \wedge \mathbf{B} = \psi' \, \mathbf{d}\eta \wedge \mathbf{B}_0.$$

Again due to (5.15):

$$\mathbf{d} \wedge \mathbf{B} = a\psi'\omega^{-1}\mathbf{d}\eta \wedge \mathbf{d}\eta \wedge \mathbf{d}\zeta = 0, \tag{5.17}$$

so the second Maxwell equation is automatically satisfied.

The third free Maxwell equation $\mathbf{d} \wedge \mathbf{H} - \partial \mathbf{D}/\partial t = 0$, by similar reasoning, is reduced to the condition

$$b1^* \, \psi' \, \mathbf{d}\eta \wedge \mathbf{d}\xi + \psi'\omega\mathbf{D}_0 = 0$$

which allows us to write down

$$\mathbf{D}_0 = -b1^*\omega^{-1}\mathbf{d}\eta \wedge \mathbf{d}\xi \quad \text{and} \quad \mathbf{D} = -\omega^{-1}\mathbf{k} \wedge \mathbf{H} = -\mathbf{u} \wedge \mathbf{H}. \tag{5.18}$$

Due to this condition, the fourth free Maxwell equation $\mathbf{d} \wedge \mathbf{D} = 0$ is automatically satisfied. It is also visible from (5.18) that $\mathbf{D} \parallel \mathbf{H}$ and $\mathbf{D} \parallel \mathbf{k}$; this is displayed in Fig. 5.4.

Fig. 5.4 Forms **H** and **D** for given **k**

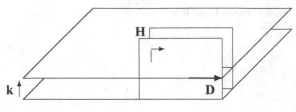

Fig. 5.5 Configuration of forms **E**, **D**, **H** and **B** for given **k**

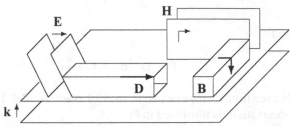

Summarizing, the found solutions of free Maxwell equations can be written in the forms:

$$\mathbf{E}(\mathbf{r}, t) = a\psi[\,\eta(\mathbf{r}) - \omega t\,]\,\mathbf{d}\zeta(\mathbf{r}), \qquad \mathbf{B}(\mathbf{r}, t) = a\omega^{-1}\psi[\,\eta(\mathbf{r}) - \omega t\,]\,\mathbf{d}\eta \wedge \mathbf{d}\zeta,$$
$$(5.19)$$

$$\mathbf{H}(\mathbf{r}, t) = b1^*\psi[\,\eta(\mathbf{r}) - \omega t\,]\,\mathbf{d}\xi(\mathbf{r}), \qquad \mathbf{D}(\mathbf{r}, t) = -b1^*\omega^{-1}\psi[\,\eta(\mathbf{r}) - \omega t\,]\,\mathbf{d}\eta \wedge \mathbf{d}\xi.$$
$$(5.20)$$

Since the forms **d**ζ and **d**ξ do not depend on time, the fields (5.19, 5.20) at a fixed point in space, have their attitudes constant in time, and only their magnitudes and orientations change in a rhythm given by the scalar function ψ; therefore the wave is linearly polarized. Formulas (5.19, 5.20) can be treated as a premetric expressions for the electromagnetic wave. We show in Fig. 5.5 the relative configuration of the four field quantities.

It is worth to consider the Poynting two-form:

$$\mathbf{S} = \mathbf{E} \wedge \mathbf{H} = ab1^*\psi(\mathbf{r}, t)^2 \mathbf{d}\zeta \wedge \mathbf{d}\xi. \qquad (5.21)$$

A direction of this twisted two-form must be recognized as the propagation direction of the wave. We repeat the previous figure with the Poynting form added—this is Fig. 5.6.

The energy densities of the electric and magnetic fields are:

$$w_e = \frac{1}{2}\mathbf{E} \wedge \mathbf{D}, \qquad\qquad w_m = \frac{1}{2}\mathbf{H} \wedge \mathbf{B},$$

Fig. 5.6 The Poynting form S added to the configuration of electromagnetic forms

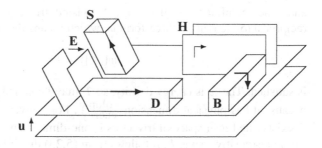

After using (5.16) and (5.18) these take the form:

$$w_e = \frac{1}{2} \mathbf{E} \wedge (\mathbf{H} \wedge \mathbf{u}), \qquad w_m = \frac{1}{2} \mathbf{H} \wedge (\mathbf{u} \wedge \mathbf{E}).$$

Therefore, the contributions of two fields are the same:

$$w_e = \frac{1}{2} (\mathbf{E} \wedge \mathbf{H}) \wedge \mathbf{u}, \qquad w_m = \frac{1}{2} (\mathbf{E} \wedge \mathbf{H}) \wedge \mathbf{u}.$$

This means, that in the running electromagnetic wave, the same amount of energy is contained in the electric field, as well as in the magnetic field. In this manner, the energy of the total electromagnetic field of the considered wave is

$$w = w_e + w_m = (\mathbf{E} \wedge \mathbf{H}) \wedge \mathbf{u} = \mathbf{S} \wedge \mathbf{u},$$

Hence an important identity arises:

$$w = \mathbf{S} \wedge \mathbf{u} \tag{5.22}$$

which relates energy density to the energy flux density. After inserting the explicit expressions (5.19), (5.20) and (5.12) we can write down

$$w(\mathbf{r}, t) = ab1^* \omega^{-1} \psi(\mathbf{r}, t)^2 \mathbf{d}\zeta(\mathbf{r}) \wedge \mathbf{d}\xi(\mathbf{r}) \wedge \mathbf{d}\eta(\mathbf{r}). \tag{5.23}$$

This is a twisted three-form. Its orientation should be positive, hence the twisted scalar 1^* should have that of the nontwisted three-form: $\mathbf{d}\zeta \wedge \mathbf{d}\xi \wedge \mathbf{d}\eta$.

The three one-forms

$$\mathbf{d}\zeta, \ \mathbf{d}\xi, \ \mathbf{d}\eta \tag{5.24}$$

are linearly independent, hence they form a basis of one-forms in each point of space. Therefore, a dual basis of vectors $\mathbf{t}_\zeta$, $\mathbf{t}_\xi$, $\mathbf{t}_\eta$ exists such that

$$\mathbf{d}\zeta[[\mathbf{t}_\zeta]] = 1, \quad \mathbf{d}\xi[[\mathbf{t}_\xi]] = 1, \quad \mathbf{d}\eta[[\mathbf{t}_\eta]] = 1,$$

and other combinations give zero. A twisted trivector T is worth introducing, reciprocal to the twisted three-form w, which means that the condition

$$w[[T]] = 1$$

is satisfied. The w is energy density, so T may be called a *volume of unit energy*, because it has physical dimension $\frac{\text{volume}}{\text{energy}}$. For a given w only one such trivector T exists (the linear space of trivectors is one-dimensional). Since the twisted three-form w is positive, so is T. It follows from (5.23) that it has the form

$$T = (ab)^{-1} 1^* \omega \psi^{-2} \mathbf{t}_\zeta \wedge \mathbf{t}_\xi \wedge \mathbf{t}_\eta. \tag{5.25}$$

We calculate the energy ΔW contained in a volume described by an twisted trivector $\Delta V = 1^* |\Delta V| \mathbf{t}_\zeta \wedge \mathbf{t}_\xi \wedge \mathbf{t}_\eta$, where $|\Delta V|$ is magnitude of the volume:

$$\Delta W = w[\Delta V] = ab1^* \omega^{-1} \psi^2 (\mathbf{d}\zeta \wedge \mathbf{d}\xi \wedge \mathbf{d}\eta)[[1^* |\Delta V| \mathbf{t}_\zeta \wedge \mathbf{t}_\xi \wedge \mathbf{t}_\eta]].$$

The value of the basis three-form $\mathbf{d}\zeta \wedge \mathbf{d}\xi \wedge \mathbf{d}\eta$ basis trivector $\mathbf{t}_\zeta \wedge \mathbf{t}_\xi \wedge \mathbf{t}_\eta$ is one; hence we obtain

$$\Delta W = ab\omega^{-1} \psi^2 |\Delta V|. \tag{5.26}$$

We apply formula (4.19), that is $\Delta \mathbf{p} = \mathbf{D} \lfloor (\Delta V \lfloor \mathbf{B})$, to calculate the momentum of the electromagnetic field in the volume (5.26). The first step is done by virtue of (5.19):

$$\Delta V \lfloor \mathbf{B} = 1^* |\Delta V| (\mathbf{t}_\zeta \wedge \mathbf{t}_\xi \wedge \mathbf{t}_\eta) \lfloor (a\omega^{-1} \psi \, \mathbf{d}\eta \wedge \mathbf{d}\zeta)$$

$$= a1^* \omega^{-1} \psi |\Delta V| \{ (\mathbf{t}_\zeta \wedge \mathbf{t}_\xi \wedge \mathbf{t}_\eta) \lfloor \mathbf{d}\eta \} \lfloor \mathbf{d}\zeta = a1^* \omega^{-1} \psi |\Delta V| (\mathbf{t}_\zeta \wedge \mathbf{t}_\xi) \lfloor \mathbf{d}\zeta$$

$$= -a1^* \omega^{-1} \psi |\Delta V| \mathbf{t}_\xi.$$

The second step follows from (5.20):

$$\Delta \mathbf{p} = \mathbf{D} \lfloor (\Delta V \lfloor \mathbf{B}) = -b1^* \omega^{-1} \psi (\mathbf{d}\eta \wedge \mathbf{d}\xi) \lfloor (-a1^* \omega^{-1} \psi |\Delta V| \mathbf{t}_\xi),$$

$$\Delta \mathbf{p} = ab \, \omega^{-2} \psi^2 |\Delta V| (\mathbf{d}\eta \wedge \mathbf{d}\xi) \lfloor \mathbf{t}_\xi = ab\omega^{-2} \psi^2 |\Delta V| \, \mathbf{d}\eta.$$

According to the notation (5.11), we may write this as

$$\Delta \mathbf{p} = ab \, \omega^{-2} \psi^2 |\Delta V| \, \mathbf{k}. \tag{5.27}$$

This result means that the electromagnetic wave brings momentum parallel to wave covector, which cannot be said about the energy flux density, described by the Poynting two-form.

After introducing the notation $h = ab\,\omega^{-2}\psi^2|\Delta V|$ one can write down the relations (5.26) and (5.27) in similar form:

$$\Delta W = h\omega, \qquad \Delta \mathbf{p} = h\mathbf{k}. \tag{5.28}$$

The striking similarity of these formulas to the *de Broglie relations* is visible. The common factor h has the physical dimension of the action.

It is possible to introduce a velocity of energy transport by the electromagnetic wave. In the traditional presentation of electrodynamics (solely by vectors and scalars), the equality

$$\mathbf{S} = w\mathbf{v}, \tag{5.29}$$

is written, defining vector $\mathbf{v}$ as the velocity of energy transport. This is analogous to the equality $\mathbf{j} = \rho\mathbf{v}$ relating the current density $\mathbf{j}$ to the charge density ρ and the charge velocity $\mathbf{v}$. Because of another type of directed quantities in our approach, we should write down relation (5.29) in the form

$$\mathbf{S} = w\lfloor\mathbf{v}, \tag{5.30}$$

where on the right-hand side the contraction of the twisted three-form w with the ordinary vector $\mathbf{v}$ is present. This formula expresses a relation inverse to (5.22). Can one calculate $\mathbf{v}$ from this formula? The mappings (3.33) and (3.34) from Sect. 3.2 are inverse to each other; if (5.30) corresponds to (3.34), its inverse should correspond to (3.33):

$$\mathbf{v} = -T\lfloor\mathbf{S}. \tag{5.31}$$

Since T is positive, the direction of $\mathbf{v}$ is the same as that of $\mathbf{S}$. We calculate the needed contraction

$$T\lfloor\mathbf{S} = (ab)^{-1}1^*\omega\,(\mathbf{t}_\zeta \wedge \mathbf{t}_\xi \wedge \mathbf{t}_\eta)\lfloor(ab1^*\,d\zeta \wedge d\xi) = \omega\,\{(\mathbf{t}_\zeta \wedge \mathbf{t}_\xi \wedge \mathbf{t}_\eta)\lfloor d\zeta\}\lfloor d\xi$$

$$= \omega\,(\mathbf{t}_\xi \wedge \mathbf{t}_\eta)\lfloor d\xi = -\omega\,\mathbf{t}_\eta\,(\mathbf{t}_\xi\lfloor d\xi) = -\omega\,\mathbf{t}_\eta.$$

Hence, according to (5.31), we find the velocity of energy transport

$$\mathbf{v} = -T\lfloor\mathbf{S} = \omega\,\mathbf{t}_\eta.$$

Let us check: what is the value of the one-form $\mathbf{u}$ on $\mathbf{v}$:

$$\mathbf{u}[[\mathbf{v}]] = \omega^{-1}\,d\eta[[\omega\,\mathbf{t}_\eta]] = d\eta[[\mathbf{t}_\eta]] = 1. \tag{5.32}$$

For this reason a desire emerges to admit the phase slowness and the energy transport velocity of the electromagnetic wave as quantities reciprocal tow each other. The found velocity is one of the vectors $\mathbf{v}$ shown in Fig. 5.2.

5.2.2   Utilizing Scalar Products

Whenever the electric and magnetic fields are present simultaneously, two scalar products $g_\varepsilon(\cdot, \cdot)$ i $g_\mu(\cdot, \cdot)$ are needed to write down the constitutive relations with the use of two operators of Hodge type N_ε i N_μ written according to (4.39):

$$\varepsilon_0 N_\varepsilon(\mathbf{E}) = \varepsilon_0 \mathbf{f}_*^{123} \lfloor \tilde{G}_\varepsilon[[\mathbf{E}]] = \mathbf{D}, \tag{5.33}$$

and according to (4.51)

$$\mu_0 N_\mu(\mathbf{H}) = \mu_0 \mathbf{f}_*^{123} \lfloor \tilde{G}_\mu[[\mathbf{H})]] = \mathbf{B}. \tag{5.34}$$

We write the dielectric scalar-product matrix for the basic one-forms (5.24):[2]

$$\tilde{g}_\varepsilon = \begin{bmatrix} \varepsilon^{\eta\eta} & \varepsilon^{\eta\zeta} & \varepsilon^{\eta\xi} \\ \varepsilon^{\zeta\eta} & \varepsilon^{\zeta\zeta} & \varepsilon^{\zeta\xi} \\ \varepsilon^{\xi\eta} & \varepsilon^{\xi\zeta} & \varepsilon^{\xi\xi} \end{bmatrix}, \tag{5.35}$$

where $\varepsilon^{\eta\eta} = \tilde{g}_\varepsilon(\mathbf{d}\eta, \mathbf{d}\eta)$, $\varepsilon^{\eta\zeta} = \tilde{g}_\varepsilon(\mathbf{d}\eta, \mathbf{d}\zeta)$ and so on. Matrix (5.35), according to (3.61) gives a mapping of one-forms into vectors:

$$\tilde{G}[[\mathbf{f}^k]] = \tilde{g}^{kj} \mathbf{e}_j. \tag{5.36}$$

From now on we shall use indices ζ, ξ, η instead of $1, 2, 3$. For the one-form $\mathbf{d}\zeta$, the prescription (5.36) yields

$$\tilde{G}_\varepsilon[[\mathbf{d}\zeta]] = \varepsilon^{\zeta\zeta} \mathbf{t}_\zeta + \varepsilon^{\zeta\xi} \mathbf{t}_\xi + \varepsilon^{\zeta\eta} \mathbf{t}_\eta \tag{5.37}$$

and analogously for $\mathbf{d}\xi$, $\mathbf{d}\eta$.

We write down the solutions (5.19), (5.20):

$$\mathbf{E} = a\psi \, \mathbf{d}\zeta, \qquad \mathbf{B} = a\omega^{-1}\psi \, \mathbf{d}\eta \wedge \mathbf{d}\zeta, \tag{5.38}$$

$$\mathbf{H} = b1^*\psi \, \mathbf{d}\xi, \qquad \mathbf{D} = -b1^*\omega^{-1}\psi \, \mathbf{d}\eta \wedge \mathbf{d}\xi, \tag{5.39}$$

---

[2] Tilde over g denotes the scalar product of one-forms. Its matrix is inverse to the matrix of the scalar product of vectors.

the constitutive relation of **D** with **E**:

$$\mathbf{D} = \varepsilon_0 f_*^{\eta\zeta\xi} \lfloor \tilde{G}_\varepsilon[[a\psi \mathbf{d}\zeta]] = a1^* \varepsilon_0 \psi f^{\eta\zeta\xi} \lfloor \tilde{G}_\varepsilon[[\mathbf{d}\zeta]] \tag{5.40}$$

and of **B** with **H**:

$$\mathbf{B} = \mu_0 f_*^{\eta\zeta\xi} \lfloor \tilde{G}_\mu[[b1^*\psi \mathbf{d}\xi]] = b\mu_0 \psi f^{\eta\zeta\xi} \lfloor \tilde{G}_\mu[[\mathbf{d}\xi]]. \tag{5.41}$$

We put $f^{\eta\zeta\xi} = \mathbf{d}\zeta \wedge \mathbf{d}\xi \wedge \mathbf{d}\eta$ and apply (5.37)

$$f^{\eta\zeta\xi} \lfloor \tilde{G}_\varepsilon[[\mathbf{d}\zeta]] = (\mathbf{d}\zeta \wedge \mathbf{d}\xi \wedge \mathbf{d}\eta) \lfloor (\varepsilon^{\zeta\zeta} \mathbf{t}_\zeta + \varepsilon^{\zeta\xi} \mathbf{t}_\xi + \varepsilon^{\zeta\eta} \mathbf{t}_\eta)$$

$$= \varepsilon^{\zeta\zeta} (\mathbf{d}\zeta \wedge \mathbf{d}\xi \wedge \mathbf{d}\eta) \lfloor \mathbf{t}_\zeta + \varepsilon^{\zeta\xi} (\mathbf{d}\zeta \wedge \mathbf{d}\xi \wedge \mathbf{d}\eta) \lfloor \mathbf{t}_\xi + \varepsilon^{\zeta\eta} (\mathbf{d}\zeta \wedge \mathbf{d}\xi \wedge \mathbf{d}\eta) \lfloor \mathbf{t}_\eta$$

$$= \varepsilon^{\zeta\zeta} (\mathbf{d}\xi \wedge \mathbf{d}\eta) \mathbf{d}\zeta[[\mathbf{t}_\zeta]] + \varepsilon^{\zeta\xi} (\mathbf{d}\eta \wedge \mathbf{d}\zeta) \mathbf{d}\xi[[\mathbf{t}_\xi]] + \varepsilon^{\zeta\eta} (\mathbf{d}\zeta \wedge \mathbf{d}\xi) \mathbf{d}\eta[[\mathbf{t}_\eta]]$$

$$= \varepsilon^{\zeta\zeta} \mathbf{d}\xi \wedge \mathbf{d}\eta + \varepsilon^{\zeta\xi} \mathbf{d}\eta \wedge \mathbf{d}\zeta + \varepsilon^{\zeta\eta} \mathbf{d}\zeta \wedge \mathbf{d}\xi.$$

We insert this into (5.40)

$$\mathbf{D} = a1^* \varepsilon_0 \psi (\varepsilon^{\zeta\zeta} \mathbf{d}\xi \wedge \mathbf{d}\eta + \varepsilon^{\zeta\xi} \mathbf{d}\eta \wedge \mathbf{d}\zeta + \varepsilon^{\zeta\eta} \mathbf{d}\zeta \wedge \mathbf{d}\xi). \tag{5.42}$$

On the other hand, according to (5.39)

$$\mathbf{D} = -b1^* \omega^{-1} \psi \, \mathbf{d}\eta \wedge \mathbf{d}\xi, \tag{5.43}$$

hence in the bracket on the right-hand side of (5.42) only one term should remain, which means that

$$\varepsilon^{\zeta\xi} = \varepsilon^{\zeta\eta} = 0. \tag{5.44}$$

These conditions imply that in the symmetric matrix $\tilde{g}_\varepsilon$, zero elements are present:

$$\tilde{g}_\varepsilon = \begin{bmatrix} \varepsilon^{\eta\eta} & 0 & \varepsilon^{\eta\xi} \\ 0 & \varepsilon^{\zeta\zeta} & 0 \\ \varepsilon^{\xi\eta} & 0 & \varepsilon^{\xi\xi} \end{bmatrix}. \tag{5.45}$$

We compare the nonzero elements from (5.42) with (5.43):

$$a1^* \varepsilon_0 \psi \varepsilon^{\zeta\zeta} \mathbf{d}\xi \wedge \mathbf{d}\eta, = -b1^* \omega^{-1} \psi \, \mathbf{d}\eta \wedge \mathbf{d}\xi,$$

which yields

$$\omega^{-1} = ab^{-1} \varepsilon_0 \varepsilon^{\zeta\zeta}. \tag{5.46}$$

Now we write down the magnetic counterpart of (5.37)

$$\tilde{G}_\mu[[\mathbf{d}\zeta]] = \mu^{\zeta\zeta}\,\mathbf{t}_\zeta + \mu^{\zeta\xi}\,\mathbf{t}_\xi + \mu^{\zeta\eta}\mathbf{t}_\eta \tag{5.47}$$

and in analogy to previous considerations, we obtain the result:

$$T\lfloor\tilde{G}_\mu[[\mathbf{d}\xi]] = \mu^{\xi\zeta}\,\mathbf{d}\xi \wedge \mathbf{d}\eta + \mu^{\xi\xi}\,\mathbf{d}\eta \wedge \mathbf{d}\zeta + \mu^{\xi\eta}\,\mathbf{d}\zeta \wedge \mathbf{d}\xi.$$

We insert this into (5.41)

$$\mathbf{B} = b\mu_0\psi\,(\mu^{\xi\zeta}\,\mathbf{d}\xi \wedge \mathbf{d}\eta + \mu^{\xi\xi}\,\mathbf{d}\eta \wedge \mathbf{d}\zeta + \mu^{\xi\eta}\,\mathbf{d}\zeta \wedge \mathbf{d}\xi). \tag{5.48}$$

Comparing this to (5.38) shows that

$$\mu^{\xi\zeta} = \mu^{\xi\eta} = 0, \tag{5.49}$$

hence the symmetric matrix $\tilde{g}_\mu$ has the shape

$$\tilde{g}_\mu = \begin{bmatrix} \mu^{\eta\eta} & \mu^{\eta\zeta} & 0 \\ \mu^{\zeta\eta} & \mu^{\zeta\zeta} & 0 \\ 0 & 0 & \mu^{\xi\xi} \end{bmatrix}. \tag{5.50}$$

We compare (5.38) with (5.48) with zero terms omitted:

$$a\omega^{-1}\psi\,\mathbf{d}\eta \wedge \mathbf{d}\zeta = b\mu_0\psi\mu^{\xi\xi}\,\mathbf{d}\eta \wedge \mathbf{d}\zeta,$$

which gives

$$\omega^{-1} = ba^{-1}\mu_0\mu^{\xi\xi}. \tag{5.51}$$

The two expressions (5.46) and (5.51) for ω^{-1} allow us to write:

$$\frac{a}{b}\,\varepsilon_0\varepsilon^{\zeta\zeta} = \frac{b}{a}\,\mu_0\mu^{\xi\xi}.$$

We find from this

$$\frac{a}{b} = \sqrt{\frac{\mu_0\mu^{\xi\xi}}{\varepsilon_0\varepsilon^{\zeta\zeta}}}. \tag{5.52}$$

This proves that the amplitudes in front of the field quantities must be interrelated.

We are interested also in mutual positioning of the fields. For this purpose we are going to find their scalar products. We begin by calculating the dielectric scalar product of two field strengths:

$$\tilde{g}_\varepsilon[[\mathbf{E}, \mathbf{H}]]\tilde{g}_\varepsilon = [[a\psi \, \mathbf{d}\zeta, 1^*b\psi \, \mathbf{d}\xi]] = 1^*ab\psi^2\tilde{g}_\varepsilon[[\mathbf{d}\zeta, \mathbf{d}\xi]] = 1^*ab\psi^2\varepsilon^{\zeta\xi}.$$

We obtain by virtue of (5.44):

$$\tilde{g}_\varepsilon[[\mathbf{E}, \mathbf{H}]] = 0. \tag{5.53}$$

Calculation of the magnetic scalar product of the same forms leads to

$$\tilde{g}_\mu[[\mathbf{H}, \mathbf{E}]] = 1^*ab\psi^2\mu^{\xi\zeta},$$

and by dint of (5.49) we obtain

$$\tilde{g}_\mu[[\mathbf{H}, \mathbf{E}]] = 0. \tag{5.54}$$

We see that the intensities of the electric and magnetic fields have to be perpendicular as one-forms, according to both scalar products.

It follows also from (5.44) that

$$\tilde{g}_\varepsilon[[\mathbf{k}, \mathbf{E}]] = \tilde{g}_\varepsilon[[\mathbf{d}\eta, a\psi \, \mathbf{d}\zeta]] = a\psi\varepsilon^{\eta\zeta} = 0, \tag{5.55}$$

and from (5.49) that

$$\tilde{g}_\mu[[\mathbf{k}, \mathbf{H}]] = \tilde{g}_\mu[[\mathbf{d}\eta, 1^*b\psi \, \mathbf{d}\xi]] = 1^*b\psi\mu^{\eta\xi} = 0, \tag{5.56}$$

We calculate the magnetic scalar product of **B** with **D** on the basis of (2.40):

$$\tilde{g}_\mu[[\mathbf{B}, \mathbf{D}]] = \tilde{g}_\mu[[\mathbf{u} \wedge \mathbf{E}, \mathbf{H} \wedge \mathbf{u}]] = \begin{vmatrix} \tilde{g}_\mu[[\mathbf{u}, \mathbf{H}]] & \tilde{g}_\mu[[\mathbf{u}, \mathbf{u}]] \\ \tilde{g}_\mu[[\mathbf{E}, \mathbf{H}]] & \tilde{g}_\mu[[\mathbf{E}, \mathbf{u}]] \end{vmatrix},$$

$$\tilde{g}_\mu[[\mathbf{B}, \mathbf{D}]] = \tilde{g}_\mu[[\mathbf{u}, \mathbf{H}]]\tilde{g}_\mu[[\mathbf{E}, \mathbf{u}]] - \tilde{g}_\mu[[\mathbf{u}, \mathbf{u}]]\tilde{g}_\mu[[\mathbf{E}, \mathbf{H}]]. \tag{5.57}$$

We have found already that $\tilde{g}_\mu[[\mathbf{E}, \mathbf{H}]] = \tilde{g}_\mu[[\mathbf{u}, \mathbf{H}]] = 0$, hence we get

$$\tilde{g}_\mu[[\mathbf{B}, \mathbf{D}]] = 0. \tag{5.58}$$

We ascertain in this manner that the two-forms **B** are **D** are electrically perpendicular. This observation is already taken into account in Fig. 5.5.

Exercise 5.1 Prove that

$$\tilde{g}_\varepsilon[[\mathbf{B}, \mathbf{D}]] = 0. \tag{5.59}$$

Table 5.1 Scalar products between the field quantities of one wave

	$\tilde{g}_\mu[[\mathbf{E},\mathbf{H}]] = 0$	$\tilde{g}_\mu[[\mathbf{B},\mathbf{D}]] = 0$	$\tilde{g}_\mu[[\mathbf{k},\mathbf{H}]] = 0$
$\tilde{g}_\varepsilon[[\mathbf{k},\mathbf{E}]] = 0$	$\tilde{g}_\varepsilon[[\mathbf{E},\mathbf{H}]] = 0$	$\tilde{g}_\varepsilon[[\mathbf{B},\mathbf{D}]] = 0$	

Table 5.1 is a summary of the found scalar products.

The conditions (5.44) and (5.49) can be written differently:

$$\tilde{g}_\varepsilon[[\mathbf{d}\zeta, \mathbf{d}\eta]] = \tilde{g}_\varepsilon[[\mathbf{d}\zeta, \mathbf{d}\xi]] = 0, \tag{5.60}$$

$$\tilde{g}_\mu[[\mathbf{d}\xi, \mathbf{d}\eta]] = \tilde{g}_\mu[[\mathbf{d}\xi, \mathbf{d}\zeta]] = 0, \tag{5.61}$$

It is seen from this that the two scalar functions ζ, ξ cannot be arbitrary—they have to be such that, their exterior derivatives are perpendicular with respect to both scalar products $\tilde{g}_\varepsilon$ and $\tilde{g}_\mu$. When $\tilde{g}_\varepsilon$ and $\tilde{g}_\mu$ are not proportional, this demand is nontrivial. It can be satisfied by specific waves, called *eigenwaves*. Moreover, the derivative $\mathbf{d}\zeta$ must be dielectrically perpendicular to $\mathbf{d}\eta$, whereas $\mathbf{d}\xi$ is magnetically perpendicular to $\mathbf{d}\eta$. Even in homogeneous media, when the functions η, ζ, ξ may be curvilinear coordinates, the attitudes of their exterior derivatives may change from point to point, so satisfying of conditions (5.60, 5.61) may be difficult in the whole region filled by a given substance.

The free Maxwell equation $\mathbf{d} \wedge \mathbf{D} = 0$ applied to $\mathbf{D}$ expressed by (5.33) yields

$$\mathbf{d} \wedge N_\varepsilon[[\psi \mathbf{d}\zeta]] = 0.$$

Since ψ is scalar function, we are allowed to write down

$$\mathbf{d} \wedge \{\psi\, N_\varepsilon[[\mathbf{d}\zeta]]\} = 0,$$

$$(\mathbf{d}\psi) \wedge N_\varepsilon(\mathbf{d}\zeta) + \psi\, \mathbf{d} \wedge N_\varepsilon[[\mathbf{d}\zeta]] = 0. \tag{5.62}$$

Due to (5.42) with one term left, the first term in (5.62) is equal to

$$-\psi'\, \mathbf{d}\eta \wedge 1^*\omega^{-1}\mathbf{d}\eta \wedge \mathbf{d}\xi.$$

This is zero because of repeated factor $\mathbf{d}\eta$. In this manner equation (5.62) assumes the shape

$$\mathbf{d} \wedge N_\varepsilon[[\mathbf{d}\zeta]] = 0,$$

which can be written also as

$$\mathbf{d} \wedge \mathbf{f}_*^{\eta\zeta\xi} \lfloor \tilde{G}_\varepsilon[[\mathbf{d}\zeta]] = 0. \tag{5.63}$$

The identity $\mathbf{p} \wedge \mathbf{f}_*^{123} \lfloor \tilde{G}[[\mathbf{s}]] = \tilde{g}[[\mathbf{p}, \mathbf{s}]] \mathbf{f}_*^{123}$ known for two one-forms $\mathbf{p}$ and $\mathbf{s}$ allows us to write

$$\tilde{g}_\varepsilon[[\mathbf{d}, \mathbf{d}]]\zeta = 0. \tag{5.64}$$

This result may be expressed as follows: the scalar function ζ satisfies the Laplace equation with a Laplacian defined through the dielectric scalar product. Functions satisfying the Laplace equation are called *harmonic*, hence we shall call ζ a *dielectrically harmonic function*.

Another free Maxwell equation $\mathbf{d} \wedge \mathbf{B} = 0$ with similar use of (5.34) leads to another Laplace equation:

$$\tilde{g}_\mu[[\mathbf{d}, \mathbf{d}]]\xi = 0. \tag{5.65}$$

Therefore we call ξ a *magnetically harmonic function*. In this way we arrive at the observation: The freedom of choice of two scalar functions ζ and ξ is limited—they must be harmonic with respect to Laplacians with appropriate scalar products.

5.2.3 Examples

We consider a homogeneous medium, such that the scalar products g_ε g_μ are proportional; then one scalar product can be used, which we denote is as g. Considerations are then the same as for an isotropic medium, where g is the ordinary scalar product of our isotropic space and the relations $\tilde{g}_\varepsilon = \varepsilon_r \tilde{g}$, $\tilde{g}_\mu = \mu_r \tilde{g}$, are fulfilled, where the scalars ε_r and μ_r are the relative permittivity and permeability, respectively. In such a case the constitutive relations (5.33) and (5.34) should be written with the same operator $\tilde{G}$:

$$\varepsilon_0 N_\varepsilon[[\mathbf{E}]] = \varepsilon \mathbf{f}_*^{123} \lfloor \tilde{G}[[\mathbf{E}]] = \mathbf{D}, \tag{5.66}$$

and

$$\mu_0 N_\mu[[\mathbf{H}]] = \mu \mathbf{f}_*^{123} \lfloor \tilde{G}[[\mathbf{H}]] = \mathbf{B}, \tag{5.67}$$

where $\varepsilon = \varepsilon_0 \varepsilon_r$, $\mu = \mu_0 \mu_r$.

The observations written after equation (5.61) mean, that three one-forms $\mathbf{d}\zeta$, $\mathbf{d}\xi$, $\mathbf{d}\eta$ are perpendicular to each other. Hence the conclusion from this is that the three curvilinear coordinates must be orthogonal. We know, besides, that two of them, ζ and ξ, have to be harmonic with respect to the Laplacian given by g.

(a) Cylindrical Coordinates

Let us take *cylindrical coordinates* as the first example:

$$\rho = \sqrt{x^2 + y^2}, \quad \alpha = \operatorname{arctg}\frac{y}{x}, \quad z = z.$$

It is well known that they are orthogonal. According to (3.86) we know the exterior derivative of the angle coordinate:

$$\mathbf{d}\alpha = \rho^{-1}\left(-\sin\alpha\,\mathbf{f}^1 + \cos\alpha\,\mathbf{f}^2\right). \tag{5.68}$$

Formula (3.87) describes another exterior derivative:

$$\mathbf{d}\rho = \cos\alpha\,\mathbf{f}^1 + \sin\alpha\,\mathbf{f}^2. \tag{5.69}$$

We have also $\mathbf{d}z = \mathbf{f}^3$, where $\mathbf{f}^i$ are dual forms to the vector basis of Cartesian coordinate $\mathbf{e}_j$. Only two cylindrical coordinates α and z are harmonic, so to begin with let us accept them in the role of the curvilinear coordinates ζ and ξ.

We write down the electromagnetic fields according to (5.19, 5.20):

$$\mathbf{E}(\mathbf{r}, t) = a\,\psi[\,\eta(\rho) - \omega t\,]\,\mathbf{d}z, \quad \mathbf{B}(\mathbf{r}, t) = a\omega^{-1}\psi[\,\eta(\rho) - \omega t\,]\,\mathbf{d}\eta \wedge \mathbf{d}z,$$
$$\tag{5.70}$$

$$\mathbf{H}(\mathbf{r}, t) = 1^*b\,\psi[\,\eta(\rho) - \omega t\,]\,\mathbf{d}\alpha, \quad \mathbf{D}(\mathbf{r}, t) = -1^*b\,\omega^{-1}\psi[\,\eta(\rho) - \omega t\,]\,\mathbf{d}\eta \wedge \mathbf{d}\alpha.$$
$$\tag{5.71}$$

In this case, η is a function of one variable ρ, which means that the wave has a constant phase on surfaces of cylinders with the common axis Z. Therefore, if a solution of shape (5.70, 5.71), exists, we ought to call it a *cylindrical wave*. We know from considerations in Chap. 1 about the attitude of the exterior product of two one-forms that the two-form $\mathbf{S} = \mathbf{E} \wedge \mathbf{H}$, parallel to $\mathbf{d}z \wedge \mathbf{d}\alpha$, has the attitude of the coordinate line ρ. This means that the wave (5.70, 5.71) propagates along the radius ρ, perpendicular to the Z-axis.

We know from previous considerations that the fields (5.70, 5.71) fulfil all Maxwell's equations, so only the constitutive relations remain to be verified. The dielectric constitutive relation (4.39), that is $\mathbf{D} = \varepsilon \mathbf{f}_*^{123} \lfloor \tilde{G}[[\mathbf{E}]]$, after reduction by scalar ψ assumes the shape

$$-1^*b\,\omega^{-1}\mathbf{d}\eta \wedge \mathbf{d}\alpha = \varepsilon \mathbf{f}_*^{123} \lfloor \tilde{G}[[a\mathbf{d}z]]. \tag{5.72}$$

We have $\tilde{G}[[dz]] = \tilde{G}[[\mathbf{f}^3]] = \mathbf{e}_3$, so the expression $\varepsilon a\, \mathbf{f}_*^{123}\lfloor \mathbf{e}_3 = \varepsilon a\, \mathbf{f}_*^{12}$ is on the right-hand side. One has $\mathbf{d}\eta = \eta'(\rho)\, \mathbf{d}\rho$ for the function η of one variable ρ, hence the formulas (5.68) and (5.69) allow us to write

$$\mathbf{d}\eta \wedge \mathbf{d}\alpha = \eta'(\rho)\left(\cos\alpha\, \mathbf{f}^1 + \sin\alpha\, \mathbf{f}^2\right) \wedge \rho^{-1}\left(-\sin\alpha\, \mathbf{f}^1 + \cos\alpha\, \mathbf{f}^2\right)$$

$$= \frac{\eta'(\rho)}{\rho}\left(\cos^2\alpha + \sin^2\alpha\right)\mathbf{f}^1 \wedge \mathbf{f}^2 = \frac{\eta'(\rho)}{\rho}\mathbf{f}^{12}.$$

We insert all these computations in (5.72), obtaining

$$-1^*b\,\frac{\eta'(\rho)}{\omega\rho}\mathbf{f}^{12} = \varepsilon a\, \mathbf{f}_*^{12},$$

that is

$$-b\,\frac{\eta'(\rho)}{\omega\rho}\mathbf{f}_*^{12} = \varepsilon a\, \mathbf{f}_*^{12},$$

The comparison of coefficients in front of the same basic two-form gives the following ordinary differential equation to be satisfied by η:

$$\eta'(\rho) = -\varepsilon\omega\,\frac{a}{b}\,\rho.$$

This has the solution

$$\eta(\rho) = -\varepsilon\omega\,\frac{a}{2b}\,\rho^2 + A \tag{5.73}$$

with the integration constant A.

The magnetic constitutive relation (4.51), i.e., $\mathbf{B} = \mu \mathbf{f}_*^{123}\lfloor \tilde{G}[[\mathbf{H}]]$ leads to

$$a\omega^{-1}\mathbf{d}\eta \wedge \mathbf{d}z = \mu \mathbf{f}_*^{123}\lfloor \tilde{G}[[1^*b\, \mathbf{d}\alpha]]. \tag{5.74}$$

We transform the right-hand side:

$$\mu b\, \mathbf{f}_*^{123}\lfloor\tilde{G}\left[\left[\rho^{-1}\left(-\sin\alpha\, \mathbf{f}_*^1 + \cos\alpha\, \mathbf{f}_*^2\right)\right]\right] = \mu b\rho^{-1}\mathbf{f}_*^{123}\lfloor(-\sin\alpha\, \mathbf{e}_1^* + \cos\alpha\, \mathbf{e}_2^*)$$

$$= \mu b\rho^{-1}\left(-\sin\alpha\, \mathbf{f}^{23} - \cos\alpha\, \mathbf{f}^{13}\right) = -\mu b\rho^{-1}\left(\cos\alpha\, \mathbf{f}^1 + \sin\alpha\, \mathbf{f}^2\right) \wedge \mathbf{f}^3.$$

We insert this in (5.74)

$$a\omega^{-1}\eta'(\rho)\left(\cos\alpha\, \mathbf{f}^1 + \sin\alpha\, \mathbf{f}^2\right) \wedge \mathbf{f}^3 = -\mu b\rho^{-1}\left(\cos\alpha\, \mathbf{f}^1 + \sin\alpha\, \mathbf{f}^2\right) \wedge \mathbf{f}^3.$$

We see from this that η has to satisfy the differential equation

$$\eta'(\rho)\,\rho = -\mu\,\omega\,\frac{b}{a},$$

which has the solution

$$\eta(\rho) = -\mu\,\omega\,\frac{b}{a}\,\ln\frac{\rho}{\rho_0} \tag{5.75}$$

with the integration constant ρ_0. Unfortunately, the solutions (5.73) and (5.75) are reconciled by a choice of appropriate constants a, b, A, ρ_0. This means that formulas (5.70, 5.71) do not describe any possible electromagnetic wave.

After interchanging the forms $\mathbf{d}z$ and $\mathbf{d}\alpha$, the two constitutive relations also lead to two different solutions for η. We conclude from this that *a running linearly polarized cylindrical electromagnetic wave propagating radially from the cylinder axis does not exist.*

A second pair of possibilities, this time with $\eta(\alpha)$ instead of $\eta(\rho)$ in the phase of the wave, leads to different solutions for η from the dielectric and magnetic constitutive relations, hence similarly the pair does not give a possible electromagnetic wave. If such a solution existed it could be called a *whirly electromagnetic wave.*

Exercise 5.2 Prove that the function $\gamma(\rho) = \ln\rho$ is harmonic.

We consider now a wave propagating along the Z-axis, which means that a function $\eta(z)$ of one variable stands in the phase. The harmonic function of the variable ρ is $\ln\rho$, hence we are seeking an electromagnetic wave in the form

$$\mathbf{E}(\mathbf{r},t) = a\,\psi[\,\eta(z) - \omega t\,]\,\mathbf{d}\ln\rho, \quad \mathbf{B}(\mathbf{r},t) = a\omega^{-1}\psi[\,\eta(z) - \omega t\,]\,\eta'(z)\mathbf{f}^3 \wedge \mathbf{d}\ln\rho, \tag{5.76}$$

$$\mathbf{H}(\mathbf{r},t) = 1^*b\,\psi[\,\eta(z) - \omega t\,]\,\mathbf{d}\alpha, \quad \mathbf{D}(\mathbf{r},t) = -1^*b\,\omega^{-1}\psi[\,\eta(z) - \omega t\,]\,\eta'(z)\mathbf{f}^3 \wedge \mathbf{d}\alpha, \tag{5.77}$$

The dielectric constitutive relation leads to the condition

$$-1^*b\,\omega^{-1}\eta'(z)\,\mathbf{f}^3 \wedge \mathbf{d}\alpha = \varepsilon\mathbf{f}_*^{123}\lfloor\tilde{G}[[a\rho^{-1}\mathbf{d}\rho]].$$

We apply (5.68) and (5.69):

$$-1^*b\,\omega^{-1}\eta'\,\mathbf{f}^3 \wedge \rho^{-1}\left(-\sin\alpha\,\mathbf{f}^1 + \cos\alpha\,\mathbf{f}^2\right) = \varepsilon\mathbf{f}_*^{123}\lfloor a\rho^{-1}(\cos\alpha\,\mathbf{e}_1 + \sin\alpha\,\mathbf{e}_2),$$

$$b\,\omega^{-1}\eta'\,\rho^{-1}\left(\sin\alpha\,\mathbf{f}_*^{31} - \cos\alpha\,\mathbf{f}_*^{32}\right) = \varepsilon a\rho^{-1}\left(-\cos\alpha\,\mathbf{f}_*^{32} + \sin\alpha\,\mathbf{f}_*^{31}\right).$$

We notice the same two-forms in the brackets of both sides, so the following equality remains:

$$b\,\omega^{-1}\eta'\rho^{-1} = \varepsilon a\rho^{-1}.$$

The functions ρ^{-1} are cancelled and we obtain the condition

$$\varepsilon\frac{a}{b} = \frac{\eta'}{\omega}. \tag{5.78}$$

The magnetic constitutive relation yields

$$a\omega^{-1}\eta'\mathbf{f}^3 \wedge \mathbf{d}\ln\rho = \mu\mathbf{f}_*^{123}\lfloor\tilde{G}[1^*b\,\mathbf{d}\alpha],$$

$$a\omega^{-1}\eta'\rho^{-1}\mathbf{f}^3 \wedge \left(\cos\alpha\,\mathbf{f}^1 + \sin\alpha\,\mathbf{f}^2\right) = \mu b\rho^{-1}\mathbf{f}_*^{123}\lfloor(-\sin\alpha\,\mathbf{e}_1^* + \cos\alpha\,\mathbf{e}_2^*),$$

$$a\omega^{-1}\eta'\rho^{-1}\left(\cos\alpha\,\mathbf{f}^{31} + \sin\alpha\,\mathbf{f}^{32}\right) = \mu b\rho^{-1}\left(-\sin\alpha\,\mathbf{f}^{32} + \cos\alpha\,\mathbf{f}^{31}\right).$$

The same two-forms are in the brackets, hence

$$a\omega^{-1}\eta'\rho^{-1} = \mu b\rho^{-1},$$

or

$$\frac{\eta'}{\omega} = \mu\frac{b}{a}. \tag{5.79}$$

After comparing this condition with (5.78), we ascertain that the equality

$$\varepsilon\frac{a}{b} = \mu\frac{b}{a},$$

must occur, hence

$$\frac{a}{b} = \pm\sqrt{\frac{\mu}{\varepsilon}}.$$

The constants ω, ε and μ positive in (5.78, 5.79), so the sign of the quotient a/b is the same as that of the derivative η'. We choose plus. We arrive at the conclusion that the constants a and b are related:

$$\frac{a}{b} = \sqrt{\frac{\mu}{\varepsilon}} \tag{5.80}$$

Inserting (5.80) into (5.79) gives

$$\eta' = \omega\sqrt{\varepsilon\mu}. \tag{5.81}$$

Notice that the derivative of $\eta(z)$ is constant, so we choose $\eta(z) = \kappa z + A$, where A is an integration constant, whereas we recognize in $\kappa = \omega\sqrt{\varepsilon\mu}$ the wave number, which is magnitude of the wave covector. The condition

$$\frac{\eta'}{\omega} = \frac{\kappa}{\omega} = \sqrt{\varepsilon\mu} \tag{5.82}$$

is the well-known relation expressing the inverse of phase velocity by the characteristics of the medium.

Summarizing the obtained results, we write down the solution (5.76, 5.77)

$$\mathbf{E}(\mathbf{r}, t) = a\,\psi(\sqrt{\varepsilon\mu}\,\omega z - \omega t)\,\mathbf{d}\ln\rho, \quad \mathbf{B}(\mathbf{r}, t) = a\sqrt{\varepsilon\mu}\,\psi(\sqrt{\varepsilon\mu}\,\omega z - \omega t)\,\mathbf{f}^3 \wedge \mathbf{d}\ln\rho, \tag{5.83}$$

$$\mathbf{H}(\mathbf{r}, t) = 1^*\sqrt{\frac{\varepsilon}{\mu}}\,a\,\psi(\sqrt{\varepsilon\mu}\,\omega z - \omega t)\,\mathbf{d}\alpha, \quad \mathbf{D}(\mathbf{r}, t) = -1^*\epsilon a\,\psi(\sqrt{\varepsilon\mu}\,\omega z - \omega t)\,\mathbf{f}^3 \wedge \mathbf{d}\alpha. \tag{5.84}$$

This is the electromagnetic wave propagating along the Z-axis. Its surfaces of constant phase are planes $z = \text{const}$, but the fields $\mathbf{E}, \mathbf{H}$ are not constant on those planes because the one-forms $\mathbf{E}$ are proportional to $\mathbf{d}\ln\rho$, therefore their geometric images are cylinders with common axis Z; whereas the one-forms $\mathbf{H}$, proportional to $\mathbf{d}\alpha$, are half-planes outgoing from the Z-axis (see Fig. 3.16). In the traditional language this means that the electric vectors $\mathbf{E}$ are parallel to lines of the radius ρ and pseudovectors $\mathbf{H}$ to lines of the angular coordinate α. A vectorial image of the fields is shown in Fig. 5.7. (Pseudovectors $\mathbf{H}$ are improperly drawn as ordinary vectors with arrows in place of directed rings—to make the image simpler.) We repeat: the phase is constant on the planes $z = \text{const}$ but the fields are not constant there. Maybe such an electromagnetic wave could be called a *semiplane wave*?

We may also consider a solution similar to (5.76), but with interchanged functions α and $\ln\rho$:

$$\mathbf{E}(\mathbf{r}, t) = a\,\psi[\eta(z) - \omega t]\,\mathbf{d}\alpha, \quad \mathbf{B}(\mathbf{r}, t) = a\omega^{-1}\psi[\eta(z) - \omega t]\,\eta'(z)\mathbf{f}^3 \wedge \mathbf{d}\alpha, \tag{5.85}$$

$$\mathbf{H}(\mathbf{r}, t) = 1^*b\,\psi[\eta(z) - \omega t]\,\mathbf{d}\ln\rho, \quad \mathbf{D}(\mathbf{r}, t) = -1^*\omega^{-1}\psi[\eta(z) - \omega t]\,\eta'(z)\mathbf{f}^3 \wedge \mathbf{d}\ln\rho, \tag{5.86}$$

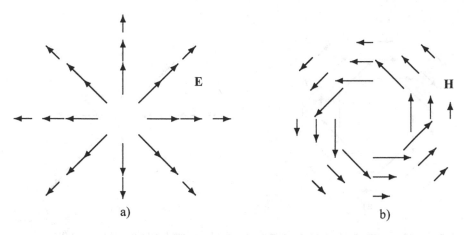

Fig. 5.7 (a) Electric field strength, (b) magnetic field strength for the wave (5.83, 5.84)

This time the electric constitutive relation leads to

$$\frac{\eta'}{\omega} = -\varepsilon \frac{a}{b},$$

whereas the magnetic relation leads to

$$\frac{\eta'}{\omega} = -\mu \frac{b}{a}. \tag{5.87}$$

Now the sign of the quotient a/b is opposite to that of the derivative η'. If we choose plus for η' the constants a and b are related to each other by

$$\frac{a}{b} = -\sqrt{\frac{\mu}{\varepsilon}} \tag{5.88}$$

Inserting (5.88) into (5.87) gives as previously

$$\eta' = \omega\sqrt{\varepsilon\mu}.$$

The derivative of $\eta(z)$ is constant, hence we choose $\eta(z) = \kappa z + A$, where A is an integration constant, and $\kappa = \omega\sqrt{\varepsilon\mu}$ is the magnitude of wave covector.

We may now write down the solution (5.85, 5.86):

$$\mathbf{E}(\mathbf{r}, t) = a\,\psi(\sqrt{\varepsilon\mu}\,\omega z - \omega t)\,\mathrm{d}\alpha, \quad \mathbf{B}(\mathbf{r}, t) = a\sqrt{\varepsilon\mu}\,\psi(\sqrt{\varepsilon\mu}\,\omega z - \omega t)\,\mathbf{f}^3 \wedge \mathrm{d}\alpha, \tag{5.89}$$

$$\mathbf{H}(\mathbf{r}, t) = -1^*\sqrt{\frac{\varepsilon}{\mu}}\,a\,\psi(\sqrt{\varepsilon\mu}\,\omega z - \omega t)\,\mathrm{d}\ln\rho, \quad \mathbf{D}(\mathbf{r}, t) = 1^*\epsilon a\,\psi(\sqrt{\varepsilon\mu}\,\omega z - \omega t)\,\mathbf{f}^3 \wedge \mathrm{d}\ln\rho.$$

$$\tag{5.90}$$

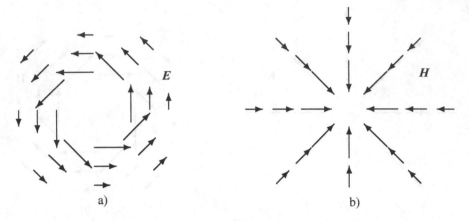

Fig. 5.8 (**a**) Electric field strength, (**b**) magnetic field strength for the wave (5.89, 5.90)

This is a second possible running electromagnetic wave propagating along the Z-axis with a polarization perpendicular to that of (5.83, 5.84). The field vectors of this wave are similar to that of Fig. 5.7, but with $\mathbf{E}$ in place of $\mathbf{H}$ and $-\mathbf{H}$ in place of $\mathbf{E}$ (see Fig. 5.8).

For both waves (5.83, 5.84) and (5.89, 5.90), the same expression for the density of energy flux:

$$\mathbf{S} = \mathbf{E} \wedge \mathbf{H} = 1^* a^2 \sqrt{\frac{\varepsilon}{\mu}}\, \psi^2 \mathbf{d} \ln \rho \wedge \mathbf{d}\alpha = \frac{a^2}{\rho^2} \sqrt{\frac{\varepsilon}{\mu}}\, \psi^2 \mathbf{f}_*^{12}$$

can be obtained. We know that the direction of the twisted two-form $\mathbf{f}_*^{12}$ is the same as that of the ordinary vector $\mathbf{e}_3$, hence we notice that the energy propagates in the positive direction of the Z-axis for both waves (5.83, 5.84) and (5.89, 5.90), but the flux density is not constant—it decreases as ρ^{-2} when going away from the Z-axis. A singularity on the Z-axis is present—the energy flux density becomes infinite.

Both waves also have the same energy density:

$$w = \frac{1}{2}\left(\mathbf{E} \wedge \mathbf{D} + \mathbf{H} \wedge \mathbf{B}\right) = \varepsilon \frac{a^2}{\rho^2}\, \psi^2 \mathbf{f}_*^{123},$$

which is singular on the Z-axis and decreases as ρ^{-2} with increasing ρ..

The wave (5.83, 5.84) may also propagate between two conducting cylinders with the same axis Z and different radii. Then the integral of the energy flux density over a ring K perpendicular to the Z-axis contained between radii ρ_1, ρ_2:

$$\int\limits_K S(\mathbf{r}, t)[[ds]] = \int\limits_{\rho_1}^{\rho_2} \int\limits_{0}^{2\pi} a^2 \sqrt{\frac{\varepsilon}{\mu}}\, \psi^2 \frac{1}{\rho^2}\, \rho\, d\rho\, d\alpha = a^2 \sqrt{\frac{\varepsilon}{\mu}}\, \psi^2(\kappa z - \omega t)\, 2\pi \int\limits_{\rho_1}^{\rho_2} \frac{d\rho}{\rho}$$

$$= 2\pi a^2 \sqrt{\frac{\varepsilon}{\mu}}\, \psi^2(\kappa z - \omega t)\, \ln \frac{\rho_2}{\rho_1}$$

gives a finite value for the power carried by the wave.

(b) Spherical Coordinates

We take *spherical coordinates* as a second example:

$$r = \sqrt{x^2 + y^2 + z^2}, \qquad \alpha = \operatorname{arctg} \frac{y}{x}, \qquad \theta = \operatorname{arc\,cos} \frac{z}{r},$$

which are known to be orthogonal. We know their exterior derivatives from formulas (3.88) (3.86) and (3.89):

$$\mathbf{d}r = \sin\theta \cos\alpha\, \mathbf{f}^1 + \sin\theta \sin\alpha\, \mathbf{f}^2 + \cos\theta\, \mathbf{f}^3. \tag{5.91}$$

$$\mathbf{d}\alpha = \rho^{-1}\left(-\sin\alpha\, \mathbf{f}^1 + \cos\alpha\, \mathbf{f}^2\right), \tag{5.92}$$

$$\mathbf{d}\theta = r^{-1}\left(\cos\theta \cos\alpha\, \mathbf{f}^1 + \cos\theta \sin\alpha\, \mathbf{f}^2 - \sin\theta\, \mathbf{f}^3\right), \tag{5.93}$$

We know already that the coordinate α is harmonic. The other coordinates are not harmonic.

Exercise 5.3 Demonstrate that function $v(r) = r^{-1}$ is harmonic.

Exercise 5.4 Demonstrate that function

$$\lambda(\theta) = -\frac{\cos\theta}{2\sin^2\theta} + \frac{1}{4}\ln\frac{1 - \cos\theta}{1 + \cos\theta}$$

is harmonic, and its external derivative is

$$\mathbf{d}\lambda(\theta) = \lambda'(\theta)\, \mathbf{d}\theta = \frac{1}{r\sin^3\theta}\left(\cos\theta \cos\alpha\, \mathbf{f}^1 + \cos\theta \sin\alpha\, \mathbf{f}^2 - \sin\theta\, \mathbf{f}^3\right).$$

It is worth verifying whether a running electromagnetic wave can exist that propagates radially from the origin of coordinates, that is, one for which surfaces of constant phase would be spheres $r = $ const. Such a wave (if it existed) could be called a *spherical wave*. We assume that η depends only on r, whereas $\zeta = \alpha$, $\xi = \lambda(\theta)$:

$$\mathbf{E}(\mathbf{r},t) = a\,\psi[\,\eta(r) - \omega t\,]\,\mathbf{d}\alpha, \quad \mathbf{B}(\mathbf{r},t) = a\omega^{-1}\psi[\,\eta(r) - \omega t\,]\,\mathbf{d}\eta \wedge \mathbf{d}\alpha, \tag{5.94}$$

$$\mathbf{H}(\mathbf{r},t) = 1^*b\,\psi[\,\eta(r) - \omega t\,]\,\mathbf{d}\lambda, \quad \mathbf{D}(\mathbf{r},t) = -1^*b\,\omega^{-1}\psi[\,\eta(r) - \omega t\,]\,\mathbf{d}\eta \wedge \mathbf{d}\lambda, \tag{5.95}$$

The dielectric constitutive relation implies the condition:

$$-1^*b\,\omega^{-1}\mathbf{d}\eta \wedge \mathbf{d}\lambda = \varepsilon \mathbf{f}_*^{123}\lfloor \tilde{G}[[a\mathbf{d}\alpha]],$$

that is

$$-1^*b\,\omega^{-1}\eta'(r)\left(\sin\theta\cos\alpha\,\mathbf{f}^1 + \sin\theta\cos\alpha\,\mathbf{f}^2 + \cos\theta\,\mathbf{f}^3\right)$$

$$\wedge \frac{1}{r\sin^3\theta}\left(\cos\theta\cos\alpha\,\mathbf{f}^1 + \cos\theta\sin\alpha\,\mathbf{f}^2 - \sin\theta\,\mathbf{f}^3\right)$$

$$= \varepsilon a\,\mathbf{f}_*^{123}\lfloor \rho^{-1}(-\sin\alpha\,\mathbf{e}_1 + \cos\alpha\,\mathbf{e}_2).$$

We utilize the dependence $\rho = r\sin\alpha$:

$$-b\,\omega^{-1}\frac{\eta'(r)}{\rho\sin^2\theta}\left(-\sin^2\theta\cos\alpha\,\mathbf{f}_*^{13} - \sin^2\theta\sin\alpha\,\mathbf{f}_*^{23} + \cos^2\theta\cos\alpha\,\mathbf{f}_*^{31} + \cos^2\theta\sin\alpha\,\mathbf{f}_*^{32}\right)$$

$$= \varepsilon a\,\rho^{-1}\left(-\sin\alpha\,\mathbf{f}_*^{23} + \cos\alpha\,\mathbf{f}_*^{31}\right),$$

hence

$$-b\,\omega^{-1}\frac{\eta'(r)}{\sin^2\theta}\left(\cos\alpha\,\mathbf{f}_*^{31} - \sin\alpha\,\mathbf{f}_*^{23}\right) = \varepsilon a\left(-\sin\alpha\,\mathbf{f}_*^{23} + \cos\alpha\,\mathbf{f}_*^{31}\right).$$

The same two-forms are in the brackets, so the following equality

$$\frac{\eta'(r)}{\sin^2\theta} = -\frac{\epsilon\omega a}{b}$$

must be satisfied. On the left-hand side, nontrivial dependence on θ is present, which is absent from the right-hand side; therefore the equality cannot be satisfied for all θ. We conclude from this that a solution of the dielectric constitutive relation does not exist.

One can similarly check that the second possibility with interchanged forms $d\alpha$ and $d\lambda$ also has no solution. In this manner we arrive at the conclusion that there is no running spherical electromagnetic wave.

The other two pairs of possibilities with spherical coordinates also do not lead to solutions.

5.2.4 Semiplane Waves

The only examples from the previous subsection, that have led us to solutions concerned a wave propagating along the Z-axis, for which the surfaces of constant phase were $z = $ const. They are solutions (5.76, 5.77) and (5.85, 5.86). There are *planar*[3] coordinates α i $\ln \rho$. Note how similar are their exterior derivatives:

$$\mathbf{d}\ln\rho = \rho^{-1}\left(\cos\alpha\, \mathbf{f}^1 + \sin\alpha\, \mathbf{f}^2\right).$$

$$\mathbf{d}\alpha = \rho^{-1}\left(-\sin\alpha\, \mathbf{f}^1 + \cos\alpha\, \mathbf{f}^2\right),$$

We see that the transition from $\mathbf{d}\ln\rho$ to $\mathbf{d}\alpha$ is the change $\mathbf{f}^1 \to \mathbf{f}^2$ and $\mathbf{f}^2 \to -\mathbf{f}^1$. This is rotation by an angle $\pi/2$ around an axis parallel to the Z-axis, in each point separately. This observation lets us introduce a more general semiplane wave in an isotropic medium.

We assume that we have any planar coordinate $\zeta(x, y)$ that is harmonic:

$$\frac{\partial^2 \zeta}{\partial x^2} + \frac{\partial^2 \zeta}{\partial y^2} = 0. \tag{5.96}$$

One may create two one-forms from it:

$$\mathbf{f} = \frac{\partial \zeta}{\partial x}\, \mathbf{f}^1 + \frac{\partial \zeta}{\partial y}\, \mathbf{f}^2, \tag{5.97}$$

$$\mathbf{h} = -\frac{\partial \zeta}{\partial y}\, \mathbf{f}^1 + \frac{\partial \zeta}{\partial x}\, \mathbf{f}^2. \tag{5.98}$$

They are perpendicular, which follows from the orthogonality of basic one-forms: $g(\mathbf{f}^i, \mathbf{f}^j) = \delta^{ij}$. The exterior derivative of (5.97) is zero because of the double action of $\mathbf{d}\wedge$:

$$\mathbf{d}\wedge\mathbf{f} = \mathbf{d}\wedge\mathbf{d}\wedge\zeta = 0.$$

---

[3] Planar means here that they are functions of only two cartesian coordinates x and y.

The exterior derivative of (5.98) is:

$$\mathbf{d} \wedge \mathbf{h} = -\frac{\partial^2 \zeta}{\partial y^2} \mathbf{f}^2 1 + \frac{\partial^2 \zeta}{\partial x^2} \mathbf{f}^1 2 = \left(\frac{\partial^2 \zeta}{\partial x^2} + \frac{\partial^2}{\partial y^2} \right) \mathbf{f}^1 2 = 0$$

by virtue of (5.96). Both one-forms $\mathbf{f}$ and $\mathbf{h}$ will be used for considering a possible wave.

The first form (5.97) is the exterior derivative of ζ: $\mathbf{f} = \mathbf{d}\zeta$, and the second (5.98) is the first one rotated by an angle $\pi/2$ separately in each point of space. The exterior derivatives of (5.97) and (5.98) are

$$\mathbf{d} \wedge \mathbf{h} = -\frac{\partial^2 \zeta}{\partial y^2} \mathbf{f}^{21} + \frac{\partial^2 \zeta}{\partial x^2} \mathbf{f}^{12} = \left(\frac{\partial^2 \zeta}{\partial x^2} + \frac{\partial^2 \zeta}{\partial y^2} \right) \mathbf{f}^{12} = 0$$

by virtue of (5.96).

We are looking for a semiplane wave, for which the electromagnetic fields have form similar to (5.83, 5.84):

$$\mathbf{E}(\mathbf{r}, t) = a\, \psi(\kappa z - \omega t)\, \mathbf{f}(\mathbf{r}), \quad \mathbf{B}(\mathbf{r}, t) = a\kappa\omega^{-1} \psi(\kappa z - \omega t)\, \mathbf{f}^3 \wedge \mathbf{f}(\mathbf{r}), \qquad (5.99)$$

$$\mathbf{H}(\mathbf{r}, t) = 1^* b\, \psi(\kappa z - \omega t)\, \mathbf{h}(\mathbf{r}), \quad \mathbf{D}(\mathbf{r}, t) = -1^* b\kappa\omega^{-1} \psi(\kappa z - \omega t)\, \mathbf{f}^3 \wedge \mathbf{h}(\mathbf{r}). \tag{5.100}$$

We use formulas (5.97), (5.98) and notations $\zeta'_x = \frac{\partial \zeta}{\partial x}$, $\zeta'_y = \frac{\partial \zeta}{\partial y}$:

$$\mathbf{E} = a\, \psi(\kappa z - \omega t) \left(\zeta'_x \mathbf{f}^1 + \zeta'_y \mathbf{f}^2 \right), \quad \mathbf{B} = a\kappa\omega^{-1} \psi(\kappa z - \omega t) \left(\zeta'_x \mathbf{f}^{31} + \zeta'_y \mathbf{f}^{32} \right), \tag{5.101}$$

$$\mathbf{H} = b\, \psi(\kappa z - \omega t) \left(-\zeta'_y \mathbf{f}^1_* + \zeta'_x \mathbf{f}^2_* \right), \quad \mathbf{D} = b\kappa\omega^{-1} \psi(\kappa z - \omega t) \left(\zeta'_y \mathbf{f}^{31}_* - \zeta'_x \mathbf{f}^{32}_* \right). \tag{5.102}$$

We verify the dielectric constitutive relation

$$\varepsilon \mathbf{f}^{123}_* \lfloor \tilde{G}[\mathbf{E}] = \varepsilon \mathbf{f}^{123}_* \lfloor \tilde{G} \left[a\psi \left(\zeta'_x \mathbf{f}^1 + \zeta'_y \mathbf{f}^2 \right) \right] = \varepsilon a \psi\, \mathbf{f}^{123}_* \lfloor \left(\zeta'_x \mathbf{e}_1 + \zeta'_y \mathbf{e}_2 \right)$$

$$= \varepsilon a \psi \left(\zeta'_x \mathbf{f}^{23}_* + \zeta'_y \mathbf{f}^{31}_* \right).$$

If this is equal to $\mathbf{D}$ the following relation must be fulfilled

$$\varepsilon a = b\kappa\omega^{-1}. \tag{5.103}$$

We check the magnetic constitutive relation:

$$\mu \mathbf{f}_*^{123} \lfloor \tilde{G}[\mathbf{H}] = \mu \mathbf{f}_*^{123} \lfloor \tilde{G} \left[b \, \psi \left(-\zeta_y' \, \mathbf{f}_*^1 + \zeta_x' \, \mathbf{f}_*^2 \right) \right] = \mu b \, \psi \, \mathbf{f}_*^{123} \lfloor \left(-\zeta_y' \, \mathbf{e}_1^* + \zeta_x' \, \mathbf{e}_2^* \right)$$

$$= \mu b \, \psi \left(-\zeta_y' \, \mathbf{f}^{23} + \zeta_x' \, \mathbf{f}^{31} \right).$$

If this is equal to $\mathbf{B}$ the following relation should be satisfied:

$$\mu b = a \kappa \omega^{-1}. \tag{5.104}$$

The equation system (5.103), (5.104) has the solution

$$\kappa = \omega \sqrt{\varepsilon \mu}, \quad \frac{b}{a} = \sqrt{\frac{\varepsilon}{\mu}}.$$

Inserting this into (5.101, 5.102) gives the following expressions for the semiplane electromagnetic wave:

$$\mathbf{E} = a \, \psi (\kappa z - \omega t) \left(\zeta_x' \, \mathbf{f}^1 + \zeta_y' \, \mathbf{f}^2 \right), \quad \mathbf{B} = a \, \sqrt{\varepsilon \mu} \psi (\kappa z - \omega t) \left(\zeta_x' \, \mathbf{f}^{31} + \zeta_y' \, \mathbf{f}^{32} \right),$$

$$\tag{5.105}$$

$$\mathbf{H} = a \, \sqrt{\frac{\varepsilon}{\mu}} \psi (\kappa z - \omega t) \left(-\zeta_y' \, \mathbf{f}_*^1 + \zeta_x' \, \mathbf{f}_*^2 \right), \quad \mathbf{D} = a \, \varepsilon \psi (\kappa z - \omega t) \left(\zeta_y' \, \mathbf{f}_*^{31} - \zeta_x' \, \mathbf{f}_*^{32} \right).$$

$$\tag{5.106}$$

Let us ponder to which physical system this wave corresponds. The electric field strength can be written down:

$$\mathbf{E}(\mathbf{r}, t) = a \, \psi (\kappa z - \omega t) \, \mathbf{d}\zeta (x, y) = \psi (\kappa z - \omega t) \, \mathbf{E}_0 (x, y),$$

where $\mathbf{E}_0(x, y) = a \, \mathbf{d}\zeta (x, y)$. We recall the relation $\mathbf{E}(\mathbf{r}) = -\mathbf{d}\Phi(\mathbf{r})$ from electrostatics, expressing the field strength by the electric potential, so we are allowed to substitute $\Phi_0(\mathbf{r}) = -a\zeta (x, y)$, and obtain

$$\mathbf{E}_0(x, y) = -\mathbf{d}\Phi_0(x, y).$$

Due to (5.96) we know that Φ satisfies the Laplace equation. In this manner, $\mathbf{E}_0(x, y)$, the electrostatic field is independent of the z-coordinate; thus it has translational symmetry along the Z-axis. Such fields are present in systems of conductors with the same translational symmetry, for instance around infinite parallel linear cables or in conducting cylinders. Therefore the expression

$$\mathbf{E}(\mathbf{r}, t) = \psi (\kappa z - \omega t) \, \mathbf{E}_0 (x, y)$$

corresponds to an electromagnetic wave propagating along the described system of conductors, which is called a *transmission line*.

Knowing this, let us look more closely at the previously found solution (5.83, 5.84). The electric induction **D** has its attitudes extending radially from the Z-axis, hence an integral over cylinder S with its axis coinciding with the Z-axis has $ds^* = -1^*\mathbf{e}_3 dz \wedge \mathbf{e}_\alpha d\alpha$ as its integration element, where $\mathbf{e}_\alpha$ is the vector dual to one-form $d\alpha$. Therefore the integral over the side surface of the above-mentioned cylinder with height dh is

$$\int_S \mathbf{D}[[ds^*]] = -1^*\varepsilon a \int_z^{z+dh} \int_0^{2\pi} \psi(\kappa z - \omega t) \left(\mathbf{f}^3 \wedge \mathbf{d}\alpha \right) [[-1^*\mathbf{e}_3 dz \wedge \mathbf{e}_\alpha d\alpha]]$$

$$= \varepsilon a \int_z^{z+dh} \int_0^{2\pi} \psi(\kappa z - \omega t) \, dz \, d\alpha.$$

The integration function does not depend on α, so

$$\int_S \mathbf{D}[[ds^*]] = 2\pi\varepsilon a \int_z^{z+dh} \psi(\kappa z - \omega t) \, dz = 2\pi \psi(\kappa z - \omega t) \, dh.$$

The integrals over the two bases of the cylinder give zero, because the field **D** is parallel to them, hence we can replace the left-hand side by a more extensive integral over the closed surface S' of S and the two bases of the cylinder:

$$\oint_{S'} \mathbf{D}[[ds^*]] = 2\pi \varepsilon a \, \psi(\kappa z - \omega t) \, dh.$$

According to Gauss's law, the left-hand side is a charge contained in the region confined by S', hence

$$dQ(S') = 2\pi \varepsilon a \, \psi(\kappa z - \omega t) \, dh.$$

This expression does not involve the radius of the cylinder, so the charge must be concentrated on its axis. As we see, it is proportional to the height dh, hence we may introduce the linear density λ of charge on the Z-axis:

$$\lambda(z, t) = \frac{dQ}{dh} = 2\pi \varepsilon a \, \psi(\kappa z - \omega t).$$

This formula says that for the found wave (5.105, 5.106), linear charge accompanies the Z-axis, and its linear density changes which the same frequency as the electromagnetic wave, because it is described by the same function ψ.

We calculate also a curvilinear integral of the magnetic field strength over a circle K surrounding the Z-axis and perpendicular to it. The integration element is expressed by the twisted one-form $dl^* = 1^* t_\alpha d\alpha$. Then for $\mathbf{H}$ taken from (5.84), the integral is

$$\oint_K \mathbf{H}[[dl^*]] = 1^* \sqrt{\frac{\varepsilon}{\mu}} a \int_0^{2\pi} \psi(\kappa z - \omega t) d\alpha [[1^* t_\alpha d\alpha]] = \sqrt{\frac{\varepsilon}{\mu}} a \psi(\kappa z - \omega t) \int_0^{2\pi} d\alpha [[t_\alpha]]$$

$$= 2\pi \sqrt{\frac{\varepsilon}{\mu}} a \psi(\kappa z - \omega t).$$

According to the Ampère-Oersted law (4.4), the left-hand side is equal to

$$I + \frac{d}{dt} \int_\Sigma \mathbf{D}[ds^*], \tag{5.107}$$

where Σ is a circle with boundary K. Since $\mathbf{D}$ is parallel to Σ, the right-hand term in (5.107) is zero and there remains

$$I(z, t) = 2\pi \sqrt{\frac{\varepsilon}{\mu}} a \psi(\kappa z - \omega t).$$

This expression does not depend on the radius of the circle, so the calculated current flows on the Z-axis. We ascertain that the found wave (5.83, 5.84) is accompanied by a current, flowing on the Z-axis with magnitude changing at the frequency of the electromagnetic wave.

Summarizing, we may claim that the found electromagnetic waves (5.105, 5.102) are present around transmission lines or cylindrical wave-guides. However, an open question remains as to whether non-planar electromagnetic waves exist that are not semiplanes.

5.3 Plane Wave

By a *plane electromagnetic wave* we understand a solution of the free Maxwell equations in which the phase contains spatial dependence only through a linear form $\mathbf{k}$, i.e., is of the form

$$\varphi = \mathbf{k}[[\mathbf{r}]] - \omega t \tag{5.108}$$

and the amplitudes of the field quantities are constant. Thus we seek the following solution:

$$E(r, t) = \psi(k[[r]] - \omega t)E_0, \tag{5.109}$$

$$B(r, t) = \psi(k[[r]] - \omega t)B_0, \tag{5.110}$$

$$H(r, t) = \psi(k[[r]] - \omega t)H_0, \tag{5.111}$$

$$D(r, t) = \psi(k[[r]] - \omega t)D_0, \tag{5.112}$$

where $\psi(\cdot)$ is a scalar function of a scalar argument, E_0, B_0, H_0, D_0 are constant forms, ω is a scalar constant and $k[[r]]$ is the value of the linear form k, i.e the wave covector on the radius vector r. Due to Exercise 3.27 we have the identity $d \wedge k[[r]] = k$. In the expected solution (5.109–5.112) all fields maintain their attitudes for all times and positions; only magnitudes and orientations may change. This means that the wave is linearly polarized. Why are these solutions called *plane*? Because the field quantities are constant where the function $k(r) = k[[r]]$ is constant, i.e., on planes.

We proceed to consider Maxwell's equations. By virtue of Exercise 3.23 we ascertain that the equation $d \wedge E + \partial B / \partial t = 0$ gives the condition

$$\psi' \, k \wedge E_0 - \psi' \omega B_0 = 0$$

(where the prime denotes the derivative with respect to the whole argument) and allows us to write

$$B_0 = \omega^{-1} k \wedge E_0 \quad \text{and} \quad B = \omega^{-1} k \wedge E = u \wedge E. \tag{5.113}$$

This equality expressed in the traditional language has the form $B = \omega^{-1} k \times E$. We see also from (5.113) that the two forms are parallel: $B \parallel E$, which in terms of vectors is written as $B \cdot E = 0$. Moreover, $B \parallel k$, which corresponds to $B \cdot k = 0$. The relation $B = u \wedge E$ is illustrated in Fig. 5.3.

The second Maxwell's equation $d \wedge B = 0$ gives the condition

$$\psi' \, k \wedge B_0 = 0.$$

This condition and this Maxwell equation are automatically satisfied by virtue of (5.113).

The third free Maxwell equation $d \wedge H - \partial D / \partial t = 0$ reduces to the condition

$$\psi' \, k \wedge H_0 + \psi' \omega D_0 = 0$$

and allows us to write

$$D_0 = -\omega^{-1} k \wedge H_0 \quad \text{and} \quad D = -\omega^{-1} k \wedge H = -u \wedge H. \tag{5.114}$$

Therefore, the fourth free Maxwell equation $d \wedge D = 0$ is automatically satisfied. The constant forms E_0, H_0, present in relations (5.113) and (5.114), for the time being, are arbitrary. It is clear from (5.114) that $D \parallel H$ and $D \parallel k$; this is shown this in Fig. 5.4. Condition (5.114) can be translated into the traditional language as $D = -\omega^{-1}k \times H$, and the parallelity conditions as $D \cdot H = 0$ and $D \cdot k = 0$.

Summarizing, we write now the plane-wave solutions of Maxwell's equations:

$$E(r, t) = \psi(k[r] - \omega t)E_0, \qquad (5.115)$$

$$B(r, t) = \psi(k[r] - \omega t)u \wedge E_0, \qquad (5.116)$$

$$H(r, t) = \psi(k[r] - \omega t)H_0, \qquad (5.117)$$

$$D(r, t) = -\psi(k[r] - \omega t)u \wedge H_0, \qquad (5.118)$$

where E_0, H_0 are arbitrary constant one-forms. This could be considered as a premetric form of the plane waves.

We consider the electromagnetic wave in a medium with different electric and magnetic properties. Then application of two scalar products is necessary: the electric one $g_\varepsilon(\cdot, \cdot)$ and the magnetic one $g_\mu(\cdot, \cdot)$. The problem has the same degree of complication, if we assume that the medium is magnetically isotropic, and only electrically anisotropic. Therefore, we assume that the magnetic scalar product g_μ coincides with the natural scalar product of the physical space. If the vector basis $\{e_1, e_2, e_3\}$ is orthonormal according to this scalar product, the one-form basis $\{f^1, f^2, f^3\}$ has the same property; thus the magnetic constitutive relation can be taken in the form $B = \mu H$, which means that we have the following relations for the form coordinates:

$$B_{23} = \mu H_1, \quad B_{31} = \mu H_2, \quad B_{12} = \mu H_3. \qquad (5.119)$$

Recall that according to formula (2.51), when $\tilde{G} = \mathcal{E}$, the dielectric scalar product of two one-forms $E = \sum_{i=1}^{3} E_i f^i$ and $F = \sum_{j=1}^{3} F_j f^j$ assumes the shape

$$\tilde{g}_\varepsilon(E, F) = \sum_{i,j=1}^{3} E_i \varepsilon^{ij} F_j. \qquad (5.120)$$

Now the constitutive relations come into play. Choose a magnetically orthonormal basis such that $k \parallel f^3$, i.e., $k = kf^3$ (k here is the coordinate of the wave covector) and

$$u = uf^3 \qquad (5.121)$$

(u is the coordinate of the slowness), and substitute this into (5.118) and (5.119):

$$\mathbf{B} = u\,\mathbf{f}^3 \wedge (E_1\mathbf{f}^1 + E_2\mathbf{f}^2 + E_3\mathbf{f}^3) = u(E_1\mathbf{f}^{31} - E_2\mathbf{f}^{23}),$$

$$\mathbf{D} = -u\,\mathbf{f}^3 \wedge (H_1\mathbf{f}_*^1 + H_2\mathbf{f}_*^2 + H_3\mathbf{f}_*^3) = u(-H_1\mathbf{f}_*^{31} + H_2\mathbf{f}_*^{23}),$$

From this the following coordinates of $\mathbf{B}$ and $\mathbf{D}$ are obtained:

$$B_{12} = 0, \quad B_{31} = uE_1, \quad B_{23} = -uE_2, \tag{5.122}$$

$$D_{12} = 0, \quad D_{31} = -uH_1, \quad D_{23} = uH_2. \tag{5.123}$$

Let us see now what the constitutive equation $\mathbf{B} = \mu[[\mathbf{H}]]$ gives. We have assumed that the medium is magnetically isotropic, which means that we may use (5.119). We unite this with (5.122) and obtain:

$$H_1 = -\frac{u}{\mu}E_2, \quad H_2 = \frac{u}{\mu}E_1, \quad H_3 = 0. \tag{5.124}$$

The constant form $\mathbf{H}_0$ is not arbitrary any more. We substitute this into (5.123):

$$D_{12} = 0, \quad D_{31} = \frac{u^2}{\mu}E_2, \quad D_{23} = \frac{u^2}{\mu}E_1. \tag{5.125}$$

In this manner equations (5.122), (5.124) and (5.125) allow us to express the fields $\mathbf{D}, \mathbf{B}, \mathbf{H}$ by $\mathbf{E}$:

$$\mathbf{E} = E_1\mathbf{f}^1 + E_2\mathbf{f}^2 + E_3\mathbf{f}^3, \tag{5.126}$$

$$\mathbf{B} = u(E_1\mathbf{f}^3 \wedge \mathbf{f}^1 - E_2\mathbf{f}^2 \wedge \mathbf{f}^3), \tag{5.127}$$

$$\mathbf{H} = \frac{u}{\mu}(-E_2\mathbf{f}_*^1 + E_1\mathbf{f}_*^2). \tag{5.128}$$

$$\mathbf{D} = \frac{u^2}{\mu}(E_2\mathbf{f}^3 \wedge \mathbf{f}_*^1 + E_1\mathbf{f}^2 \wedge \mathbf{f}_*^3), \tag{5.129}$$

We know from Eqs. (5.113) and (5.114) that $\mathbf{B}, \mathbf{D}$ are parallel to the planes of $\mathbf{k}$, and it is clear from (5.128) that $\mathbf{H}$ is magnetically perpendicular to $\mathbf{k}$ (since it does not contain $\mathbf{f}_*^3 \parallel \mathbf{k}$), but $\mathbf{E}$ is not magnetically perpendicular to $\mathbf{k}$, if $E_3 \neq 0$. The configuration of the named quantities is depicted in Fig. 5.5.

5.3.1 Eigenwaves

We write the dielectric constitutive equations:

$$D_{23} = \varepsilon_0(\varepsilon^{11}E_1 + \varepsilon^{12}E_2 + \varepsilon^{13}E_3),$$

$$D_{31} = \varepsilon_0(\varepsilon^{21}E_1 + \varepsilon^{22}E_2 + \varepsilon^{23}E_3),$$

$$D_{12} = \varepsilon_0(\varepsilon^{31}E_1 + \varepsilon^{32}E_2 + \varepsilon^{33}E_3).$$

Condition (5.125) allows us to change the left-hand sides:

$$\frac{u^2}{\varepsilon_0\mu}E_1 = \varepsilon^{11}E_1 + \varepsilon^{12}E_2 + \varepsilon^{13}E_3, \tag{5.130}$$

$$\frac{u^2}{\varepsilon_0\mu}E_2 = \varepsilon^{21}E_1 + \varepsilon^{22}E_2 + \varepsilon^{23}E_3, \tag{5.131}$$

$$0 = \varepsilon^{31}E_1 + \varepsilon^{32}E_2 + \varepsilon^{33}E_3. \tag{5.132}$$

We conclude from this that the coordinates of **E** can not be arbitrary—they have to fulfill the above system of linear equations. The last equation yields

$$E_3 = -\frac{\varepsilon^{31}}{\varepsilon^{33}}E_1 - \frac{\varepsilon^{32}}{\varepsilon^{33}}E_2. \tag{5.133}$$

We substitute this relation in Eqs. (5.130) and (5.131):

$$\left(\varepsilon^{11} - \frac{\varepsilon^{13}\varepsilon^{31}}{\varepsilon^{33}}\right)E_1 + \left(\varepsilon^{12} - \frac{\varepsilon^{13}\varepsilon^{32}}{\varepsilon_{33}}\right)E_2 = \frac{u^2}{\varepsilon_0\mu}E_1, \tag{5.134}$$

$$\left(\varepsilon^{21} - \frac{\varepsilon^{23}\varepsilon^{31}}{\varepsilon^{33}}\right)E_1 + \left(\varepsilon^{22} - \frac{\varepsilon^{23}\varepsilon^{32}}{\varepsilon^{33}}\right)E_2 = \frac{u^2}{\varepsilon_0\mu}E_2, \tag{5.135}$$

Notice that a matrix $A = \{a^{ij}\}$ with elements

$$a^{ij} = \varepsilon^{ij} - \frac{\varepsilon^{i3}\varepsilon^{3j}}{\varepsilon^{33}}$$

is present on the left-hand side. We call this the *reduced permittivity matrix*. Since the matrix $\mathcal{E} = \{\varepsilon^{ij}\}$ is symmetric, A has the same property. Equations (5.134), (5.135) constitute the following eigenvalue equation for A:

$$A \begin{pmatrix} E_1 \\ E_2 \end{pmatrix} = \frac{u^2}{\varepsilon_0 \mu} \begin{pmatrix} E_1 \\ E_2 \end{pmatrix}. \tag{5.136}$$

Matrix A is symmetric; hence its eigenvalues and eigenvectors exist. If it has two distinct eigenvalues $a_{(1)}$ and $a_{(2)}$, we obtain two distinct conditions for the phase slowness:

$$\frac{u^2}{\varepsilon_0 \mu} = a_{(1)}, \quad \text{or} \quad \frac{u^2}{\varepsilon_0 \mu} = a_{(2)},$$

hence

$$u = \sqrt{a_{(1)} \varepsilon_0 \mu} \quad \text{or} \quad u = \sqrt{a_{(2)} \varepsilon_0 \mu}.$$

This gives two different values of phase slownesses for two solutions of (5.136):

$$u^{(i)} = \sqrt{a_{(i)} \varepsilon_0 \mu} \text{ for } i \in \{1, 2\}.$$

Assumption (5.121) concerns both eigenwaves:

$$\mathbf{u}^{(i)} = u^{(i)} \mathbf{f}^3. \tag{5.137}$$

Suitably, with given circular frequency ω two magnitudes of the wave covector $k^{(i)} = \omega u^{(i)}$ occur. If the electric permittivity does not depend on the frequency of the electromagnetic field, the phase slowness coordinate is the same for various frequencies and is characteristic of the given medium and of a chosen direction in it. The eigenequation (5.136) can also be written as:

$$\sum_{m=1}^{2} a^{jm} E_m^{(i)} = a^{(i)} E_j^{(i)}. \tag{5.138}$$

In this manner we ascertain that only pairs $(E_1, E_2)^T$ (T denotes transposition, because we should write columns) satisfying the eigenequation (5.136) give solutions of the free Maxwell's equations with specific wave covectors $\mathbf{k}^{(i)} = \omega \mathbf{u}^{(i)}$; we call them *eigenwaves*. We rewrite formulas (5.126)–(5.129) for the eigenwaves with the third coordinate E_3 of the electric field substituted from (5.133):

$$\mathbf{E}^{(i)} = E_1^{(i)} \mathbf{f}^1 + E_2^{(i)} \mathbf{f}^2 - \varepsilon_{33}^{-1} (\varepsilon_{31} E_1^{(i)} + \varepsilon_{32} E_2^{(i)}) \mathbf{f}^3, \tag{5.139}$$

$$\mathbf{B}^{(i)} = u^{(i)} (E_1^{(i)} \mathbf{f}^{31} - E_2^{(i)} \mathbf{f}^{23}), \tag{5.140}$$

$$\mathbf{D}^{(i)} = \frac{(u^{(i)})^2}{\mu_0} (E_1^{(i)} \mathbf{f}_*^{23} + E_2^{(i)} \mathbf{f}_*^{31}), \tag{5.141}$$

$$\mathbf{H}^{(i)} = \frac{u^{(i)}}{\mu_0} (-E_2^{(i)} \mathbf{f}_*^1 + E_1^{(i)} \mathbf{f}_*^2), \tag{5.142}$$

where the index $i \in \{1, 2\}$ enumerates the eigenwaves.

Table 5.1 is still valid—we should only add superscripts (i). Orthogonality of $\mathbf{B}^{(i)}$ and $\mathbf{D}^{(i)}$ in both metrics proves that their attitudes lie on the principal directions of the ellipse obtained by the intersection of the ellipsoid (quasi-sphere) of the dielectric scalar product with the plane of $\mathbf{k}$. Therefore, it is not strange that only two such attitudes are possible and—by this—only two linearly independent eigenwaves.

5.3.2 Relations Between Eigenwaves

We calculate magnetic scalar products between forms of the same kind, corresponding to different indices (i). To find the scalar product $g_\mu(\mathbf{E}^{(1)}, \mathbf{E}^{(2)})$ we use (5.133):

$$g_\mu(\mathbf{E}^{(1)}, \mathbf{E}^{(2)}) = E_1^{(1)} E_1^{(2)} + E_2^{(1)} E_2^{(2)}$$

$$+ (\varepsilon^{33})^{-2} [(\varepsilon^{31})^2 E_1^{(1)} E_1^{(2)} + (\varepsilon^{32})^2 E_2^{(1)} E_2^{(2)} + \varepsilon^{31} \varepsilon^{32} (E_1^{(1)} E_2^{(2)} + E_2^{(1)} E_1^{(2)})].$$

In other cases we apply (5.122), (5.124) and (5.125):

$$g_\mu(\mathbf{B}^{(1)}, \mathbf{B}^{(2)}) = u_{(1)} u_{(2)} (E_1^{(1)} E_1^{(2)} + E_2^{(1)} E_2^{(2)}),$$

$$g_\mu(\mathbf{D}^{(1)}, \mathbf{D}^{(2)}) = \frac{u_{(1)}^2 u_{(2)}^2}{\mu^2} (E_1^{(1)} E_1^{(2)} + E_2^{(1)} E_2^{(2)}),$$

$$g_\mu(\mathbf{H}^{(1)}, \mathbf{H}^{(2)}) = \frac{u_{(1)} u_{(2)}}{\mu^2} (E_1^{(1)} E_1^{(2)} + E_2^{(1)} E_2^{(2)}).$$

Two eigenvectors $(E_1^{(1)}, E_2^{(1)})^T$ and $(E_1^{(2)}, E_2^{(2)})^T$ of one symmetric matrix A are perpendicular columns, hence we obtain

$$g_\mu(\mathbf{E}^{(1)}, \mathbf{E}^{(2)}) = \frac{1}{(\varepsilon^{33})^2} \{[(\varepsilon^{31})^2 - (\varepsilon^{32})^2] E_1^{(1)} E_1^{(2)} + \varepsilon^{31} \varepsilon^{32} (F_1^{(1)} F_2^{(2)} + E_2^{(1)} E_1^{(3)})\},$$

$$g_\mu(\mathbf{B}^{(1)}, \mathbf{B}^{(2)}) = 0, \quad g_\mu(\mathbf{D}^{(1)}, \mathbf{D}^{(2)}) = 0, \quad g_\mu(\mathbf{H}^{(1)}, \mathbf{H}^{(2)}) = 0.$$

Let us elaborate on the dielectric scalar product of the two one-forms $\mathbf{E}^{(i)}$ starting from (5.120):

$$g_\varepsilon(\mathbf{E}^{(1)}, \mathbf{E}^{(2)}) = \sum_{i,j=1}^{3} E_i^{(1)} \varepsilon^{ij} E_j^{(2)}.$$

We apply (5.133):

$$g_\varepsilon(\mathbf{E}^{(1)}, \mathbf{E}^{(2)}) = E_1^{(1)} \varepsilon^{11} E_1^{(2)} + E_1^{(1)} \varepsilon^{12} E_2^{(2)} - E_1^{(1)} \frac{\varepsilon^{13}}{\varepsilon^{33}} (\varepsilon^{31} E_1^{(2)} + \varepsilon^{32} E_2^{(2)})$$

$$+ E_2^{(1)} \varepsilon^{11} E_1^{(2)} + E_2^{(1)} \varepsilon^{12} E_2^{(2)} - E^{(1)} \frac{\varepsilon^{23}}{\varepsilon^{33}} (\varepsilon^{31} E_1^{(2)} + \varepsilon^{32} E_2^{(2)})$$

$$- (\varepsilon^{33})^{-1} (\varepsilon^{31} E_1^{(1)} + \varepsilon^{32} E_2^{(1)}) \varepsilon^{31} E_1^{(2)} - (\varepsilon^{33})^{-1} (\varepsilon^{31} E_1^{(1)} + \varepsilon^{32} E_2^{(1)}) \varepsilon^{32} E_2^{(2)}$$

$$+ (\varepsilon^{33})^{-1} (\varepsilon^{31} E_1^{(1)} + \varepsilon^{32} E_2^{(1)}) (\varepsilon^{31} E_1^{(2)} + \varepsilon^{32} E_2^{(2)}).$$

After reduction of similar terms, we are left with

$$g_\varepsilon(\mathbf{E}^{(1)}, \mathbf{E}^{(2)}) = E_1^{(1)} \left(\varepsilon^{11} - \frac{\varepsilon^{13} \varepsilon^{31}}{\varepsilon^{33}} \right) E_1^{(2)} + E_1^{(1)} \left(\varepsilon^{12} - \frac{\varepsilon^{13} \varepsilon^{32}}{\varepsilon^{33}} \right) E_2^{(2)}$$

$$+ E_2^{(1)} \left(\varepsilon^{21} - \frac{\varepsilon^{23} \varepsilon^{31}}{\varepsilon^{33}} \right) E_1^{(2)} + E_2^{(1)} \left(\varepsilon^{22} - \frac{\varepsilon^{23} \varepsilon^{32}}{\varepsilon^{33}} \right) E_2^{(2)}.$$

We recognize elements of A, hence

$$g_\varepsilon(\mathbf{E}^{(1)}, \mathbf{E}^{(2)}) = \sum_{i,j=1}^{2} E_i^{(1)} a^{ij} E_j^{(2)}.$$

Since $(E_1^{(2)}, E_2^{(2)})^T$ is the eigenvector to the eigenvalue $a_{(2)}$, we obtain

$$g_\varepsilon(\mathbf{E}^{(1)}, \mathbf{E}^{(2)}) = a_{(2)} (E_1^{(1)} E_1^{(2)} + E_2^{(1)} E_2^{(2)}) = 0 \tag{5.143}$$

because eigenvectors corresponding to different eigenvalues are orthogonal. In this manner we see that the forms $\mathbf{E}^{(i)}$ are dielectrically orthogonal.

The relation (5.16) concerns both eigenwaves:

$$\mathbf{B}^{(i)} = \mathbf{u}^{(i)} \wedge \mathbf{E}^{(i)}.$$

Table 5.2 Perpendicularity relations between eigenwaves

	$g_\mu(\mathbf{B}^{(1)}, \mathbf{B}^{(2)}) = 0$	$g_\mu(\mathbf{D}^{(1)}, \mathbf{D}^{(2)}) = 0$	$g_\mu(\mathbf{H}^{(1)}, \mathbf{H}^{(2)}) = 0$
$g_\varepsilon(\mathbf{E}^{(1)}, \mathbf{E}^{(2)}) = 0$	$g_\varepsilon(\mathbf{B}^{(1)}, \mathbf{B}^{(2)}) = 0$	$g_\varepsilon(\mathbf{D}^{(1)}, \mathbf{D}^{(2)}) = 0$	

It helps us to calculate the dielectric scalar product of both two-forms $\mathbf{B}^{(i)}$:

$$\tilde{g}_\varepsilon(\mathbf{B}^{(1)}, \mathbf{B}^{(2)}) = \tilde{g}_\varepsilon(\mathbf{u}^{(1)} \wedge \mathbf{E}^{(1)}, \mathbf{u}^{(2)} \wedge \mathbf{E}^{(2)}) = \begin{vmatrix} \tilde{g}_\varepsilon(\mathbf{u}^{(1)}, \mathbf{u}^{(2)}) & \tilde{g}_\varepsilon(\mathbf{u}^{(1)}, \mathbf{E}^{(2)}) \\ \tilde{g}_\varepsilon(\mathbf{E}^{(1)}, \mathbf{u}^{(2)}) & \tilde{g}_\varepsilon(\mathbf{E}^{(1)}, \mathbf{E}^{(2)}) \end{vmatrix}$$

$$= \tilde{g}_\varepsilon(\mathbf{u}^{(1)}, \mathbf{u}^{(2)}) \tilde{g}_\varepsilon(\mathbf{E}^{(1)}, \mathbf{E}^{(2)}) - \tilde{g}_\varepsilon(\mathbf{u}^{(1)}, \mathbf{E}^{(2)}) \tilde{g}_\varepsilon(\mathbf{E}^{(1)}, \mathbf{u}^{(2)}).$$

Both eigenwaves have phase slownesses with equal attitudes, the same as their phase densities $\mathbf{k}^{(i)}$; hence, recalling (5.55) we have

$$\tilde{g}_\varepsilon(\mathbf{u}^{(1)}, \mathbf{E}^{(2)}) = \tilde{g}_\varepsilon(\mathbf{u}^{(2)}, \mathbf{E}^{(1)}) = 0.$$

By uniting this with (5.143) we obtain

$$\tilde{g}_\varepsilon(\mathbf{B}^{(1)}, \mathbf{B}^{(2)}) = 0. \tag{5.144}$$

The two-forms $\mathbf{B}^{(1)}$ and $\mathbf{B}^{(2)}$ are perpendicular in both magnetic and dielectric metrics.

Exercise 5.5 Prove the following orthogonality

$$g_\varepsilon(\mathbf{D}^{(1)}, \mathbf{D}^{(2)}) = 0.$$

All the perpendicularity relations between quantities with different (i)s can be summarized in Table 5.2.

5.3.3 Densities of Energy, Momentum and Energy Flux

The energy densities of the electric and magnetic fields

$$w_e = \frac{1}{2} \mathbf{E} \wedge \mathbf{D}, \qquad\qquad w_m = \frac{1}{2} \mathbf{H} \wedge \mathbf{B}$$

after using (5.114) and (5.113) are

$$w_e = \frac{1}{2} \mathbf{E} \wedge (\mathbf{H} \wedge \mathbf{u}), \qquad\qquad w_m = \frac{1}{2} \mathbf{H} \wedge (\mathbf{u} \wedge \mathbf{E}),$$

that is, the contributions of the two fields to the energy are exactly the same

$$w_e = \frac{1}{2}\mathbf{u} \wedge \mathbf{E} \wedge \mathbf{H}, \qquad\qquad w_m = \frac{1}{2}\mathbf{u} \wedge \mathbf{E} \wedge \mathbf{H}.$$

The energy of the whole electromagnetic field for the considered plane waves is thus

$$w = w_e + w_m = \mathbf{u} \wedge \mathbf{E} \wedge \mathbf{H} = \mathbf{u} \wedge \mathbf{S} \tag{5.145}$$

We obtained an interesting identity relating the energy density with the energy flux density $\mathbf{S} = \mathbf{E} \wedge \mathbf{H}$. For the fields (5.139–5.142) we get

$$w = \frac{u^2}{\mu}(E_1{}^2 + E_2{}^2)\mathbf{f}_*^{123}. \tag{5.146}$$

Let us consider the momentum of the electromagnetic wave contained in a region of volume ΔV according to the formula (4.19), i.e.,

$$\Delta\mathbf{p} = \mathbf{D}\lfloor(\Delta V \lfloor \mathbf{B}). \tag{5.147}$$

We insert in this $\mathbf{B} = \mathbf{u} \wedge \mathbf{E}$, $\mathbf{D} = -\mathbf{u} \wedge \mathbf{H}$, and as the volume twisted trivector we choose the exterior product of three vectors (one of them, namely $\mathbf{a}$, being twisted):

$$\Delta V = \mathbf{a} \wedge \mathbf{b} \wedge \Delta\mathbf{l} \tag{5.148}$$

which satisfy the conditions:

$$\mathbf{a} \parallel \mathbf{u}, \quad \mathbf{b} \parallel \mathbf{u}, \tag{5.149}$$

$$\mathbf{a} \parallel \mathbf{E}, \quad \mathbf{b} \parallel \mathbf{H}, \quad \Delta\mathbf{l} \parallel \mathbf{E}, \quad \Delta\mathbf{l} \parallel \mathbf{H}. \tag{5.150}$$

See Fig. 5.9.

Fig. 5.9 Appropriately chosen $\mathbf{a}$, $\mathbf{b}$ and $\Delta\mathbf{l}$

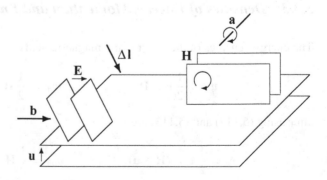

We calculate first the contraction needed for (5.147)

$$\Delta V \lfloor \mathbf{B} = (\mathbf{a} \wedge \mathbf{b} \wedge \Delta \mathbf{l}) \lfloor (\mathbf{u} \wedge \mathbf{E}) = \{(\mathbf{a} \wedge \mathbf{b} \wedge \Delta \mathbf{l}) \lfloor \mathbf{u}\} \lfloor \mathbf{E}.$$

The contractions of $\mathbf{a}$ and $\mathbf{b}$ with $\mathbf{u}$ vanish because of (5.149); hence only

$$\Delta V \lfloor \mathbf{B} = \mathbf{u}[[\Delta \mathbf{l}]] (\mathbf{a} \wedge \mathbf{b}) \lfloor \mathbf{E} = \mathbf{u}[[\Delta \mathbf{l}]] \, \mathbf{E}[[\mathbf{b}]] \, \mathbf{a}$$

remains. We calculate the second contraction for (5.147)

$$\Delta \mathbf{p} = \mathbf{D} \lfloor (\Delta V \lfloor \mathbf{B}) = -(\mathbf{u} \wedge \mathbf{H}) \lfloor (\mathbf{u}[[\Delta \mathbf{l}]] \, \mathbf{E}[[\mathbf{b}]] \, \mathbf{a}) = -\mathbf{u}[[\Delta \mathbf{l}]] \, \mathbf{E}[[\mathbf{b}]] \, (\mathbf{u} \wedge \mathbf{H}) \lfloor \mathbf{a}$$

$$\Delta \mathbf{p} = -\mathbf{u}[[\Delta \mathbf{l}]] \, \mathbf{E}[[\mathbf{b}]] \, \mathbf{H}[[\mathbf{a}]] \, \mathbf{u} \tag{5.151}$$

We see that the momentum one-form is proportional to the slowness one-form.

An interesting question arises: what is the sign of the coefficient standing in front of the slowness $\mathbf{u}$? In order to answer this we take from Fig. 5.6 the relevant one-forms $\mathbf{E}$, $\mathbf{H}$, $\mathbf{u}$ and choose vectors $\mathbf{a}$, $\mathbf{b}$, $\Delta \mathbf{l}$ such that $\mathbf{E}[[\mathbf{b}]]$, $\mathbf{H}[[\mathbf{a}]]$ are positive but $\mathbf{u}[[\Delta \mathbf{l}]]$ is negative. At the same time, the volume (5.148) is positive (see Fig. 5.9). Now the whole coefficient in front of the last $\mathbf{u}$ in (5.151) is positive. In this manner the nontwisted one-forms $\Delta \mathbf{p}$ and $\mathbf{u}$ have the same direction.

Exercise 5.6 Show that the scalar coefficient in front of $\mathbf{u}$ in (5.151) is a value of the three-form $\mathbf{u} \wedge \mathbf{E} \wedge \mathbf{H}$ on the volume trivector (5.148)

$$-\mathbf{u}[[\Delta \mathbf{l}]] \, \mathbf{E}[[\mathbf{b}]] \, \mathbf{H}[[\mathbf{a}]] = (\mathbf{u} \wedge \mathbf{E} \wedge \mathbf{H})[[\Delta V]].$$

Thus we can write relation (5.151) as

$$\Delta \mathbf{p} = \mathbf{u} \, (\mathbf{u} \wedge \mathbf{E} \wedge \mathbf{H})[[\Delta V]].$$

Therefore, we can introduce a *density of momentum* of the plane electromagnetic wave as the tensor

$$T = \mathbf{u} \otimes (\mathbf{u} \wedge \mathbf{E} \wedge \mathbf{H})$$

which by virtue of (5.145) can be written as

$$T = \mathbf{u} \otimes w. \tag{5.152}$$

The one-dimensional attitude of the Poynting twisted two-form as the outer product $\mathbf{S} = \mathbf{E} \wedge \mathbf{H}$ of one-forms is the intersection of the planes of the two factors. If $\mathbf{E}$ and $\mathbf{H}$ are as in Fig. 5.6, the attitude of $\mathbf{S}$ is oblique with respect to the planes of $\mathbf{k}$. The direction of $\mathbf{S}$ must be conceded as the direction of wave propagation. We have now settled the dilemma considered immediately after

Fig. 5.2—a phase velocity is not given uniquely, hence it should be abandoned in the anisotropic medium. On the other hand, the propagation direction of the plane wave is determined by the energy flux density, i.e., by the Poynting twisted two-form.

The energy flux density $\mathbf{S} = \mathbf{E} \wedge \mathbf{H}$ for fields (5.139–5.142) is

$$\mathbf{S} = \frac{u}{\mu} [(E_1^2 + E_2^2)\mathbf{f}_*^{12} - E_3 E_1 \mathbf{f}_*^{23} - E_3 E_2 \mathbf{f}_*^{31}]. \tag{5.153}$$

The last two terms are parallel to $\mathbf{e}_1$ and $\mathbf{e}_2$, respectively, hence $\mathbf{S}$ is not magnetically perpendicular to the planes of $\mathbf{k}$ if $E_3 \neq 0$. The parallel part, i.e., $\mathbf{S}_\| = \frac{-u\bar{E}_3}{\mu}(E_1 \mathbf{f}^2 \wedge \mathbf{f}_*^3 + E_2 \mathbf{f}^3 \wedge \mathbf{f}_*^1)$, is proportional to $\mathbf{D}$, as is visible from (5.141). The common factor h has the physical dimension of the action. Thus $\mathbf{S}$ has its direction inclined to $\mathbf{D}$. After using (5.133) we rewrite (5.153) as

$$\mathbf{S} = \frac{u}{\mu} \left[(E_1^2 + E_2^2)\mathbf{f}_*^{12} + \frac{1}{\varepsilon^{33}}(\varepsilon^{31} E_1^2 + \varepsilon^{32} E_1 E_2)\mathbf{f}_*^{23} \right. $$
$$\left. + \frac{1}{\varepsilon^{33}}(\varepsilon^{31} E_1 E_2 + \varepsilon^{32} E_2^2)\mathbf{f}_*^{31} \right]. \tag{5.154}$$

Up to now our basis was fitted to the wave covector by the condition $\mathbf{k} = k\mathbf{f}^3$. We may adjust it further to the eigenwaves. Choose the basis vectors and one-forms such that $E_2^{(1)} = 0$ and $E_1^{(2)} = 0$; then expressions (5.139–5.142) and (5.153) assume the form:

$$\mathbf{E}^{(1)} = E_1^{(1)}\mathbf{f}^1 + E_3^{(1)}\mathbf{f}^3 = E_1^{(1)} \left(\mathbf{f}^1 - \frac{\varepsilon^{31}}{\varepsilon^{33}} \mathbf{f}^3 \right), \tag{5.155}$$

$$\mathbf{B}^{(1)} = u_{(1)} E_1^{(1)}\mathbf{f}^{31}, \quad \mathbf{H}^{(1)} = \frac{u_{(1)}^2}{\mu} E_1^{(1)}\mathbf{f}_*^2, \quad \mathbf{D}^{(1)} = \frac{u_{(1)}^2}{\mu} E_1^{(1)}\mathbf{f}_*^{23}, \tag{5.156}$$

$$\mathbf{S}^{(1)} = \frac{u_{(1)}}{\mu} E_1^{(1)2} \left(\mathbf{f}_*^{12} + \frac{\varepsilon^{31}}{\varepsilon^{33}} \mathbf{f}_*^{23} \right), \tag{5.157}$$

for the first eigenwave, and

$$\mathbf{E}^{(2)} = E_2^{(2)}\mathbf{f}^2 + E_3^{(2)}\mathbf{f}^3 = E_2^{(2)} \left(\mathbf{f}^2 - \frac{\varepsilon^{32}}{\varepsilon^{33}} \mathbf{f}^3 \right), \tag{5.158}$$

$$\mathbf{B}^{(2)} = -u_{(2)} E_2^{(2)}\mathbf{f}^{23}, \quad \mathbf{H}^{(2)} = -\frac{u_{(2)}^2}{\mu} E_2^{(2)}\mathbf{f}_*^1, \quad \mathbf{D}^{(2)} = \frac{u_{(2)}^2}{\mu} E_2^{(2)}\mathbf{f}_*^{31}, \tag{5.159}$$

$$\mathbf{S}^{(2)} = \frac{u_{(2)}}{\mu} E_2^{(2)2} \left(\mathbf{f}_*^{12} + \frac{\varepsilon^{32}}{\varepsilon^{33}} \mathbf{f}_*^{31} \right) \tag{5.160}$$

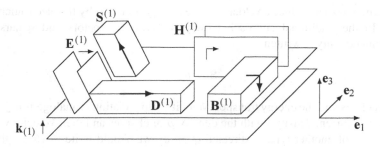

Fig. 5.10 Relevant forms for the eigenwave (5.155, 5.156)

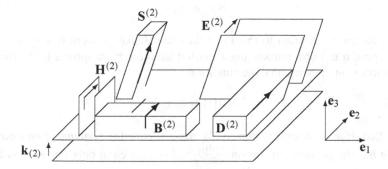

Fig. 5.11 Relevant forms for the eigenwave (5.158, 5.159)

for the second eigenwave. In the wave (5.155), (5.156) the nontwisted two-form $\mathbf{B}^{(1)}$ is parallel to $\mathbf{e}_2$, and the twisted two-form $\mathbf{D}^{(1)}$ to $\mathbf{e}_1$, whereas in (5.158), (5.159), the two-form $\mathbf{B}^{(2)}$ is parallel to $\mathbf{e}_1$, and the two-form $\mathbf{D}^{(2)}$ to $\mathbf{e}_2$.

For such a choice of basis vectors, we may think that Fig. 5.6 shows the first eigenwave. We repeat it here as Fig. 5.10 with the basis vectors and the Poynting form added. In such a case the second eigenwave should be as in Fig. 5.11.

The Poynting twisted two-forms, corresponding to the two eigenwaves, may have different directions. In such a case two plane waves with the same frequency and the same planes of constant phase propagate, i.e., transport the energy, in different directions and have distinct phase velocities, and therefore different phase velocities.

It is worth pondering why the eigenwaves are not needed in an isotropic medium. The matrix $\mathcal{E}$ is then diagonal, $\varepsilon^{i3} = \varepsilon^{3j} = 0$ and the elements of A given by $a^{ij} = \varepsilon^{ij} - \frac{\varepsilon^{i3}\varepsilon^{3j}}{\varepsilon^{33}}$ should be written as $a^{ij} = \varepsilon^{ij} = \varepsilon_r \delta^{ij}$. Thus in the isotropic medium matrix A becomes a multiple of the unit matrix; hence each column $(E_1, E_2)^T$ is its eigenvector and no two of them are distinguished. Also one eigenvalue exists and in this connection only one phase slowness occurs in the given medium $u = \sqrt{\varepsilon_r \varepsilon_0 \mu_0}$. Then formula (5.133) for arbitrary E_1, E_2 gives $E_3 = 0$, which means that $\mathbf{E}$ is also magnetically perpendicular to the wave covector $\mathbf{k}$.

One may also introduce a velocity of the energy transport by the electromagnetic wave. In the traditional approach (solely in terms of vectors and scalars) the following equality is written

$$\mathbf{S} = w\mathbf{v}, \tag{5.161}$$

which defines $\mathbf{v}$ as the energy transport velocity. This relation is analogous to $\mathbf{j} = \rho\mathbf{v}$ linking the current density $\mathbf{j}$ with the density ρ of charges and their velocity $\mathbf{v}$.

Because of another type of directed quantity we should write (5.161) rather as the contraction

$$\mathbf{S} = w\lfloor\mathbf{v}, \tag{5.162}$$

This is an inverse relation to (5.145). Can one calculate $\mathbf{v}$ from this formula? In this purpose it is worth introducing a twisted trivector W reciprocal to the twisted three-form w in the sense that the condition

$$w[[W]] = 1$$

is satisfied. If w is the energy density, W may be called the *volume of unit energy*, since it has the physical dimension $\frac{\text{volume}}{\text{energy}}$. For a given w only one such twisted trivector W exists. If $w = |w|\mathbf{f}_*^{123}$, then $W = |w|^{-1}\mathbf{e}_{123}^*$.

We may look at (5.162) as a mapping of a vector into a twisted two-form similar to (3.34). We know that the inverse mapping (3.33) is the contraction of the twisted two-form with the appropriate twisted trivector; hence we have

$$\mathbf{v} = -W\lfloor\mathbf{S}.$$

Since the twisted trivector W is positive, the direction of $\mathbf{v}$ is the same as that of $\mathbf{S}$. In this way we obtain the sought formula for the *velocity of the energy transport by the electromagnetic field*:

$$\mathbf{v} = -|w|^{-1}\mathbf{e}_{123}^*\lfloor\mathbf{S}. \tag{5.163}$$

After substitution of (5.146) and (5.154) we obtain

$$\mathbf{v} = \frac{1}{u(E_1^2 + E_2^2)}\{(E_1^2 + E_2^2)\mathbf{e}_3 + \frac{1}{\varepsilon^{33}}(\varepsilon^{31}E_1^2 + \varepsilon^{32}E_1E_2)\mathbf{e}_1 + \frac{1}{\varepsilon^{33}}(\varepsilon^{31}E_1E_2 + \varepsilon^{32}E_2^2)\mathbf{e}_2\}$$

$$= u^{-1}\left\{\mathbf{e}_3 + \frac{E_1(\varepsilon_{31}E_1 + \varepsilon^{32}E_2)}{\varepsilon^{33}(E_1^2 + E_2^2)}\mathbf{e}_1 + \frac{E_2(\varepsilon^{31}E_1 + \varepsilon^{32}E_2)}{\varepsilon^{33}(E_1^2 + E_2^2)}\mathbf{e}_2\right\}.$$

Since our basis was adjusted to the wavity such that $\mathbf{k} \parallel \mathbf{f}^3$ and $\mathbf{u} = u\mathbf{f}^3$, we may calculate the value of $\mathbf{u}$ on $\mathbf{v}$:

$$\mathbf{u}[[\mathbf{v}]] = uu^{-1}\mathbf{f}^3 \left[\left[\mathbf{e}_3 + \frac{E_1(\epsilon_{31}E_1 + \varepsilon^{32}E_2)}{\varepsilon^{33}(E_1^2 + E_2^2)}\mathbf{e}_1 + \frac{E_2(\varepsilon^{31}E_1 + \varepsilon^{32}E_2)}{\epsilon_{33}(E_1^2 + E_2^2)}\mathbf{e}_2 \right] \right],$$

$$\mathbf{u}[[\mathbf{v}]] = \mathbf{f}^3[[\mathbf{e}_3]] = 1.$$

In this sense we may claim that the phase slowness and the velocity of the energy transport by the plane electromagnetic wave are mutually inverse quantities. The energy transport velocity is one of the vectors depicted in Fig. 5.2.

Chapter 6
Interfaces

This chapter introduces notions useful for considering interfaces between two anisotropic media. In Sect. 6.1, directed quantities in two-dimensional space are defined.
It turns out there are twelve of them: six nontwisted and twisted multivectors, and
six (of the same type) exterior forms. An interesting question is the passage from
spatial to planar quantities. In order to preserve the information, for instance, the
spatial vector has to be replaced by a pair: a planar scalar and a planar vector—
all three are respectively nontwisted or twisted quantities. Similar replacements of
spatial quantity by pairs of planar quantities are needed for other types. For the
exterior forms, so-called restrictions to plane are necessary.

The integral Maxwell equations serve in Sect. 6.2 to derive conditions on
interfaces for the field quantites. For **E** and **B**, they yield equality of their
restrictions from two sides to the interface, whereas for **H** and resp. **D**, differences
between the restrictions are proportional to surface current and charge, respectively.
Considerations of this section are metric-free.

Section 6.3 is devoted to finding fields generated by infinite planes with surface
charges or currents. Since they are solutions to Maxwell's equations, the medium
metric is needed, similarly as in Sects. 4.2 and 4.3. The following cases are
considered: electric field of a single plane with uniform surface charge, of a pair
of planes with the same surface charge, and a pair of planes with opposite surface
charges. The last case is the ideal electric capacitor. Then magnetic field of single
plane with uniform surface current, of pair of planes with the same surface current,
and of pair of planes with opposite surface currents. The last case may be called an
ideal magnetic capacitor.

If a region, filled with the electromagnetic field, is enclosed between a pair of
parallel plates, the field exerts forces on them. The plane electric capacitor is an
example of such a situation—its plates are attracted to each other. Considerations of
this type are presented in Sect. 6.4. An electric stress tensor characterizes forces that
could act on three possible pairs of plates arranged in three independent directions.

© Springer Nature Switzerland AG 2021
B. Jancewicz, *Directed Quantities in Electrodynamics*,
https://doi.org/10.1007/978-3-030-90471-5_6

A similar role is played by a magnetic stress tensor. Their sum constitutes an electromagnetic stress tensor. This section is metric-free.

Section 6.5 is devoted to a plane electromagnetic wave impeding on an interface between two different anisotropic dielectrics each with given metrics. The refracted wave, penetrating the second medium, becomes a superposition of two eigenwaves—this phenomenon is known as birefrigence.

6.1   Directed Quantities of a Plane

We introduce now the directed quantities possible in a two-dimensional space, i.e., on a plane. The definition of vectors and bivectors is the same as before, because their orientations are defined on the attitudes; hence they can be contained on the plane. Only one attitude is possible for bivectors, which is the plane on which the considerations are performed. For this reason the linear space of bivectors is one-dimensional.

For a twisted vector we assume that the attitude is a straight line, but the orientation is the straight arrow intersecting the line of the attitude. All arrows intersecting the line in the same side represent the same orientation (see Fig. 6.1).

It should be obvious from this definition that for a given attitude only two orientations are possible (see Fig. 6.2)—we call them *opposite*.

Twisted vectors can be added by juxtaposing appropriate segments such that the intersecting arrows pass continuously through their junction. Then by joining the free ends by a third segment we obtain their sum (see Fig. 6.3), in which the addition $c + d = e$ is depicted.

The exterior product of two vectors gives the bivector according to the same prescription, as given in Sect. 1.2 and illustrated in Fig. 1.16. To obtain the exterior product of two twisted vectors we should first juxtapose the arrows compatibly, as in addition (intersecting arrows pass continuously through the junction), but instead, joining the free ends, we draw the two missing sides of the parallelogram. Then—

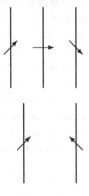

Fig. 6.1 Three images of the same orientation of twisted vectors

Fig. 6.2 Two opposite orientations of twisted vectors

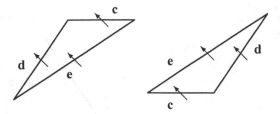

Fig. 6.3 Addition **c** + **d** of two twisted vectors

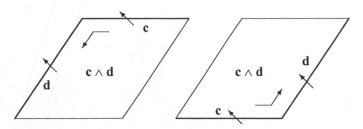

Fig. 6.4 Exterior product **c** ∧ **d** of two twisted vectors **c**, **d**

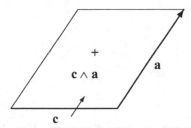

Fig. 6.5 Exterior product **c** ∧ **a** of twisted vector **c** and nontwisted vector **a**

similarly as in Fig. 1.21—the curved arrow has its orientation from the first to the second twisted vector. Figure 6.4 illustrates this prescription for the product **c** ∧ **d**.

The twisted bivector differs from the nontwisted bivector only by the orientation; we assume this orientation as the sign plus or minus. The linear space of twisted bivectors is one-dimensional. A twisted bivector can be obtained as the exterior product of vector **a** and twisted vector **c**. The product **c** ∧ **a** is shown in Fig. 6.5, and **a** ∧ **c** on Fig. 6.6. The sign of the result is positive if the arrow intersecting **c** enters the parallelogram, and minus if it goes out. It is clear from the figures that this product is anticommutative.

The definition of the scalar does not depend on the dimension of the vector space; hence it is the same as before. Only for the twisted scalar do we assume another orientation, namely the curved arrow lying on the plane. For the two distinct orientations of the twisted scalar, descriptions "right" or "left" are not proper,

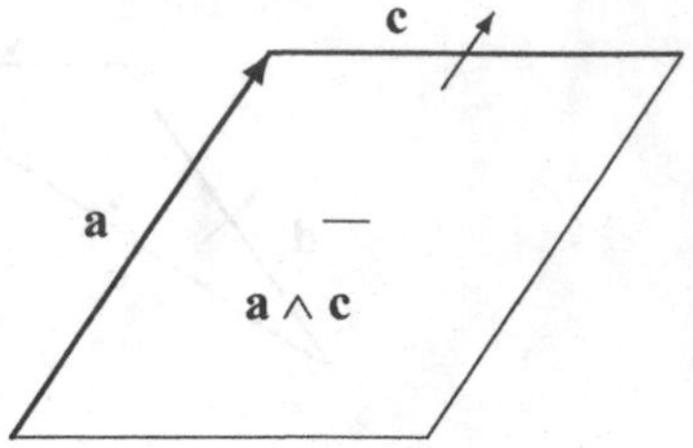

Fig. 6.6 Exterior product
$a \wedge c$ of nontwisted vector **a**
and twisted vector **c**

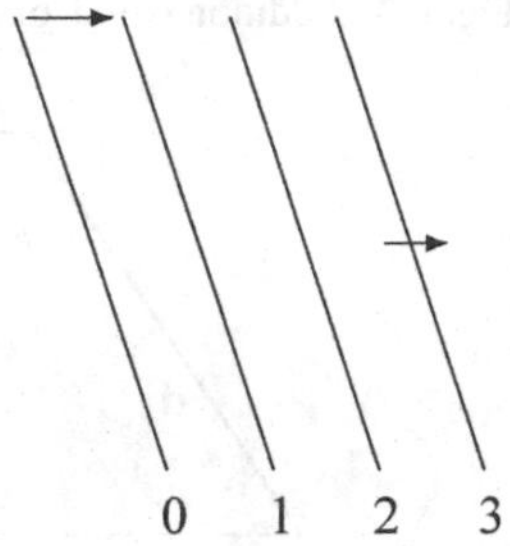

Fig. 6.7 Nontwisted planar
one-form

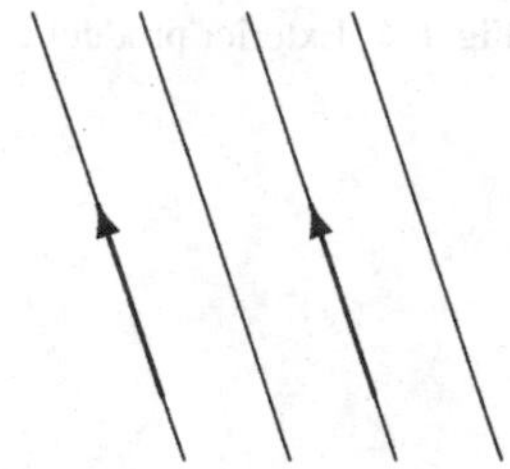

Fig. 6.8 Twisted planar
one-form

because these words refer rather to three dimensions. If we look at the plane from a
fixed side, we may describe the orientation as *clockwise* or *anticlockwise*.

The one-form, as always, is the linear mapping of vectors into scalars. The kernel
of the mapping is now one-dimensional; hence the attitude of a one-form is a straight
line and its geometric image is now a family of parallel and equidistant lines with
an arrow between neighbouring lines, showing the orientation. The arrow may also
pierce one of the lines to make the image similar to that of the twisted vector (see
Fig. 6.7). The linear space of one-forms is two-dimensional—one may easily pick
up a basis consisting of two elements.

Twisted one-forms differ from nontwisted one-forms only by orientation, which
is now represented by arrows lying on the lines (see Fig. 6.8).

A nontwisted two-form is imaged geometrically as a family of cells of equal
areas, tightly filling the plane. Its orientation is the same as that of bivectors, i.e., the
curved arrow. The nontwisted two-form is illustrated in Fig. 6.9. Twisted two-forms
differ from nontwisted ones only by orientation, which is now the sign. Figure 6.10
shows a twisted two-form.

The exterior product of two nontwisted or two twisted one-forms gives the
nontwisted two-form according to the prescription illustrated in Figs. 6.11 and 6.12.

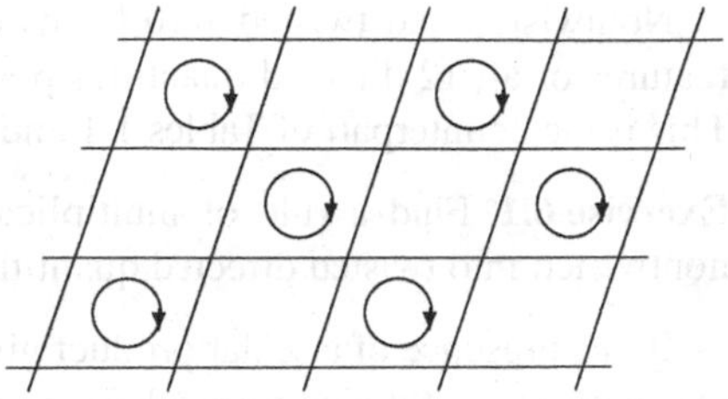

Fig. 6.9 Nontwisted planar two-form

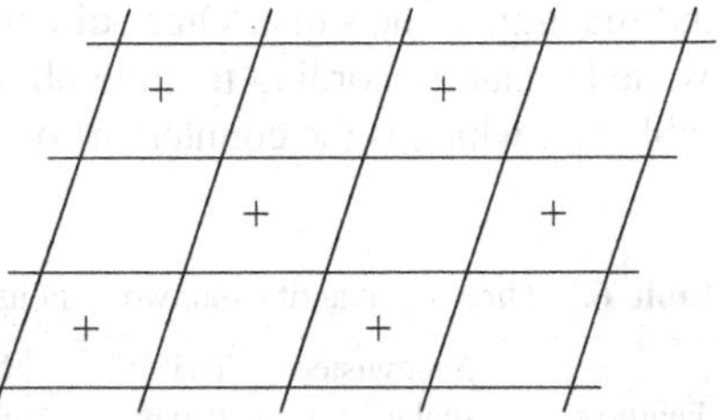

Fig. 6.10 Twisted planar two-form

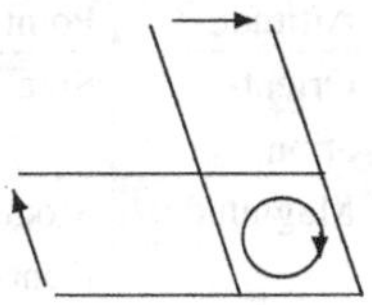

Fig. 6.11 Exterior product of two nontwisted one-forms

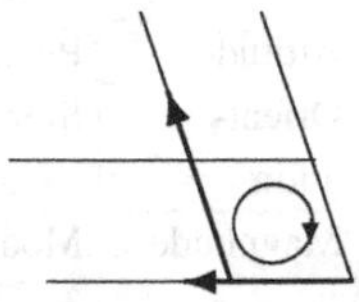

Fig. 6.12 Exterior product of two twisted one-forms

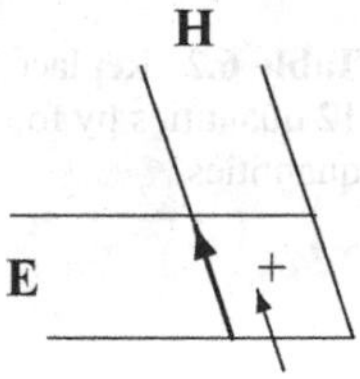

Fig. 6.13 Exterior product $E \wedge H$ of nontwisted one-form E with twisted one-form H

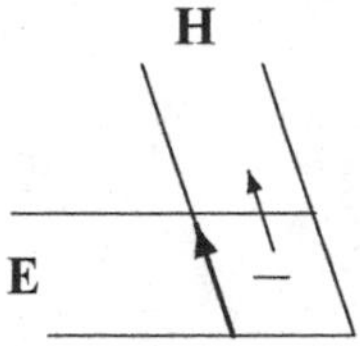

Fig. 6.14 Exterior product $H \wedge E$ of twisted one-form H with nontwisted one-form E

On the other hand, the exterior product of nontwisted one-form E with twisted one-form H in the order $E \wedge H$ is shown in Fig. 6.13, whereas in order $H \wedge E$— in Fig. 6.14.

Nontwisted and twisted zero-forms can be also introduced. Table 6.1 lists the features of all 12 directed quantities possible in the two-dimensional vector space. This is the counterpart of Tables 1.1 and 1.2.

Exercise 6.1 Find a rule of multiplication by the unit twisted scalar, changing nontwisted into twisted directed quantities and vice versa.

In the presence of a scalar product given on our plane, we can change one-forms into vectors of attitudes perpendicular to that of the forms, orientation carried over, and magnitude the same. Other directed quantities are changed into nontwisted or twisted scalars according to quite obvious rules. All the changes are summed up in Table 6.2, which is the counterpart of Tables 2.1 and 2.2.

Table 6.1 Directed quantities in two dimensions

Features	Nontwisted scalar	Twisted scalar	Nontwisted vector	Twisted vector	Nontwisted bivector	Twisted bivector
Attitude	Point	Point	Line	Line	Plane	Plane
Orient-ation	Sign	Curved arrow	Lying arrow	Piercing arrow	Curved arrow	Sign
Magnitude	Module	Module	Length	Length	Area	Aarea
	Nontwisted zero-form	Twisted zero-form	Nontwisted one-form	Twisted one-form	Nontwisted two-form	Twisted two-form
Attitude	Plane	Plane	Line	Line	Point	Point
Orient-ation	Sign	Curved arrow	Piercing arrow	Lying arrow	Curved arrow	Sign
Magnitude	Module	Module	Linear density	Linear density	Surface density	Surface density

Table 6.2 Replacement of 12 quantities by four quantities

Previous quantity	New quantity
Nontwisted scalar	Scalar
Twisted scalar	Pseudoscalar
Nontwisted zero-form	Scalar
Twisted zero-form	Pseudoscalar
Nontwisted vector	Vector
Twisted vector	Pseudovector
Nontwisted one-form	Vector
Twisted one-form	Pseudovector
Nontwisted bivector	Pseudoscalar
Twisted bivector	Scalar
Nontwisted two-form	Pseudoscalar
Twisted two-form	Scalar

6.1.1   From 3D to 2D

Let a distinguished plane Σ exist in three-dimensional space. If spatial nontwisted vectors and bivectors are parallel to the plane, we can by a translation put them on a distinguished plane. Then we consider them as elements belonging to the geometry of the plane. It is important that the whole direction—the attitude and the orientation—of such a multivector is contained in the plane.

The situation with twisted quantities is more complicated, since twisted spatial multivectors never lie fully on a plane. Consider a twisted vector $\mathbf{l}$ with its attitude lying on a plane Σ. The ring or parallelogram can not be parallel to Σ. We now try to find some orientation appropriate for a twisted vector in Σ. We care that the parallelogram representing the orientation of $\mathbf{l}$ has two sides parallel to Σ. When seen from above, one side of this parallelogram has its orientation in one direction; the other, seen from below, is in the opposite direction. We see in Fig. 6.15 one of the two sides parallel to Σ. At this point we give an external orientation to Σ by choosing an arrow piercing it, and now, for a while, we lift the twisted vector $\mathbf{l}$ (along with the parallelogram of its orientation) above Σ in the direction shown by the piercing arrow. One side of the parallelogram lies on Σ as an arrow, see Fig. 6.16. We leave this arrow on Σ as the arrow piercing the segment $\mathbf{l}$ when it returns to the plane (see Fig. 6.17). The planar twisted vector obtained in this way is denoted as $\mathbf{l}_\Sigma$ and called the *flattening* of the spatial twisted vector. If the arrow piercing Σ had the opposite orientation, the flattening would also have the opposite orientation (see Fig. 6.18).

The situation with twisted bivectors is simpler. If we have a twisted bivector $\mathbf{S}$ with its attitude parallel to Σ, we compare its orientation with that of Σ and define its *flattening* $\mathbf{S}_\Sigma$ with the plus sign if either orientation is the same (see Figs. 6.19 and 6.20), and with the minus sign if opposite.

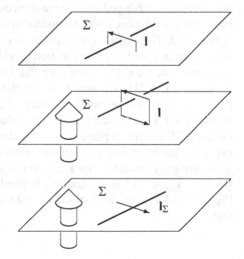

Fig. 6.15 One side of the parallelogram defining the orientation of $\mathbf{l}$ is below Σ

Fig. 6.16 The parallelogram defining the orientation of $\mathbf{l}$ is is moved in the external direction of Σ

Fig. 6.17 One side of the parallelogram defining the orientation of $\mathbf{l}$ is left on Σ

Fig. 6.18 One side of the parallelogram defining the orientation of **l** is left on Σ for the opposite orientation of Σ

Fig. 6.19 Spatial twisted bivector **S** put on surface Σ

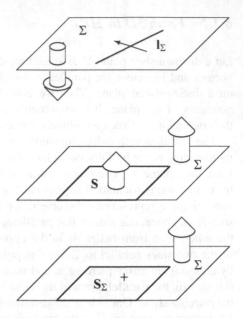

Fig. 6.20 Flattering S_Σ of **S** to surface Σ

Up to now in the considerations about means of transition from three-dimensional directed quantities to planar ones we have discussed only multivectors parallel to a chosen plane Σ and forms nonparallel to Σ. Some information, however, is lost in such a procedure, because there is no room on a plane for multivectors nonparallel to it, and also after the restriction any trace of parallel forms falls into oblivion. We now describe the means of preserving full information about all the spatial directed quantities.

There is no need for any action in the case of scalars, because a spatial scalar can also be considered as a planar scalar. The same is true of zero-forms, only their attitude is changed from three-dimensional space to the plane. (This may be treated as a restriction.) When considering other quantities a vector **a** nonparallel to Σ is needed. This vector distinguishes an external orientation of Σ. In the case of a twisted scalar ν with the combination of a straight arrow and directed ring showing its orientation, we rotate them such that the straight arrow has the same direction as **a** (see Fig. 6.21). We next translate the oriented ring on Σ and thus we have the two-dimensional orientation which we concede as the orientation of the planar twisted scalar ν_Σ obtained from ν. The values of both twisted scalars are obviously the same. If we have a planar twisted scalar we may uniquely reconstruct a spatial twisted scalar by adding to its planar orientation the straight arrow of **a**.

An arbitrary spatial vector **r** can be uniquely decomposed into two components: $\mathbf{r} = \mathbf{r}_\| + \mathbf{r}_a$, of which the first, $\mathbf{r}_\|$, is parallel to Σ and the second, $\mathbf{r}_a$—to **a** (see Fig. 6.22). Moreover, the formula $\mathbf{r}_a = \lambda\mathbf{a}$ is satisfied, where λ is a scalar. We have thus

$$\mathbf{r} = \mathbf{r}_\| + \lambda\mathbf{a}. \tag{6.1}$$

Fig. 6.21 Spatial twisted scalar ν determines planar twisted scalar ν_Σ

Fig. 6.22 For a spatial nontwisted vector $\mathbf{r}$, finding component $\mathbf{r}_\parallel$ parallel to Σ

Fig. 6.23 For a spatial twisted vector $\mathbf{l}$, finding component $\mathbf{l}_\parallel$ parallel to Σ

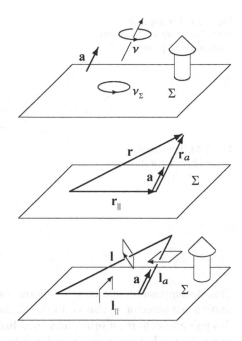

If $\mathbf{a}$ is fixed, then by changing $\mathbf{r}$ we change the vector $\mathbf{r}_\parallel$ and the scalar λ, and these quantities belong to the directed quantities of the plane. In this manner the spatial vector $\mathbf{r}$ is represented on the plane as the pair $\{\mathbf{r}_\parallel, \lambda\}$, i.e., we determine the mapping $\mathbf{r} \to \{\mathbf{r}_\parallel, \lambda\}$. With fixed $\mathbf{a}$ this pair allows to reconstruct the spatial vector according to (6.1). The set of such pairs is three-dimensional: two dimensions stand on the first place and one on the second.

The same vector $\mathbf{a}$ allows us to change a spatial twisted vector $\mathbf{l}$ nonparallel to Σ into planar quantities. Now, we write it in the form

$$\mathbf{l} = \mathbf{l}_\parallel + \nu\mathbf{a}, \tag{6.2}$$

where $\mathbf{l}_a = \nu\mathbf{a}$ is its component parallel to $\mathbf{a}$ and $\mathbf{l}_\parallel$—parallel to Σ (see Fig. 6.23). Here ν is a spatial twisted scalar. The vector $\mathbf{a}$ distinguishes some external orientation of Σ; thus in the described manner we may obtain the flattening $\mathbf{l}_{\parallel\Sigma}$ of $\mathbf{l}_\parallel$ to Σ. The spatial twisted scalar ν is changed into a planar twisted scalar ν_Σ according to the prescription given above. In this manner we determined the mapping $\mathbf{l} \to \{\mathbf{l}_{\parallel\Sigma}, \nu_\Sigma\}$ changing a spatial twisted vector into the pair of planar quantities.

We decompose an arbitrary spatial bivector $\mathbf{S}$ on the sum $\mathbf{S} = \mathbf{S}_\parallel + \mathbf{S}_a$ of the component $\mathbf{S}_\parallel$ parallel to Σ and the component $\mathbf{S}_a$ parallel to $\mathbf{a}$. We show these in Fig. 6.24. The component $\mathbf{S}_a$ can be factorized into the external product $\mathbf{S}_a = \mathbf{a} \wedge \mathbf{b}$ with $\mathbf{b}$ parallel to Σ, which yields

$$\mathbf{S} = \mathbf{S}_\parallel + \mathbf{a} \wedge \mathbf{b}. \tag{6.3}$$

Fig. 6.24 For a spatial
nontwisted bivector **S**, finding
component $\mathbf{S}_{||}$ parallel to Σ

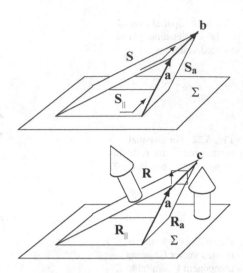

Fig. 6.25 For a spatial
twisted bivector **R**, finding
component $\mathbf{R}_{||}$

This decomposition allows us to choose two planar quantities lying on Σ, repre-
senting the bivector **S**, that is, to define the mapping $\mathbf{S} \rightarrow \{\mathbf{S}_{||}, \mathbf{b}\}$. With **a** fixed
this pair allows to reconstruct the spatial bivector according to (6.3). Notice that the
orientation of **b** depends on the order of factors in the exterior product in (6.3). The
set of pairs $\{\mathbf{S}_{||}, \mathbf{b}\}$ is three-dimensional, because one dimension is in the first place,
and two are in the second.

We similarly decompose a spatial twisted bivector **R** on the sum $\mathbf{R} = \mathbf{R}_{||} + \mathbf{R_a}$
of the component $\mathbf{R}_{||}$ parallel to Σ and the component $\mathbf{R_a}$ parallel to **a**. Figure 6.25
shows this. The component parallel to **a** may be factorized onto the exterior product
$\mathbf{R_a} = \mathbf{a} \wedge \mathbf{c}$ with the twisted vector **c** parallel to Σ, which gives

$$\mathbf{R} = \mathbf{R}_{||} + \mathbf{a} \wedge \mathbf{c}. \tag{6.4}$$

This decomposition allows us to choose two planar quantities lying on Σ, namely
the flattening $\mathbf{R}_{||\Sigma}$ of $\mathbf{R}_{||}$ and $\mathbf{c}_\Sigma$, representing the twisted bivector **R**, i.e., define the
mapping $\mathbf{R} \rightarrow \{\mathbf{R}_{||\Sigma}, \mathbf{c}_\Sigma\}$.

The third grade quantities can also be represented on the plane. Trivector **T** would
be factorized into the exterior product $\mathbf{T} = \mathbf{a} \wedge \mathbf{S}$ with the bivector **S** parallel to Σ
and just this bivector represents **T** on Σ. A twisted trivector **V** should be represented
as $\mathbf{V} = \mathbf{a} \wedge \mathbf{R}$ with a twisted bivector **R** parallel to Σ. The flattened twisted bivector
$\mathbf{R}_\Sigma$ represents **V** on Σ. This prescription ensures that the positive twisted trivector
V is represented by the positive twisted bivector $\mathbf{R}_\Sigma$, since the orientation of **a**
determines the external orientation of Σ and in the decomposition of the positive
$\mathbf{V} = \mathbf{a} \wedge \mathbf{R}$ the orientations of **a** and **R** must be compatible, and then the flattening
$\mathbf{R}_\Sigma$ must be positive. The dimension of the set of these quantities remains the same,
since the set of spatial twisted trivectors is one-dimensional and the same is true for
the set of planar twisted bivectors.

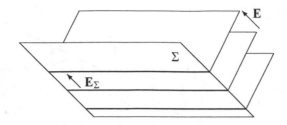

Fig. 6.26 Spatial one-form $\mathbf{E}$ determines planar one-form $\mathbf{E}_\Sigma$

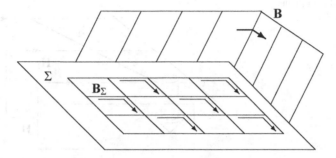

Fig. 6.27 Spatial two-form $\mathbf{B}$ determines planar two-form $\mathbf{B}_\Sigma$

In the summary we can state that the multivectors are decomposed into two components: parallel to Σ and parallel to $\mathbf{a}$. The component parallel to Σ is left without changes or, for twisted quantities, its flattening should be taken, whereas the component parallel to $\mathbf{a}$ is changed into a quantity of one grade lower.

For a nontwisted one-form $\mathbf{E}$ defined in the three-dimensional vector space $\mathbf{R}^3$ one may ask: what does its limitation to the plane, i.e., to two-dimensional subspace Σ, look like? The answer is: it is the *restriction* $\mathbf{E}_\Sigma$ of the mapping $\mathbf{E}$ to vectors from Σ. It should be a one-form in Σ of the kind shown in Fig. 6.7. The lines of $\mathbf{E}_\Sigma$ are obtained simply as intersections of the planes of $\mathbf{E}$ with Σ (see Fig. 6.26). The orientation of $\mathbf{E}_\Sigma$ is depicted as the arrow lying on Σ directed from one line to another. If Σ is parallel to the planes of $\mathbf{E}$, then $\mathbf{E}_\Sigma = 0$, since in such a case $\mathbf{E}(\mathbf{r}) = 0$ for each $\mathbf{r} \parallel \Sigma$.

Exercise 6.2 Show that the restriction $\mathbf{E}_\Sigma$ of $\mathbf{E}$ to Σ after a change of one-forms into vectors corresponds to the projection $\mathbf{E}_\Sigma$ of the vector $\mathbf{E}$ on Σ.

Limitation of a two-form $\mathbf{B}$ to a plane Σ is its *restriction* $\mathbf{B}_\Sigma$ only to bivectors lying on Σ. Its geometric image is a family of equal cells obtained from intersections of tubes of $\mathbf{B}$ with Σ (see Fig. 6.27). If Σ is parallel to the attitude of $\mathbf{B}$, we obtain $\mathbf{B}_\Sigma = 0$.

Exercise 6.3 Show that the restriction $\mathbf{B}_\Sigma$ of $\mathbf{B}$ to Σ after a change of two-forms into twisted scalars corresponds to the choice of the component $\mathbf{B}_\perp$ of the twisted vector $\mathbf{B}$ perpendicular to Σ.

Fig. 6.28 Value of spatial **H**
on spatial **l**

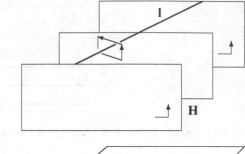

Fig. 6.29 Value of planar
H_Σ on planar l_Σ

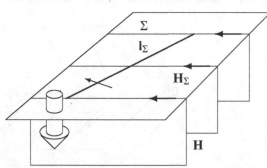

If we have a spatial twisted one-form **H** nonparallel to Σ, we define its *restriction* H_Σ in such a way that its value **H** on **l** is the same as the value of H_Σ on l_Σ:

$$H[[l]] = H_\Sigma[[l_\Sigma]].$$

The value of $H[[l]]$ is illustrated in Fig. 6.28, (the segment **l** fills exactly two layers between the planes and the orientations are compatible, hence $H[[l]] = +2$), whereas the value $H_\Sigma[[l_\Sigma]]$ is shown in Fig. 6.29 (the segment l_Σ fills exactly two strips between the lines and the orientations are compatible, hence $H_\Sigma[[l_\Sigma]] = +2$).

Exercise 6.4 Show that the restriction H_Σ of **H** to Σ after a change of twisted one-forms into twisted vectors corresponds to the projection H_Σ of **H** on Σ.

We proceed similarly with twisted two-forms . The restriction D_Σ of a twisted two-form **D** to Σ has the positive sign if the orientations **D** and Σ are the same (see Fig. 6.30), and negative in the opposite case. This ensures that for a twisted bivector **S** with attitude lying on Σ we obtain the identity

$$D[[S]] = D_\Sigma[[S_\Sigma]].$$

Exercise 6.5 Show that the restriction D_Σ of a twisted two-form **D** to Σ after a change of twisted two-forms into scalars corresponds to choosing of the component $D_\perp$ of vector **D** perpendicular to Σ.

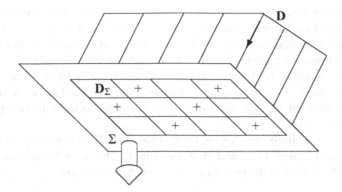

Fig. 6.30 Spatial two-form **D** determines planar two-form $\mathbf{D}_\Sigma$

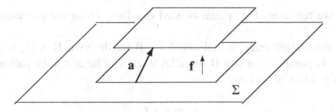

Fig. 6.31 Nontwisted one-form **f** discriminated by **a** and Σ

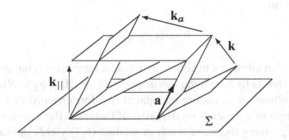

Fig. 6.32 For a spatial nontwisted one-form **k**, finding component $\mathbf{k}_{\|}$ parallel to Σ

In the case of exterior forms a nontwisted one-form **f** parallel to Σ should be discriminated. It is natural to connect it with the vector **a** by the condition $\mathbf{f}[[\mathbf{a}]] = 1$ (see Fig 6.31). Let us call it a *one-form dual to* **a**. Then an arbitrary spatial one-form **k** can be decomposed into two components: $\mathbf{k_a}$—parallel to **a**, i.e., such that $\mathbf{k_a}[[\mathbf{a}]] = 0$, and $\mathbf{k}_{\|}$—parallel to Σ (see Fig. 6.32). The component parallel to Σ may be written as $\mathbf{k}_{\|} = \lambda\mathbf{f}$ with a scalar λ which can be found from formula $\lambda = \mathbf{k}\lfloor\mathbf{a} = \mathbf{k}[[\mathbf{a}]]$. Thus we have

$$\mathbf{k} = \mathbf{k_a} + \lambda\mathbf{f}. \tag{6.5}$$

The component parallel to **a** has an interesting property: its restriction to Σ is the same as that of the original form: $\mathbf{k}_{\mathbf{a}\Sigma} = \mathbf{k}_\Sigma$. Therefore, we represent the spatial one-form **k** on the plane as the pair $\{\mathbf{k}_\Sigma, \lambda\}$. One might expect some troubles with a reconstruction of the spatial one-form because there exist a plenty of forms with the same restriction to Σ. But among them only one is parallel to **a**. Hence a unique prescription exists of reconstructing **k** from the pair $\{\mathbf{k}_\Sigma, \lambda\}$. The set of such pairs is three-dimensional, because two dimensions are obtained from the first place and one from the second.

We proceed similarly with spatial twisted one-form **H**. We decompose it on two components: $\mathbf{H}_a$ parallel to **a** and $\mathbf{H}_{||}$ parallel to Σ. The second component can be written as $\mathbf{H}_{||} = \nu \mathbf{f}$ with the twisted scalar ν, whence we obtain

$$\mathbf{H} = \mathbf{H}_a + \nu \mathbf{f}. \tag{6.6}$$

Therefore, we represent the spatial twisted one-form **H** on the plane as the pair: $\{\mathbf{H}_\Sigma, \nu_\Sigma\}$.

We decompose arbitrary spatial two-form **B** on the sum $\mathbf{B} = \mathbf{B}_a + \mathbf{B}_{||}$ of the component $\mathbf{B}_a$ parallel to **a** and $\mathbf{B}_{||}$ parallel to Σ. The latter can be factorized onto the exterior product of two one-forms:

$$\mathbf{B}_{||} = \mathbf{g} \wedge \mathbf{f}, \tag{6.7}$$

whereby we obtain

$$\mathbf{B} = \mathbf{B}_a + \mathbf{g} \wedge \mathbf{f}. \tag{6.8}$$

This decomposition allows us to choose two planar quantities lying on Σ, representing bivector **B**, that is to determine the mapping $\mathbf{B} \to \{\mathbf{B}_\Sigma, \mathbf{g}_\Sigma\}$. When **a** and **f** are fixed, this pair allows us to restore the spatial two-form according to (6.8). Notice that the orientation of **g** depends on the order of factors in the exterior product (6.7). Each one-form **g** giving the same result in product (6.7) yields the same restriction $\mathbf{g}_\sigma$. Now we assume $\mathbf{g} \parallel \mathbf{a}$ and calculate the contraction of (6.8) with **a**:

$$\mathbf{B}\lfloor\mathbf{a} = \mathbf{B}_a\lfloor\mathbf{a} + (\mathbf{g} \wedge \mathbf{f})\lfloor\mathbf{a}.$$

The first term vanishes, because $\mathbf{B}_a \parallel \mathbf{a}$, hence

$$\mathbf{B}\lfloor\mathbf{a} = (\mathbf{g} \wedge \mathbf{f})\lfloor\mathbf{a} = \mathbf{g}\,\mathbf{f}[[\mathbf{a}]] - \mathbf{f}\,\mathbf{g}[[\mathbf{a}]] = \mathbf{g},$$

since $\mathbf{g}[[\mathbf{a}]] = 0$ and $\mathbf{f}[[\mathbf{a}]] = 1$. In this way we arrived at the prescription for how to obtain the second element in the pair $\{\mathbf{B}_\Sigma, \mathbf{g}_\Sigma\}$ representing **B**:

$$\mathbf{g}_\Sigma = (\mathbf{B}\lfloor\mathbf{a})_\Sigma. \tag{6.9}$$

Fig. 6.33 Decomposition of
spatial nontwisted two-form
$\mathbf{B}_\|$ onto factors: $\mathbf{g}$ parallel to
$\mathbf{a}$, and $\mathbf{f}$ parallel to Σ

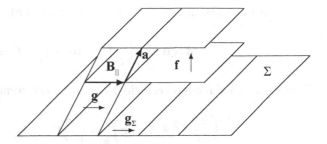

Figure 6.33 illustrates the factorization (6.7) of $\mathbf{B}_\|$. The set of pairs $\{\mathbf{B}_\Sigma, \mathbf{g}_\Sigma\}$ is three-dimensional, because one dimension is in the first place and two on the second.

We similarly decompose the spatial twisted two-form $\mathbf{D}$ on the sum $\mathbf{D} = \mathbf{D}_a + \mathbf{D}_\|$ of the component $\mathbf{D}_a$ parallel to $\mathbf{a}$ and $\mathbf{D}_\|$ parallel to Σ. The latter is factorized into the exterior product $\mathbf{D}_\| = \mathbf{f} \wedge \mathbf{h}$ in which $\mathbf{h} = \mathbf{D} \lfloor \mathbf{a}$ is a twisted one-form; hence we obtain

$$\mathbf{D} = \mathbf{D}_a + \mathbf{h} \wedge \mathbf{f}. \tag{6.10}$$

Now we define the mapping $\mathbf{D} \to \{\mathbf{D}_\Sigma, \mathbf{h}_\Sigma\}$ changing the spatial twisted two-form $\mathbf{D}$ into the pair of planar twisted forms.

The three-form τ is factorized into the exterior product $\tau = \mathbf{B} \wedge \mathbf{f}$, then the planar two-form $\mathbf{B}_\Sigma$ represents τ on Σ. The twisted three-form ρ is factorized as $\rho = \mathbf{D} \wedge \mathbf{f}$ and we treat $\mathbf{D}_\Sigma$ as the twisted two-form representing ρ on Σ.

Summarizing, spatial forms are decomposed on two components: parallel to $\mathbf{a}$ and parallel to Σ. We take from the former its restriction to Σ, and the latter the change into the quantity one grade higher (recall that forms have negative grade).

Exercise 6.6 Show graphically that for spatial nontwisted forms M, N of grades 0 or 1 the identity $(M \wedge N)_\Sigma = M_\Sigma \wedge N_\Sigma$ holds.

Let us consider what happens to differential equations when the discussed decompositions are performed. We do this with the example of formula (6.8). Choose a vector basis $\{\mathbf{e}_i\}$ such that $\mathbf{e}_1$ and $\mathbf{e}_2$ lie on Σ, and $\mathbf{e}_3$ plays the role of $\mathbf{a}$. Then $\mathbf{f}^3 \parallel \Sigma$ and $\mathbf{f}^{12} \parallel \mathbf{e}_3$; hence we write (6.8) as

$$\mathbf{B} = \mathbf{B}_{e_3} + \mathbf{g} \wedge \mathbf{f}^3 = \mathscr{B}\mathbf{f}^{12} + \mathbf{g} \wedge \mathbf{f}^3.$$

Calculate the exterior derivative:

$$\mathbf{d} \wedge \mathbf{B} = \frac{\partial \mathscr{B}}{\partial x^3}\mathbf{f}^3 \wedge \mathbf{f}^{12} + (\mathbf{d} \wedge \mathbf{g}) \wedge \mathbf{f}^3.$$

Because of the presence of $\mathbf{f}^3$ in front of the second term, only two partial derivatives remain in the bracket. Thus we introduce the notation $\mathbf{d}_2 = \mathbf{f}^1 \frac{\partial}{\partial x^1} + \mathbf{f}^2 \frac{\partial}{\partial x^2}$ and equate the above to zero because of the Maxwell equation (4.6):

$$\mathbf{d} \wedge \mathbf{B} = \left(\frac{\partial \mathcal{B}}{\partial x^3} \mathbf{f}^{12} + \mathbf{d}_2 \wedge \mathbf{g} \right) \wedge \mathbf{f}^3 = 0.$$

This is a spatial three-form equation. The planar two-form, corresponding to this is

$$\left(\frac{\partial \mathcal{B}}{\partial x^3} \mathbf{f}^{12} + \mathbf{d}_2 \wedge \mathbf{g} \right)_\Sigma = \frac{\partial \mathcal{B}}{\partial x^3} \mathbf{f}^{12}_\Sigma + \mathbf{d}_\Sigma \wedge \mathbf{g}_\Sigma = 0.$$

We see that the two planar quantities $\{ \mathbf{B}_\Sigma, \mathbf{g}_\Sigma \}$, corresponding to the spatial quantity $\mathbf{B}$, are interconnected in a common equation. Moreover, the x^3 coordinate appears here as a scalar quantity, because no direction corresponds to it in Σ.

Typical situations in which a passage to two dimensions is performed, are fields with a translational symmetry along one linear dimension, for instance the magnetic field produced by current I flowing in an infinite straight line. Let us consider such a field given by (4.55):

$$\mathbf{H}(\mathbf{r}) = \frac{1^* I}{2\pi} \, d\alpha, \tag{6.11}$$

where 1^* is the twisted unit scalar with the handedness of the basis. It is useful to consider its counterpart on the plane Σ perpendicular (in the medium scalar product) to the straight line of the current (see Fig. 6.34) left. A natural candidate for $\mathbf{a}$, needed to find the decomposition (6.6), is any vector lying on the conductor and showing the direction of the flowing current. We choose $\mathbf{m}_3$ for this role. Since the differential twisted one-form (6.11) is parallel to $\mathbf{m}_3$, the second term in (6.6) is absent, so we need to find only the restriction $\mathbf{H}_\Sigma$ of $\mathbf{H}$. According to Fig. 6.29, this restriction should have its orientation directed to the current. Figure 6.34 (right) shows the restricted field as the radially incoming lines. When we reconstruct the

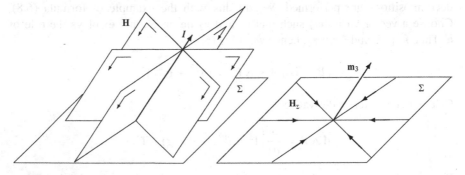

Fig. 6.34 The spatial twisted differential one-form $\mathbf{H}$ determines the planar twisted differential one-form $\mathbf{H}_\Sigma$

spatial field, we shall juxtapose these arrows with the arrow of **a** to obtain the orientation of the half-planes of **H**.

In order to make a similar procedure for the magnetic induction, we need first to calculate it. By virtue of (3.86) we obtain from (6.11)

$$\mathbf{H}(\mathbf{r}) = \frac{1^*I}{2\pi}\mathrm{d}\alpha = \frac{1^*I}{2\pi\rho}(-\sin\alpha\,\mathbf{h}^1 + \cos\alpha\,\mathbf{h}^2) = \frac{I}{2\pi\rho}(-\sin\alpha\,\mathbf{h}^1_* + \cos\alpha\,\mathbf{h}^2_*).$$

We change this into the twisted vector

$$\mathbf{H}'(\mathbf{r}) = \frac{I}{2\pi\rho}(-\sin\alpha\,\mathbf{m}^*_1 + \cos\alpha\,\mathbf{m}^*_2).$$

We apply the mapping (4.51)

$$\mathbf{B}(\mathbf{r}) = \mu_0\mathbf{f}^{123}_*\lfloor\mathbf{H}'(\mathbf{r}) = \mu_0(\det\mathcal{M})^{1/2}\mathbf{h}^{123}_*\lfloor\mathbf{H}'(\mathbf{r})$$

$$= \frac{\mu_0 I\,(\det\mathcal{M})^{1/2}}{2\pi\rho}(-\sin\alpha\mathbf{h}^{23} + \cos\alpha\mathbf{h}^{31}) = -\frac{\mu_0 I\,(\det\mathcal{M})^{1/2}}{2\pi\rho}(\sin\alpha\mathbf{h}^2 + \cos\alpha\mathbf{h}^1) \wedge \mathbf{h}^3.$$

This field is illustrated in Fig. 6.35 (left).

This should be compared with the decomposition (6.8). We see that the first term in that sum is absent—the whole two-form **B** is parallel to Σ: $\mathbf{B} = \mathbf{g} \wedge \mathbf{f}$ and the role of **f** is played by $\mathbf{h}^3$, which as dual to $\mathbf{m}_3$, is parallel to Σ. The other factor is the spatial twisted one-form

$$\mathbf{g} = -\frac{\mu_0 I\,(\det\mathcal{M})^{1/2}}{2\pi\rho}(\cos\alpha\,\mathbf{h}^1 + \sin\alpha\,\mathbf{h}^2).$$

Thus the planar twisted one-form

$$\mathbf{g}_\Sigma = -\frac{\mu_0 I\,(\det\mathcal{M})^{1/2}}{2\pi\rho}(+\cos\alpha\,\mathbf{h}^1_\Sigma + \sin\alpha\,\mathbf{h}^2_\Sigma)$$

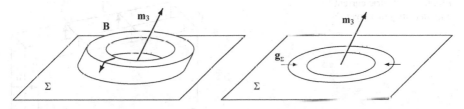

Fig. 6.35 The spatial nontwisted differential two-form **B** parallel to Σ determines the planar nontwisted differential one-form $\mathbf{g}_\Sigma$

is the quantity that describes the magnetic induction on plane Σ. Figure 6.35 (right) depicts its two lines as quasi-circles orthogonal (in the medium metric) to the lines of Fig. 6.34 (right).

If we choose Σ not perpendicular, in the medium metric, to the current, the spatial magnetic induction $\mathbf{B}$ would not be parallel to Σ; therefore, the first term in (6.8) would also be present. Thus, in addition to the twisted one-form $\mathbf{g}_\Sigma$, the nontwisted two-form $\mathbf{B}_\Sigma$ should be used to describe the same magnetic field.

6.2   Conditions on an Interface

There exist physical problems in which charges and currents are distributed in layers so thin that their thickness is negligible, and as an approximation we consider them to be surfaces, not layers. Let us give some thought to how surface charges and currents should be introduced.

Let current with density $\mathbf{j}$ flow in the layer W. The tubes of $\mathbf{j}$ must be parallel to the layer, which is shown in Fig. 6.36. The width of W is described by a vector $\mathbf{a}$. We factorize the twisted two-form $\mathbf{j}$ in the exterior product:

$$\mathbf{j} = \mathbf{f} \wedge \mathbf{J} \tag{6.12}$$

where $\mathbf{J}$ is a twisted one-form, and $\mathbf{f}$ a nontwisted one-form such that

$$\mathbf{f}[[\mathbf{a}]] = 1 \quad \text{and} \quad \mathbf{J}[[\mathbf{a}]] = 0. \tag{6.13}$$

This means that two neighbouring planes of $\mathbf{f}$ coincide with two boundaries of W, and the planes of $\mathbf{J}$ are parallel to $\mathbf{a}$ (see Fig. 6.37). (To properly mark the orientation of $\mathbf{J}$, one should look at Fig. 1.52.) We use the decomposition (6.12) to find the contraction

$$\mathbf{a} \lrcorner \mathbf{j} = \mathbf{a} \lrcorner (\mathbf{f} \wedge \mathbf{J}) = \mathbf{f}[[\mathbf{a}]] \, \mathbf{J} - \mathbf{J}[[\mathbf{a}]] \, \mathbf{f} = \mathbf{J} \tag{6.14}$$

Fig. 6.36 Electric current with density $\mathbf{j}$ in a layer W

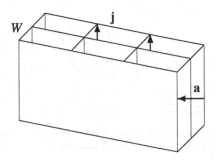

Fig. 6.37 Factorizing current density as the exterior product $\mathbf{j} = \mathbf{f} \wedge \mathbf{J}$

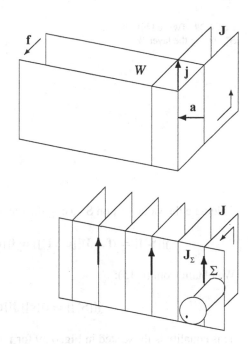

Fig. 6.38 The situation after shrinking the layer W to the plane Σ

We used (6.13) in the last transition. Notice that the change of orientation of $\mathbf{a}$ while maintaining (6.13) demands a change of orientation of $\mathbf{f}$. Then preserving (6.12) demands a change of orientation of $\mathbf{J}$.

We assume now that $\mathbf{a}$ is very small, which means that $\mathbf{f}$ and $\mathbf{j}$ are very large, and the layer W has a negligible width; we treat it now as a plane Σ. The situation looks as like Fig. 6.38. Notice that we do not pass to the limit $\mathbf{a} \to 0$.[1] From the negligibly small vector $\mathbf{a}$ only an arrow piercing Σ remains giving it an external orientation. This is needed to find the restriction of $\mathbf{J}$ to Σ. The orientation of $\mathbf{J}$ is marked by the broken arrow; its piercing part is opposite to the orientation of Σ, the other part, parallel, yields the orientation of $\mathbf{J}_\Sigma$ (see Fig. 6.38). We know that the orientation of $\mathbf{J}$ depends on that of $\mathbf{a}$, but the orientation of $\mathbf{J}_\Sigma$ does not. (The reversion of $\mathbf{a}$ reverses $\mathbf{h}$, but also reverses the external orientation of Σ.) In this manner the whole transition from the spatial twisted two-form $\mathbf{j}$ to the planar twisted one-form $\mathbf{J}_\Sigma$ does not depend on the vector $\mathbf{a}$ joining the two boundaries of the initial layer W.

Consider now the value of the twisted two-form $\mathbf{j}$ on the twisted bivector $\mathbf{S}^*$ representing a surface cutting the layer W, but completely contained in it (see Fig. 6.39). We factorize $\mathbf{S}^* = \mathbf{c} \wedge \mathbf{l}^*$ in such a manner that

$$\mathbf{f}[[\mathbf{l}^*]] = 0 \quad \text{and} \quad \mathbf{f}[[\mathbf{c}]] = 1. \tag{6.15}$$

---

[1] If we decrease the width of W, that is if $\mathbf{a} \to 0$ with fixed current flowing through W, the current density $\mathbf{j}$ would be increasing, that is $\mathbf{j} \to \infty$. The contraction $\mathbf{a} \rfloor \mathbf{j} = \mathbf{J}$ can be kept constant during this limiting transition. Therefore its restriction $\mathbf{J}_\Sigma$ to Σ can be kept constant.

Fig. 6.39 Twisted bivector
$\mathbf{S}^*$ cutting the layer W

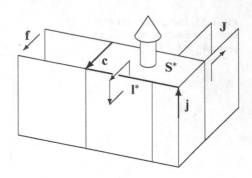

Calculate the value of $\mathbf{j}$ on $\mathbf{S}^*$ using the identity (2.12):

$$\mathbf{j}[[\mathbf{S}^*]] = (\mathbf{f} \wedge \mathbf{J})[[\mathbf{c} \wedge \mathbf{l}^*]] = \mathbf{f}[[\mathbf{c}]]\,\mathbf{J}[[\mathbf{l}^*]] - \mathbf{f}[[\mathbf{l}^*]]\,\mathbf{J}[[\mathbf{c}]]\ .$$

We obtain from (6.15):

$$\mathbf{j}[[\mathbf{S}^*]] = \mathbf{f}[[\mathbf{c}]]\,\mathbf{J}[[\mathbf{l}]]^* = \mathbf{J}[[\mathbf{l}^*]]. \tag{6.16}$$

This equality is illustrated in Fig. 6.39 for $\mathbf{j}$ and $\mathbf{S}^*$ with the same orientation; hence both sides of (6.16) should be positive. After replacement of $\mathbf{J}$ by its restriction to Σ and of $\mathbf{l}^*$ by its flattening we obtain the identity

$$I = \mathbf{j}[[\mathbf{S}^*]] = \mathbf{J}_\Sigma[[\mathbf{l}_\Sigma^*]]. \tag{6.17}$$

Due to this, the twisted one-form $\mathbf{J}_\Sigma = \mathbf{i}$ can be called the *linear density of the surface current* on Σ. Notice that the direction of $\mathbf{i}$ coincides with that of $\mathbf{j}$. The equality

$$I = \mathbf{i}[[\mathbf{l}_\Sigma^*]], \tag{6.18}$$

rewritten from (6.17), can be interpreted as follows. If the flat twisted vector $\mathbf{l}_\Sigma^*$ corresponds to linear *extension* of a gate on Σ, then $\mathbf{i}[[\mathbf{l}_\Sigma^*]]$ is the current flowing though the gate. The right-hand side of (6.18) is illustrated in Fig. 6.40.

We now pass to the electric field. Let in a layer W, characterized by $\mathbf{a}$, a spatial charge, be present with density ρ. Using the same one-form $\mathbf{f}$, we factorize the twisted three-form ρ

$$\rho = \tau \wedge \mathbf{f},$$

where τ is an twisted two-form such that $\tau \lfloor \mathbf{a} = 0$. This means that the lines representing the attitude of τ must be parallel to $\mathbf{a}$ (see Fig. 6.41). By virtue of (3.12) we calculate the contraction

$$\rho \lfloor \mathbf{a} = (\tau \wedge \mathbf{f}) \lfloor \mathbf{a} = \tau\,\mathbf{f}[[\mathbf{a}]] - (\tau \lfloor \mathbf{a}) \wedge \mathbf{f} = \tau.$$

Fig. 6.40 The situation after shrinking the layer W to the plane Σ

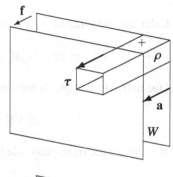

Fig. 6.41 Factorizing charge density as the exterior product $\rho = \tau \wedge \mathbf{f}$

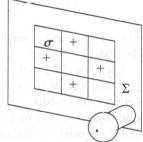

Fig. 6.42 The situation after shrinking the layer W to the plane Σ

The orientation of τ need not be the same as that of $\mathbf{a}$—it depends on the sign of ρ. It is the same only for positive ρ; such a situation is shown in Fig. 6.41. For negligibly small $\mathbf{a}$ we change W into plane Σ with the external orientation taken from $\mathbf{a}$. Now the restriction $\tau_\Sigma = \sigma$ is the *density of the surface charge*, illustrated in Fig. 6.42. The sign of the contraction is the same as that of ρ and does not depend on the vector $\mathbf{a}$ joining the two boundaries of W.

We need to consider the value of ρ on the twisted trivector V representing the volume inside W (see Fig. 6.43). Factorize $V = \mathbf{R}^* \wedge \mathbf{a}$ with twisted bivector $\mathbf{R}^*$ such that $\mathbf{R}^* \| \mathbf{f}$, which means that

$$\mathbf{f} \rfloor \mathbf{R}^* = 0. \tag{6.19}$$

Fig. 6.43 Volume
$V = \mathbf{R}^* \wedge \mathbf{a}$ inside the layer
W

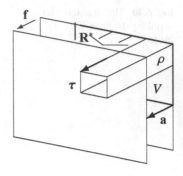

Calculate the value of ρ on V using Exercise 3.18:

$$\rho[V] = (\tau \wedge \mathbf{f})[[\mathbf{R}^* \wedge \mathbf{a}]] = \tau[[\mathbf{R}^*]]\mathbf{f}[[\mathbf{a}]] - \tau[[(\mathbf{f}\rfloor \mathbf{R}^*) \wedge \mathbf{a}]].$$

Due to (6.19) and (6.13) we obtain

$$\rho[[V]] = \tau[[\mathbf{R}^*]]\,\mathbf{f}[[\mathbf{a}]] = \tau[[\mathbf{R}^*]].$$

We replace the right-hand side by its restriction to Σ:

$$\rho[[V]] = \sigma[[\mathbf{R}_\Sigma^*]]. \tag{6.20}$$

Since $\mathbf{f}$ is absent from this, the dependence on $\mathbf{a}$, that is on the width of W, disappears.

It happens frequently that the physical properties of a medium described by ϵ and μ change rapidly in the vicinity of some surface Σ, which is treated as an interface between two media. As an idealization, the functions ϵ or μ are assumed to be discontinuous. One may expect that the fields $\mathbf{E}, \mathbf{D}, \mathbf{B}, \mathbf{H}$ are also discontinuous, and the spatial distributions ρ and $\mathbf{j}$ should be supplemented by surface ones σ and $\mathbf{i}$.

We proceed to consider the conditions imposed on the field by Maxwell's equations at an interface between two media.

For Faraday's law, we choose the directed surface S as the parallelogram $ABCD$ intersected by the interface Σ as shown in Fig. 6.44. Assume that S is stationary; hence we may write down Faraday's law with the time derivative under the integration sign:

$$\int_S \dot{\mathbf{B}}[[d\mathbf{S}]] = -\oint_{\partial S} \mathbf{E}[[d\mathbf{l}]].$$

Now the sequences DA and BC tend to zero. Then the area of S also tends to zero. If the time derivative of $\mathbf{B}$ is bounded, the integral on the left-hand side tends to zero.

Fig. 6.44 Directed surface S
with internal orientation cut
by the interface Σ

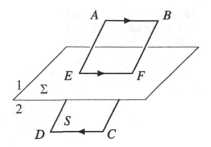

If, moreover, $\mathbf{E}$ is bounded, the linear integrals over DA and BC also tend to zero.
In this manner we get in the limit:

$$\int\limits_{AB} \mathbf{E}_1[[d\mathbf{l}]] + \int\limits_{CD} \mathbf{E}_2[[d\mathbf{l}]] = 0,$$

where the indices 1 and 2 refer to the media above and below Σ. Both integrals are
in essence performed over the same segment EF, only in the opposite directions on
the two sides of Σ, so we have:

$$\int\limits_{EF} (\mathbf{E}_1 - \mathbf{E}_2)[[d\mathbf{l}]] = 0.$$

The segment EF is arbitrary on Σ; hence the restriction of $\mathbf{E}_1 - \mathbf{E}_2$ to Σ must
vanish: $(\mathbf{E}_1 - \mathbf{E}_2)_\Sigma = 0$, whence

$$\mathbf{E}_{1\Sigma} = \mathbf{E}_{2\Sigma}. \tag{6.21}$$

In this manner we arrive at the conclusion that the restrictions of the electric field
intensities on both sides of the interface must be equal. This condition is illustrated
in Fig. 6.45. If $\mathbf{E}_1$ is parallel to Σ, its restriction to Σ is zero. In such a case (6.21)
implies that $\mathbf{E}_2$ has the same zero restriction, hence is also parallel to Σ.

For the next Maxwell equation we choose the directed surface S with the external
orientation as the parallelogram $ABCD$ intersecting Σ as shown in Fig. 6.46. We
assume that the arrow piercing Σ is directed to the medium 1. We write the Ampère-
Oersted law for stationary S as

$$\oint\limits_{\partial S} \mathbf{H}[[d\mathbf{l}^*]] = \int\limits_{S} \mathbf{j}[[d\mathbf{S}^*]] + \int\limits_{S} \mathbf{D}[[d\mathbf{S}^*]].$$

Fig. 6.45 Two fields $\mathbf{E}_1$ and $\mathbf{E}_2$ with the same restrictions to Σ

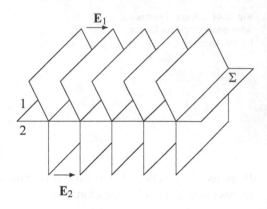

Fig. 6.46 Directed surface S with external orientation cut by the interface Σ

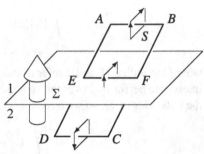

When AE, BF, DE and CF tend to zero with $\mathbf{D}$ and $\mathbf{H}$ bounded, the following terms remain:

$$\int_{AB} \mathbf{H}_1[[d\mathbf{l}^*]] + \int_{CD} \mathbf{H}_2[[d\mathbf{l}^*]] = \int_{S} \mathbf{j}[[d\mathbf{S}^*]].$$

If surface currents flow over Σ, then due to (6.17) the right-hand side can be replaced by $\int_{EF} \mathbf{i}[d\mathbf{l}^*_\Sigma]$. The two integrals on the left-hand side are in essence performed over the same segment EF; hence we use the appropriate restrictions to Σ. The orientation of $d\mathbf{l}^*$ on the segment CD is opposite to that on EF, so the second integral has a minus sign:

$$\int_{EF} \mathbf{H}_{1\Sigma}[[d\mathbf{l}^*_\Sigma]] - \int_{EF} \mathbf{H}_{2\Sigma}[[d\mathbf{l}^*_\Sigma]] = \int_{EF} \mathbf{i}[[d\mathbf{l}^*_\Sigma]].$$

We write everything under one integral sign:

$$\int_{EF} (\mathbf{H}_{1\Sigma} - \mathbf{H}_{2\Sigma} - \mathbf{i}) [[d\mathbf{l}^*_\Sigma]] = 0.$$

Fig. 6.47 Two fields H_1 and H_2 close to Σ with a surface current

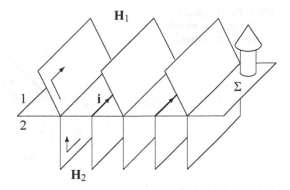

Fig. 6.48 Two fields H_1 and H_2 close to Σ without a surface current

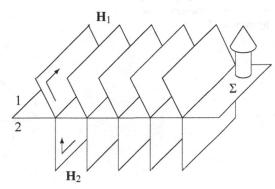

The segment EF is arbitrary on Σ, whence:

$$H_{1\Sigma} - H_{2\Sigma} = i. \tag{6.22}$$

We arrive at the conclusion that the restrictions of the magnetic field to Σ on the two sides of Σ differ by the surface current density. Figure 6.47 illustrates the smaller value (broader layers) of H_1 than that of H_2 because of the presence of a surface current on Σ.

When the surface current is absent, the restriction of the magnetic field on both sides must be equal:

$$H_{1\Sigma} = H_{2\Sigma}. \tag{6.23}$$

This condition is illustrated in Fig. 6.48.

It is possible that the magnetic field is absent on one side of Σ, for example $H_1 = 0$. Then (6.22) implies

$$H_{2\Sigma} = -i. \tag{6.24}$$

Fig. 6.49 Magnetic field present only on one side of Σ with a surface current

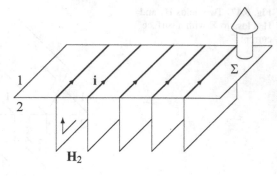

Fig. 6.50 Elliptic cylinder S with external orientation cut by interface Σ

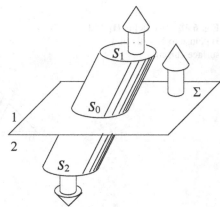

This situation is shown in Fig. 6.49. We may say that the layers of magnetic field end on surface currents.

To obtain a condition for the electric induction, we choose the closed surface S for Gauss's law as an oblique elliptic cylinder with the bases S_1 and S_2 parallel to Σ and with the side surface denoted as S_s. The interface Σ cuts the cylinder along the ellipse S_0 as shown in Fig. 6.50. We endow Σ with an external orientation going from medium 2 to 1.

The integral of Gauss's law over the closed surface S encompassing region V can be divided into three parts:

$$\int_{S_1} \mathbf{D}_1[[d\mathbf{S}^*]] + \int_{S_2} \mathbf{D}_2[[d\mathbf{S}^*]] + \int_{S_s} \mathbf{D}[[d\mathbf{S}^*]] = \int_V \rho[[dV]].$$

The bases S_1 and S_2 tend to S_0. If $\mathbf{D}$ is bounded, the integral over S_s tends to zero and there remains

$$\int_{S_1} \mathbf{D}_1([[\mathbf{S}^*]] + \int_{S_2} \mathbf{D}_2[[d\mathbf{S}^*]] = \int_V \rho[[dV]].$$

Surfaces S_1 and S_2 coincide with S_0 in the limit, only their orientations are opposite. We assume that S_0 and S_1 have the same orientation; then S_0 and S_2 are opposite, and hence we get

$$\int_{S_0} \mathbf{D}_1[[d\mathbf{S}^*]] - \int_{S_0} \mathbf{D}_2[[d\mathbf{S}^*]] = \int_V \rho[[dV]].$$

We apply now identity (6.20) and replace all other quantities by their restrictions to Σ:

$$\int_{S_0} (\mathbf{D}_1 - \mathbf{D}_2 - \sigma)_\Sigma [[d\mathbf{S}_\Sigma^*]] = 0.$$

Ellipse S_0 is arbitrary on Σ, hence the integrand has to vanish: $(\mathbf{D}_1 - \mathbf{D}_2 - \sigma)_\Sigma = 0$, which is written as

$$D_{1\Sigma} - D_{2\Sigma} = \sigma. \tag{6.25}$$

We arrive at the conclusion that the restrictions of the electric induction from either sides of the interface differ by the density of surface charges σ. We show in Fig. 6.51 a situation in which $\mathbf{D}_1$ has a bigger value (more dense tubes) than $\mathbf{D}_2$ because of some negative charges on Σ.

If $\sigma = 0$, the restrictions must be equal:

$$D_{1\Sigma} = D_{2\Sigma}. \tag{6.26}$$

This condition is illustrated in Fig. 6.52.

If the twisted two-form $\mathbf{D}_1$ is parallel to Σ and the surface charges are absent, $\mathbf{D}_2$ also must be parallel, because condition (6.26) demands zeros on either sides.

Fig. 6.51 Two fields $\mathbf{D}_1$ and $\mathbf{D}_2$ close to Σ with surface charge

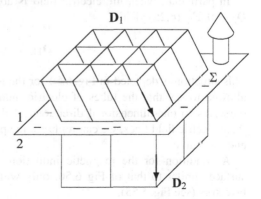

Fig. 6.52 Two fields $\mathbf{D}_1$ and $\mathbf{D}_2$ close to Σ without surface charge

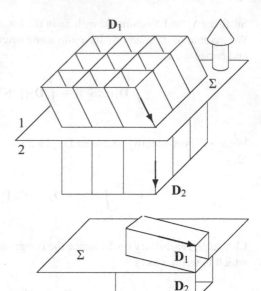

Fig. 6.53 Possible situation for fields $\mathbf{D}_1$ and $\mathbf{D}_2$ parallel to Σ without surface charge

In such a case the attitudes of $\mathbf{D}_1$ and $\mathbf{D}_2$ do not need be parallel. This is shown in Fig. 6.53.

Such a situation occurs simultaneously with that shown on Fig. 6.48. We know that the twisted two-forms $\mathbf{D}_i$ must be perpendicular to the one-forms $\mathbf{E}_i$ according to the metrics of the respective media. If the two media are identical, the tubes of $\mathbf{D}_i$ lying on the interface Σ would be parallel. We conclude from this that the situation depicted in Fig. 6.53 may occur only when media 1 and 2 have different permittivity tensors.

If $\mathbf{E}_1$ is parallel to Σ, by virtue of (6.21), $\mathbf{E}_2$ also is parallel. Then the electric inductions $\mathbf{D}_1$ are $\mathbf{D}_2$ perpendicular to Σ, each according to its scalar product. Figure 6.52 may be an illustration of such a situation.

In particular, when the electric field is absent on one side of Σ, for instance $\mathbf{D}_2 = 0$, Eq. (6.25) yields

$$\mathbf{D}_{1\Sigma} = \sigma.$$

This situation is depicted in Fig. 6.54 for the negative surface charge on Σ. We are allowed to say that the tubes of electric induction end on charges. If we restrict ourselves to one-dimensional directions of the nontwisted two-forms $\mathbf{D}$ and call them electric field lines, we can rephrase the previous sentence—electric field lines end on charges.

A condition for the magnetic induction remains. We choose the integration surface similar to that of Fig. 6.50, only with the orientation appropriate for the bivectors (see Fig. 6.55).

Fig. 6.54 Electric field present only on one side of Σ with surface charge

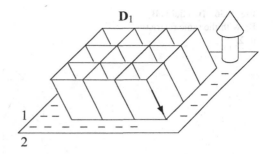

Fig. 6.55 Elliptic cylinder S with internal orientation cut by interface Σ

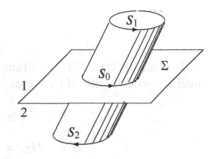

The magnetic Gauss law assumes the form

$$\int_{S_1} \mathbf{B}_1[[d\mathbf{S}]] + \int_{S_2} \mathbf{B}_2[[d\mathbf{S}]] + \int_{S_b} \mathbf{B}[[d\mathbf{S}]] = 0.$$

In the limit of S_1 and S_2 tending to S_0 we obtain

$$\int_{S_0} (\mathbf{B}_1 - \mathbf{B}_2)_\Sigma [[d\mathbf{S}_\Sigma]] = 0.$$

The arbitrariness of S_0 on Σ implies the condition

$$\mathbf{B}_{1\Sigma} = \mathbf{B}_{2\Sigma}, \tag{6.27}$$

stating that the restrictions of the magnetic induction from either sides of the interface have to be equal. We illustrate this condition in Fig. 6.56.

Media 1 and 2 may have different anisotropies of the permeability, that is, distinct magnetic metrics. Let the interface be free of surface currents—then condition (6.23) is valid. If $\mathbf{B}_1$ is parallel to Σ, by virtue of (6.27), $\mathbf{B}_2$ also must be parallel. Then $\mathbf{H}_1$ and $\mathbf{H}_2$ are perpendicular to Σ, each according to its scalar product. Figure 6.49 may be an illustration of such a situation.

Fig. 6.56 Two fields $\mathbf{B}_1$ and $\mathbf{B}_2$ with the same restrictions to Σ

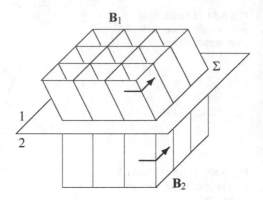

Formulas (6.21), (6.22), (6.26) and (6.27) jointly constitute the conditions on the interface, as in the title of the present section. We collect them now:

$$\mathbf{E}_{1\Sigma} = \mathbf{E}_{2\Sigma}, \qquad \mathbf{B}_{1\Sigma} = \mathbf{B}_{2\Sigma}, \qquad (6.28)$$

$$\mathbf{H}_{1\Sigma} - \mathbf{H}_{2\Sigma} = \mathbf{i}, \qquad \mathbf{D}_{1\Sigma} - \mathbf{D}_{2\Sigma} = \sigma. \qquad (6.29)$$

The first two can be called *continuity conditions* for $\mathbf{E}$ and $\mathbf{B}$ *on the interface*, and the other two in the presence of surface charges and currents may be called *discontinuity conditions* for $\mathbf{H}$ and $\mathbf{D}$ *on the interface*. In the language of spatial vectors the left-hand equalities impose conditions on components parallel to Σ (see Exercises 6.2 and 6.4), and the right-hand ones impose conditions on perpendicular components (Exercises 6.3 and 6.5).

6.3   Fields of Infinite Plates with Charges or Currents

We consider in this section static electric and magnetic fields produced by flat infinite plates with uniform surface charges or currents. Such sources have natural two-dimensional translational symmetry and as a main tool the following rule will be used: *the symmetries of causes should also be present in the effects*, called the *Curie principle* by Pérez, Calmes and Fleckinger [32]. Therefore, we shall assume that the fields also have the same translational symmetry.

6.3.1   Electric Field of a Uniformly Charged Plate

Let us find the electric field produced by an infinite plane Σ with a homogeneous surface charge of density σ. On the two sides of it two distinct anisotropic dielectric media fill the half spaces.

Two-dimensional translational symmetry parallel to Σ occurs in the posed problem. The fields $\mathbf{E}$ and $\mathbf{D}$ should also have this symmetry; hence they must be constant on planes parallel to Σ. Does the named symmetry also determine the attitudes of the fields? A natural assumption is that the attitude of $\mathbf{E}$ is determined, since the planes parallel to Σ are distinguished. Therefore we assume that the attitudes of $\mathbf{E}_i$ are the same on both sides of Σ—they are parallel to Σ. It is also natural to assume that the orientations of $\mathbf{E}_i$ are opposite: out of Σ when $\sigma > 0$, or towards Σ, when $\sigma < 0$. Now a question arises about the magnitudes: should they be the same, or may they be different? At the moment we do not have data to answer this.

The electric induction fields $\mathbf{D}_i$ have already obtained directions—perpendicular to $\mathbf{E}_i$ (each according to its scalar product) and orientations compatible with that of $\mathbf{E}_i$. The magnitudes of $\mathbf{D}_i$ in the close vicinity of Σ are determined by the discontinuity condition (6.25)

$$\mathbf{D}_{1\Sigma} - \mathbf{D}_{2\Sigma} = \sigma. \tag{6.30}$$

The orientations $\mathbf{D}_i$ close to Σ are not strictly opposite, because their attitudes may not be parallel, but their restrictions $\mathbf{D}_{i_\Sigma}$ have opposite orientations; this is illustrated in Fig. 6.57 for $\sigma > 0$.

Only now can we make assumptions about the magnitudes of the field. To not discriminate between the media on either side of Σ, assume that the magnitudes of the restrictions $\mathbf{D}_{i_\Sigma}$ are the same, which means that they are opposite:

$$\mathbf{D}_{2\Sigma} = -\mathbf{D}_{1\Sigma}.$$

Then the equality (6.30) yields $2\mathbf{D}_{1\Sigma} = \sigma$, whence we obtain

$$\mathbf{D}_{1\Sigma} = \frac{1}{2}\sigma, \qquad \mathbf{D}_{2\Sigma} = -\frac{1}{2}\sigma.$$

Fig. 6.57 Two fields $\mathbf{D}_1$, $\mathbf{D}_2$ with opposite restrictions to the positively charged plane Σ

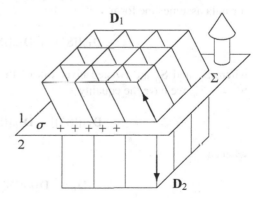

Fig. 6.58 Orientations of $\mathbf{E}_1$ and $\mathbf{E}_2$ are opposite when the plane Σ is charged

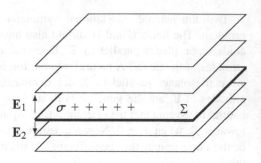

Thus we have the answer to the last question from the previous paragraph: the electric field intensities $\mathbf{E}_i$ close to Σ generally have distinct magnitudes. We illustrate this in Fig. 6.58.

We know already what the fields are close to Σ on its both sides. We know also that they can not change when shifting along Σ. But can they change when shifting away from Σ? To answer this question, let us calculate the integral of $\mathbf{D}$ over the surface of prism G situated completely in one medium, say 1, with both bases G_a and G_b parallel to Σ, and the other sides parallel to $\mathbf{D}_1$. From Gauss's theorem we have

$$\int_G \mathbf{D}_1[[d\mathbf{S}^*]] = 0.$$

Since the lateral sides are parallel to $\mathbf{D}_1$, the integral over them vanishes and the integral over the two bases remains:

$$\int_{G_a} \mathbf{D}_1[[d\mathbf{S}^*]] + \int_{G_b} \mathbf{D}_1[[d\mathbf{S}^*]] = 0.$$

Field $\mathbf{D}_1$ is constant on these bases: let them be $\mathbf{D}_{1a}$ and $\mathbf{D}_{1b}$, respectively. The constancy of the field allows us to integrate inside the argument of the form and our formula assumes the form

$$\mathbf{D}_{1a}[[\mathbf{S}_a^*]] + \mathbf{D}_{1b}[[\mathbf{S}_b^*]] = 0,$$

where $\mathbf{S}_a^*$ and $\mathbf{S}_b^*$ are twisted bivectors of the two bases. Because of the identity $\mathbf{S}_a^* = -\mathbf{S}_b^*$, we have the equality

$$\mathbf{D}_{1a}[[\mathbf{S}_a^*]] - \mathbf{D}_{1b}[[\mathbf{S}_a^*]] = 0,$$

whence

$$(\mathbf{D}_{1a} - \mathbf{D}_{1b})[[\mathbf{S}_a^*]] = 0,$$

The twisted bivector $\mathbf{S}_a^*$ is arbitrary (only parallel to Σ); hence we get

$$\mathbf{D}_{1a} - \mathbf{D}_{1b} = 0, \quad \text{i.e.} \quad \mathbf{D}_{1a} = \mathbf{D}_{1b}.$$

This means that the electric induction is constant in the whole half space filled by medium 1. It follows from the constitutive relation between $\mathbf{E}$ and $\mathbf{D}$ that the electric field intensity is also constant in the whole half space.

It is worthwhile to find the scalar potential of this electric field. If we introduce a basis $\mathbf{m}_i$ with the origin on Σ such that $\mathbf{m}_1, \mathbf{m}_2 \, || \, \Sigma$, then Φ can be a function of the third coordinate of $\mathbf{r} = \xi^i \mathbf{m}_i$ only:

$$\Phi(\mathbf{r}) = \alpha \xi^3 + \beta.$$

The constants α can be distinct on two sides of Σ, distinct in sign and absolute value. Notice that the potential has the same symmetry as the sources of the field—it is constant on planes parallel to Σ.

6.3.2 Electric Field of Two Plates—Electric Capacitor

Consider two parallel planes Σ_a and Σ_b uniformly charged with the surface densities σ_a and σ_b, respectively. Such a system is called an *ideal plane electric capacitor*. The planes divide the space into three parts: half space 1, layer 2 and half space 3—distinct anisotropic dielectric in each of them. We give to both planes the same external orientation towards medium 1. One-forms of the electric field $\mathbf{E}_1$, $\mathbf{E}_2$ are $\mathbf{E}_3$ parallel to the charged planes and to themselves. Thus the twisted two-forms $\mathbf{D}_i$ are perpendicular to one-forms $\mathbf{E}_i$ in each part of the space according to another scalar product. By making use of the superposition principle we may consider two fields of the electric induction. One comes from the charge collected on Σ_a, having three values:

$$\mathbf{D}_a = \begin{cases} \mathbf{D}_{1a} & \text{in region 1} \\ \mathbf{D}_{2a} & \text{in region 2} \\ \mathbf{D}_{3a} & \text{in region 3.} \end{cases}$$

The field coming from the charge on plane Σ_b is:

$$\mathbf{D}_b = \begin{cases} \mathbf{D}_{1b} & \text{in region 1} \\ \mathbf{D}_{2b} & \text{in region 2} \\ \mathbf{D}_{3b} & \text{in region 3.} \end{cases}$$

The field is their sum: $\mathbf{D} = \mathbf{D}_a + \mathbf{D}_b$.

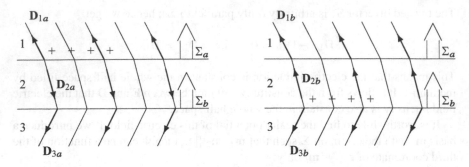

Fig. 6.59 Directions of **D** fields when plane Σ_b is positively charged or Σ_b is positively charged

We know from previous considerations that on Σ_a the following conditions are satisfied

$$\mathbf{D}_{1a\Sigma_a} = \frac{\sigma_a}{2}, \quad \mathbf{D}_{2a\Sigma_a} = -\frac{\sigma_a}{2}. \tag{6.31}$$

From the constancy of $\mathbf{D}_a$ in region 2 and the two planes Σ having compatible orientations, it follows that this induction is the same on the two boundaries of layer 2: $\mathbf{D}_{2a\Sigma_a} = \mathbf{D}_{2a\Sigma_b} = -\frac{\sigma_a}{2}$. For the field coming only from charges lying on Σ_a, we treat Σ_b as not charged; hence the electric induction has the same restriction on its two sides:

$$\mathbf{D}_{2a\Sigma_b} = \mathbf{D}_{3a\Sigma_b} = -\frac{\sigma_a}{2}. \tag{6.32}$$

This field is illustrated in Fig. 6.59 (left), made in a section cutting two planes Σ for $\sigma_a > 0$. In a similar manner we arrive at the conclusion that the field coming only from charges collected on Σ_b fulfills the conditions

$$\mathbf{D}_{1b\Sigma_a} = \mathbf{D}_{2b\Sigma_a} = \frac{\sigma_b}{2}. \tag{6.33}$$

$$\mathbf{D}_{2b\Sigma_b} = \frac{\sigma_b}{2}, \quad \mathbf{D}_{3b\Sigma_b} = -\frac{\sigma_b}{2}. \tag{6.34}$$

This is shown in Fig. 6.59 (right), made in the same section for $\sigma_b > 0$.
Joining conditions (6.31–6.34), we may write such formulas for the restriction of the net field $\mathbf{D} = \mathbf{D}_a + \mathbf{D}_b$ to both planes Σ on their two sides:

$$\mathbf{D}_{1\Sigma_a} = \frac{\sigma_a + \sigma_b}{2}, \quad \mathbf{D}_{2\Sigma_a} = \frac{\sigma_b - \sigma_a}{2},$$

$$\mathbf{D}_{2\Sigma_b} = \frac{\sigma_b - \sigma_a}{2}, \quad \mathbf{D}_{3\Sigma_b} = -\frac{\sigma_b + \sigma_a}{2}.$$

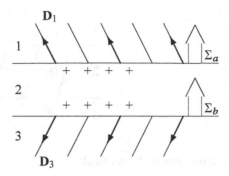

Fig. 6.60   Directions of **D** fields when two planes Σ_a and Σ_b are positively charged

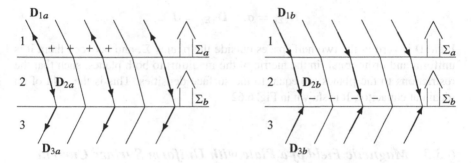

Fig. 6.61   Directions of **D** fields when plane Σ_b is positively charged or Σ_b is negatively charged

In particular, when the two planes have the surface densities: $\sigma_a = \sigma_b = \sigma$, the following restrictions are obtained:

$$\mathbf{D}_{1\Sigma_a} = \sigma, \quad \mathbf{D}_{2\Sigma_a} = 0,$$

$$\mathbf{D}_{2\Sigma_b} = 0, \quad \mathbf{D}_{3\Sigma_b} = -\sigma.$$

Field **D** vanishes between the plates and outside them, orthogonal to the planes separately according to the respective scalar products and such that their restrictions to the planes are equal to the surface densities. They are shown in Fig. 6.60. We may consider this as the electric field of two conductive charged plates, connected to each other somewhere far away. Since this is now already one conductor, it is not strange that the field between the plates is zero.

In another particular case, when the two planes have the opposite surface densities, $\sigma_a = -\sigma_b = \sigma$, the field $\mathbf{D}_a$ remains the same, and $\mathbf{D}_b$ changes the orientation (see Fig. 6.61).

Fig. 6.62 Directions of **D** fields when two planes Σ_a and Σ_b are oppositely charged

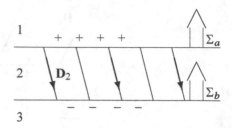

Then we obtain such restrictions of the net field:

$$\mathbf{D}_{1\Sigma_a} = 0, \quad \mathbf{D}_{2\Sigma_a} = \sigma,$$

$$\mathbf{D}_{2\Sigma_b} = \sigma, \quad \mathbf{D}_{3\Sigma_b} = 0.$$

Field **D** is zero in the two half spaces outside the planes Σ, and between them it is uniform and orthogonal (in the metric of the medium) to both planes, such that the restrictions to the planes are equal to the surface densities. This is the field of an ideal flat capacitor. It is shown in Fig. 6.62.

6.3.3 Magnetic Field of a Plate with Uniform Surface Current

We address now the magnetic field produced by an infinite plane Σ with uniform surface current of linear density **i**. Two anisotropic magnetic media fill the half spaces on either side of Σ. A two-dimensional translational symmetry parallel to Σ occurs in the posed problem. Fields **H** and **B** should also have this symmetry; hence they must be constant on planes parallel to Σ. Does the named symmetry also determine the attitudes of the fields? At the moment we cannot decide this.

We recall Sect. 4.3 in which the Biot-Savart law was considered. We established there that the magnetic induction **B** is perpendicular (with respect to the medium scalar product) to the currents in their vicinity (see Fig. 4.11). Thus we state that the magnetic induction in each medium separately must be orthogonal to **i** on each side of Σ according to the different scalar products. But there are plenty of directions orthogonal to **i**. We invoke now the reflection symmetry of the sources and the medium in the plane perpendicular (always in the medium metric) to Σ and passing through **i**—this plane is determined by the vectors $\mathbf{m}_1$ and $\mathbf{m}_3$ of the orthonormal basis depicted in Fig. 6.63. The named reflection symmetry implies that only **B** parallel to Σ is admissible (see Fig. 6.64).

The magnetic field intensities $\mathbf{H}_i$ are more directly connected with **i**. Their directions are already determined—they should be perpendicular to $\mathbf{B}_i$, hence parallel to **i**. To not discriminate between either side of Σ, we assume that their

Fig. 6.63 Vector basis
chosen for plane Σ with
surface current

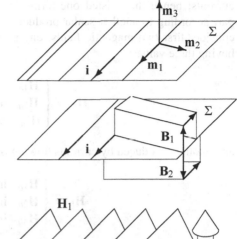

Fig. 6.64 The fields $\mathbf{B}_1$, $\mathbf{B}_2$
must be parallel to Σ

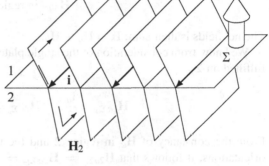

Fig. 6.65 Two fields $\mathbf{H}_1$, $\mathbf{H}_2$
with opposite restrictions to
plane Σ with surface current

restrictions to Σ are opposite: $\mathbf{H}_{2\Sigma} = -\mathbf{H}_{1\Sigma}$; then we obtain from (6.22):

$$\mathbf{H}_{1\Sigma} = \frac{\mathbf{i}}{2}.$$

Therefore only the restriction $\mathbf{H}_{1\Sigma}$ has the same orientation as $\mathbf{i}$, whereas $\mathbf{H}_{2\Sigma}$ is opposite. A glance at Fig. 6.29 helps us to conclude that this situation can be depicted as in Fig. 6.65.

We see now that the fields $\mathbf{H}_i$ have orientations compatible with their source, that is, with $\mathbf{i}$.

6.3.4   Magnetic Field of Two Plates—Magnetic Capacitor

Let us consider parallel planes Σ_a and Σ_b with uniform surface currents with densities $\mathbf{i}_a$ and $\mathbf{i}_b$, respectively. The planes divide the space onto three parts: half space 1, layer 2 and half space 3—in each of them there is a different anisotropic magnetic medium. We give the same external orientation to medium 1 to either plane. The nontwisted two-forms $\mathbf{B}_1$, $\mathbf{B}_2$ and $\mathbf{B}_3$ are parallel to the planes with currents; hence the twisted one-forms $\mathbf{H}_i$ are perpendicular to the Σ's in each part according to another scalar product. Applying the superposition principle we consider first two magnetic fields: one produced by currents flowing on Σ_a and having three values:

$$\mathbf{H}_a = \begin{cases} \mathbf{H}_{1a} & \text{in region 1} \\ \mathbf{H}_{2a} & \text{in region 2} \\ \mathbf{H}_{3a} & \text{in region 3 ,} \end{cases}$$

and another, produced by currents flowing on Σ_b:

$$\mathbf{H}_b = \begin{cases} \mathbf{H}_{1b} & \text{in region 1} \\ \mathbf{H}_{2b} & \text{in region 2} \\ \mathbf{H}_{3b} & \text{in region 3.} \end{cases}$$

The net fields is their sum: $\mathbf{H} = \mathbf{H}_a + \mathbf{H}_b$.

We know from consideration of the single plate that the following conditions are fulfilled on Σ_a:

$$\mathbf{H}_{1a\Sigma_a} = \frac{i_a}{2}, \qquad \mathbf{H}_{2a\Sigma_a} = -\frac{i_a}{2}. \tag{6.35}$$

From the constancy of $\mathbf{H}_a$ in region 2 and the two planes Σ having compatible orientations, it follows that $\mathbf{H}_{2a\Sigma_a} = \mathbf{H}_{2a\Sigma_b} = -\frac{i_a}{2}$. For the field produced only by the currents flowing only over Σ_a, we treat Σ_b as devoid of currents; hence the magnetic field intensity has its two restrictions from both sides equal:

$$\mathbf{H}_{2a\Sigma_b} = \mathbf{H}_{3a\Sigma_b} = -\frac{i_a}{2}. \tag{6.36}$$

This field is illustrated in Fig. 6.66 (left). In a similar manner we arrive at the conclusion that the field produced only by charges flowing on Σ_b fulfils the conditions

$$\mathbf{H}_{1b\Sigma_a} = \mathbf{H}_{2b\Sigma_a} = \frac{i_b}{2}. \tag{6.37}$$

$$\mathbf{H}_{2b\Sigma_b} = \frac{i_b}{2}, \qquad \mathbf{H}_{3b\Sigma_b} = -\frac{i_b}{2}. \tag{6.38}$$

This has been shown in Fig. 6.66 (right), made for the surface current density i_b of the same direction as that of i_a.

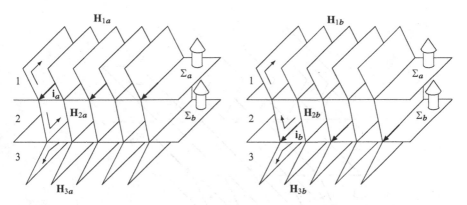

Fig. 6.66 Directions of **H** fields when plane Σ_b has surface current or Σ_b has surface current

Joining conditions (6.35–6.38) we may write the following formulas for the restrictions of the net field $\mathbf{H} = \mathbf{H}_a + \mathbf{H}_b$ to either planes Σ on their two sides:

$$\mathbf{H}_{1\Sigma_a} = \frac{\mathbf{i}_a + \mathbf{i}_b}{2}, \quad \mathbf{H}_{2\Sigma_a} = \frac{\mathbf{i}_b - \mathbf{i}_a}{2},$$

$$\mathbf{H}_{2\Sigma_b} = \frac{\mathbf{i}_b - \mathbf{i}_a}{2}, \quad \mathbf{H}_{3\Sigma_b} = -\frac{\mathbf{i}_b + \mathbf{i}_a}{2}.$$

In particular, when the two planes have the same current densities: $\mathbf{i}_a = \mathbf{i}_b = \mathbf{i}$, we obtain the following restrictions:

$$\mathbf{H}_{1\Sigma_a} = \mathbf{i}, \quad \mathbf{H}_{2\Sigma_a} = 0,$$

$$\mathbf{H}_{2\Sigma_b} = 0, \quad \mathbf{H}_{3\Sigma_b} = -\mathbf{i}.$$

Field **H** vanishes between the planes, and is uniform outside, perpendicular to the planes separately according to the appropriate scalar products and such that their restrictions to the planes are equal to the surface current densities. They are shown in Fig. 6.67.

In another particular case, when the two planes have the opposite current densities: $\mathbf{i}_a = -\mathbf{i}_b = \mathbf{i}$, the field $\mathbf{H}_a$ is the same, whereas $\mathbf{H}_b$ changes its orientation (see Figs. 6.68 (left) (the same as Figs. 6.66 (left)) and 6.68 (right)) (different from Fig. 6.66 (right) by opposite $\mathbf{i}_b$). If the two plates have the described surface currents simultaneously, the magnetic field is the superposition of both cases. Then one obtains the following restrictions of the net field:

$$\mathbf{H}_{1\Sigma_a} = 0, \quad \mathbf{H}_{2\Sigma_a} = -\mathbf{i}, \tag{6.39}$$

$$\mathbf{H}_{2\Sigma_b} = -\mathbf{i}, \quad \mathbf{H}_{3\Sigma_b} = 0. \tag{6.40}$$

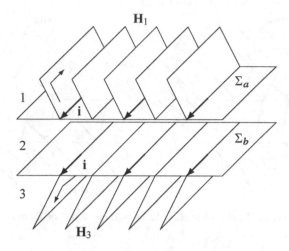

Fig. 6.67 Directions of **H** fields when two planes Σ_a and Σ_b have equal surface currents

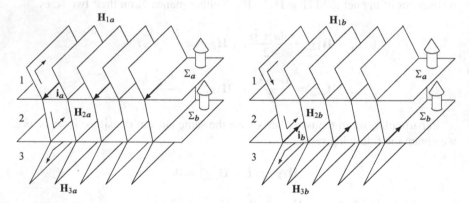

Fig. 6.68 Directions of **H** fields when plane Σ_b has a surface current or Σ_b has the opposite surface current

H $= 0$ in both half spaces outside the planes Σ, and between them **H** is uniform, perpendicular to the planes and such that its restrictions to the planes are equal to the surface current densities. This is the field of a system, which could be called an *ideal plane magnetic capacitor*. This field is shown in Fig. 6.69.

The same plane magnetic capacitor is depicted in the next two figures as seen from another side. Figure 6.70 shows the tubes of the magnetic induction **B**, and Fig. 6.71 shows the planes of the magnetic field **H**.

If media 1 and 3 are the same, we may consider the magnetic field of two conducting plates with currents, connected with each other in such a way that the current flows from Σ_a to Σ_b on one side, and conversely on the other (see Fig. 6.72). This can be considered as an *ideal solenoid* with a cross-section in the shape of a parallelogram, which contains a different medium inside than outside. Then one obtains the magnetic field, confined inside and absent outside.

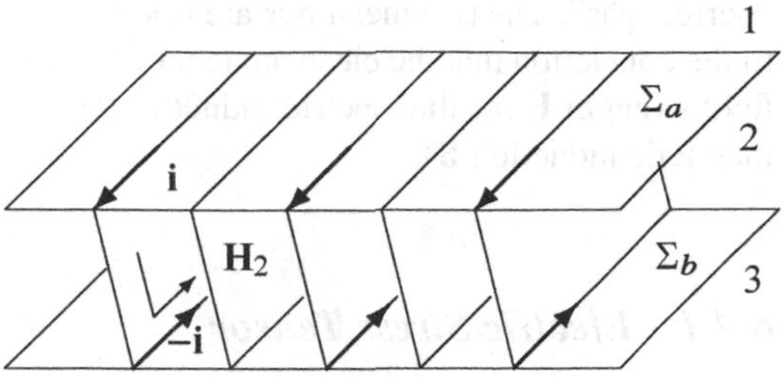

Fig. 6.69 Magnetic field **H** of an ideal plane magnetic capacitor

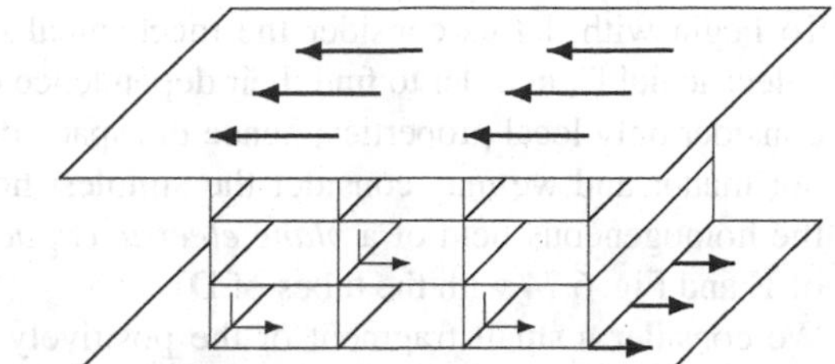

Fig. 6.70 Magnetic induction **B** of an ideal plane magnetic capacitor

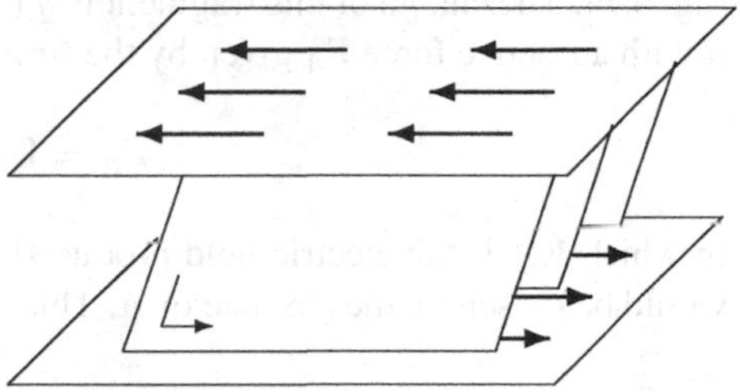

Fig. 6.71 Magnetic field **H** of an ideal plane magnetic capacitor

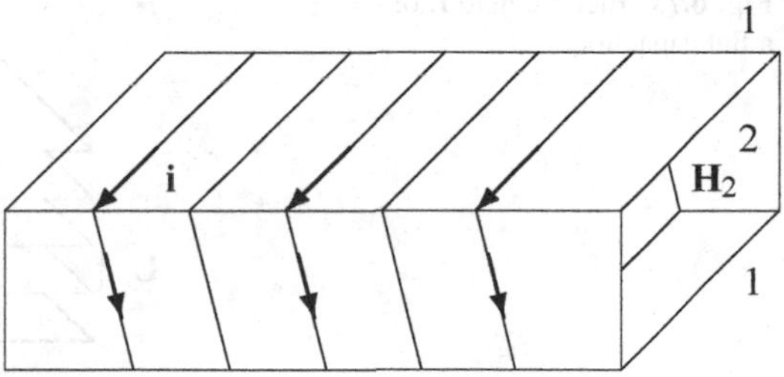

Fig. 6.72 Magnetic field **H** of an ideal angular solenoid

6.4   Electromagnetic Stress Tensor

The inspiration for this section stems from an article by F. Herrmann [26].

The word "field" appears in the common use among the physicists in two different meanings. On the one hand, it is used in the description of a distribution of a local physical quantity in space, e.g., temperature or pressure. On the other hand, in modern field theory, the word "field" is applied as the name of a specific physical system. The *electromagnetic field* is the name of a system, similar to "rigid body" or "perfect gas". The two meanings are not always distinguished, which leads students to the conviction that the electromagnetic field is only the distribution of the electric field strength **E** (or the electric induction **D**) and the magnetic field strength **H** (or magnetic induction **B**).

6.4.1   Electric Stress Tensor

To begin with, let us consider the mechanical properties of the system called an "electric field", in order to find their dependence on the field quantities **E** and **D**. We consider only local properties; hence the space distribution of these quantities does not matter and we may consider the simplest homogeneous distribution. Let it be the homogeneous field of a *plane electric capacitor* (see Fig. 6.73 with the planes of **E** and Fig. 6.74 with the tubes of **D**).

We consider a small fragment of the positively charged plate A described by the twisted bivector **Q** with the orientation to the outside of the capacitor, marked in Fig. 6.74. The charge of this fragment is q_A. The negatively charged plate A' acts on it with attractive force $\mathbf{F}_A$ given by the formula

$$\mathbf{F}_A = \mathbf{E}_{A'} q_A,$$

in which $\mathbf{E}_{A'}$ is the electric field produced by the plate A' alone, i.e., the field that would be present in the absence of A. This situation is shown in Figs. 6.75 and 6.76.

Fig. 6.73 Electric field **E** of a flat capacitor

Fig. 6.74 Electric induction **D** of a flat capacitor

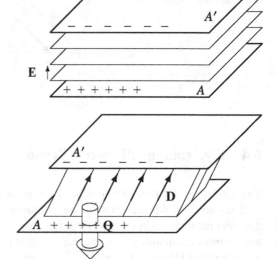

Fig. 6.75 Electric field **E** of single plane A'

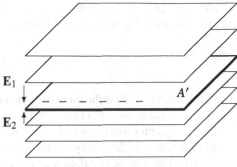

Fig. 6.76 Electric induction **D** of single plane A'

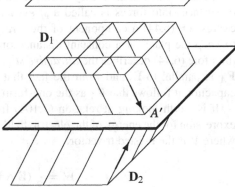

It is well known from considerations about the flat capacitor, that the field strength $\mathbf{E}_{A'}$ produced by a single plate is twice as small as the field strength $\mathbf{E}$ present in the capacitor. In this manner we have

$$\mathbf{F}_A = \frac{1}{2}\mathbf{E}\,q_A. \tag{6.41}$$

We know, that the electric charge collected on the plate A is connected with the electric induction **D** in the capacitor by the formula

$$q_A = -\mathbf{D}[[\mathbf{Q}]], \tag{6.42}$$

where the minus sign is necessary because the orientations of **D** and **Q** are opposite and we know that the charge q_A is positive. We insert (6.42) into (6.41):

$$\mathbf{F}_A = -\frac{1}{2}\mathbf{E}\,\mathbf{D}[[\mathbf{Q}]]. \tag{6.43}$$

Since $\mathbf{D}[[\mathbf{Q}]] < 0$ on A (see Fig. 6.74), the directions of $\mathbf{F}_A$ and **E** are the same.

We see in this formula that the linear mapping of the bivector **Q** into the one-form $\mathbf{F}_A$, in the terminlology of Sect. 3.3, is the simple tensor

$$\kappa_A = -\frac{1}{2}\,\mathbf{E} \otimes \mathbf{D} \tag{6.44}$$

The expression (7.104) is the force by which the plate A' acts on the chosen fragment of A. This force does not act on a distance, but is transmitted by the electric field filling the capacitor, similarly to how a tensed rubber band attracts two hooks stuck in opposite walls. In the theory of elastic media the linear mapping of the fragments of surfaces into forces is called a *stress tensor*. Therefore, we shall consider the expression (6.44) a component of the *stress tensor of the electric field*. In order to keep the plates at a constant distance one has to tense them from the outside, therefore (6.44) describes the *attractive stress*. It is clear from (7.104) that the force $\mathbf{F}_A$ is parallel to **E**, and from the fact that the bivector **Q** is parallel to **E** in the capacitor, it follows that $\mathbf{F}_A$ as the one-form is parallel to **Q**.

It is worth looking at relation (7.104) from another side. Let us start from an expression for the energy of the electric field stored in the volume V of the capacitor, where V is the twisted trivector:

$$W = \frac{1}{2}\,(\mathbf{E} \wedge \mathbf{D})[[V]].$$

Let the distance between the plates of the capacitor be described by a vector **l** with the orientation from A' to A, because we consider the force acting on A. We represent V as the exterior product of the twisted bivector $\mathbf{Q} \,\|\, \mathbf{E}$ of the plate surface and the nontwisted vector $\mathbf{l} \,\|\, \mathbf{D}$ connecting the plates: $V = \mathbf{l} \wedge \mathbf{Q}$ (see Fig. 6.77). The orientations of **l** and **Q** fit that of the positive twisted trivector V. By virtue of Exercise 3.16

$$(\mathbf{E} \wedge \mathbf{D})[[\mathbf{l} \wedge \mathbf{Q}]] = \mathbf{E}[[\mathbf{l}]]\,\mathbf{D}[[\mathbf{Q}]] + \mathbf{D}[[(\mathbf{E}\lrcorner\mathbf{Q}) \wedge \mathbf{l}]].$$

Fig. 6.77 Quantities **l**, **u** and **v** added to Fig. 6.74

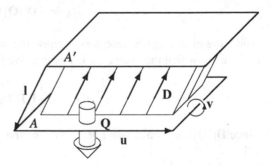

In our case $\mathbf{E} \rfloor \mathbf{Q} = 0$, so we get

$$W = \frac{1}{2} \mathbf{E}[[\mathbf{l}]] \, \mathbf{D}[[\mathbf{Q}]]. \tag{6.45}$$

Both scalar factors $\mathbf{E}[[\mathbf{l}]]$ and $\mathbf{D}[[\mathbf{Q}]]$ are negative, so W is positive.

We consider the parallelepiped-shaped sector of the capacitor with the surface of the plates given by the twisted bivector $\mathbf{Q}$ with orientation opposite to $\mathbf{D}$. We now consider how this energy changes when the plate A is pulled away from A', so we treat the right-hand side of (6.45) as a linear function of $\mathbf{l} = \mathbf{r}$ (a change of notation). This corresponds to a new physical problem—as the distance between the plates grows, the energy of the electric field increases:

$$W_A(\mathbf{r}) = \frac{1}{2} \mathbf{E}[[\mathbf{r}]] \, \mathbf{D}[[\mathbf{Q}]].$$

The force acting on A is the exterior derivative (taken with the minus sign) of this function:

$$\mathbf{F}_A = -\mathbf{d} \wedge W_A(\mathbf{r}) = -\frac{1}{2} \mathbf{E} \, \mathbf{D}[[\mathbf{Q}]].$$

The derivative must be taken with the minus sign to reproduce formula (7.104). In this manner we have found the possibility of finding forces from the expression for the energy of the capacitor.

Let us consider now other components of the force, not necessarily parallel to the plates of the capacitor. We factorize the bivector $\mathbf{Q} = \mathbf{u} \wedge \mathbf{v}$, where the nontwisted vector $\mathbf{u}$ is one edge of the plate and the twisted vector $\mathbf{v}$ is the other (these vectors are depicted in Fig. 6.77). We factorize similarly the two-form $\mathbf{D} = \mathbf{g} \wedge \mathbf{h}$, where $\mathbf{g}$ is a nontwisted one-form parallel to $\mathbf{v}$, whereas $\mathbf{h}$ is a twisted one-form parallel to $\mathbf{u}$ (see Fig. 6.78). Then by virtue of (2.12)

$$\mathbf{D}[[\mathbf{Q}]] = (\mathbf{g} \wedge \mathbf{h})[[\mathbf{u} \wedge \mathbf{v}]] = \mathbf{g}[[\mathbf{u}]] \, \mathbf{h}[[\mathbf{v}]] - \mathbf{g}[[\mathbf{v}]] \, \mathbf{h}[[\mathbf{u}]].$$

But, from the assumptions about parallelism, $\mathbf{g}[[\mathbf{v}]] = \mathbf{h}[[\mathbf{u}]] = 0$ and there remains

$$\mathbf{D}[[\mathbf{Q}]] = \mathbf{g}[[\mathbf{u}]] \, \mathbf{h}[[\mathbf{v}]].$$

We insert this into (6.45):

$$W = \frac{1}{2} \mathbf{E}[[\mathbf{l}]] \, \mathbf{g}[[\mathbf{u}]] \, \mathbf{h}[[\mathbf{v}]]. \tag{6.46}$$

It is worth considering now a capacitor, the plates of which can be stretched to increase their surfaces in such a way that their density of charges is constant. Let us consider, at the right-hand boundary of the capacitor, a hypothetical wall C

Fig. 6.78 Representing **D** as the exterior product $\mathbf{g} \wedge \mathbf{h}$

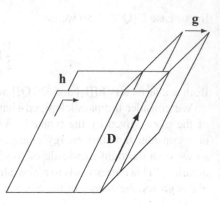

Fig. 6.79 Exterior product $\mathbf{R} = -\mathbf{l} \wedge \mathbf{v}$

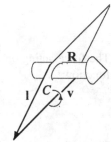

described by the twisted bivector $\mathbf{R} = -\mathbf{l} \wedge \mathbf{v}$ with orientation outside the capacitor (see Fig. 6.79) which is the orientation of stretching. Then the energy W becomes a function of the nontwisted vector $\mathbf{u}$ determining the size of the plates in the direction of stretching. After a change of notation $\mathbf{u} \to \mathbf{r}$ we write down:

$$W_C(\mathbf{r}) = \frac{1}{2}\, \mathbf{E}[[\mathbf{l}]]\, \mathbf{g}[[\mathbf{r}]]\, \mathbf{h}[[\mathbf{v}]]. \tag{6.47}$$

Then force $\mathbf{F}_C$ acting on the wall C is (with the minus sign) the exterior derivative of the function (6.47):

$$\mathbf{F}_C = -\mathbf{d} \wedge W_C = -\frac{1}{2}\, \mathbf{g}\, \mathbf{E}[[\mathbf{l}]]\, \mathbf{h}[[\mathbf{v}]]. \tag{6.48}$$

Due to (2.12) and the assumed parallelism $\mathbf{E} \parallel \mathbf{v}$, $\mathbf{h} \parallel \mathbf{l}$ one can write down

$$\mathbf{E}[[\mathbf{l}]]\, \mathbf{h}[[\mathbf{v}]] = \mathbf{E}[[\mathbf{l}]]\, \mathbf{h}[[\mathbf{v}]] - \mathbf{E}[[\mathbf{v}]]\, \mathbf{h}[[\mathbf{l}]] = (\mathbf{E} \wedge \mathbf{h})[[\mathbf{l} \wedge \mathbf{v}]]$$

with the use of the twisted two-form $\mathbf{E} \wedge \mathbf{h}$ depicted in Fig. 6.80. This implies

$$\mathbf{F}_C = -\frac{1}{2}\, \mathbf{g}\, (\mathbf{E} \wedge \mathbf{h})[[\mathbf{l} \wedge \mathbf{v}]] = \frac{1}{2}\, \mathbf{g}\, (\mathbf{E} \wedge \mathbf{h})[[\mathbf{R}]], \tag{6.49}$$

Fig. 6.80 Introducing the twisted two-form $\mathbf{E} \wedge \mathbf{h}$

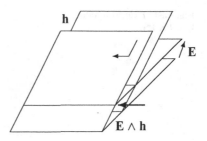

and we see that the force $\mathbf{F}_C$ is proportional to the bivector of the wall C moved during stretching of the capacitor. This force is parallel to the one-form $\mathbf{g}$ (one of the factors of $\mathbf{D}$), and hence also parallel to $\mathbf{D}$. But this is only one of the components parallel to $\mathbf{D}$. On the right-hand side of (6.49), there is a linear mapping of the twisted bivector $\mathbf{R}$ into the nontwisted one-form $\mathbf{F}_C$, which can be written as the simple tensor

$$\kappa_C' = \frac{1}{2}\, \mathbf{g} \otimes (\mathbf{E} \wedge \mathbf{h}). \tag{6.50}$$

Expression (6.50) is not yet the quantity that we are seeking, because of the assumption of constant density of charge during stretching the plates. We shall return to this point later.

Similar reasoning about stretching the capacitor parallel to $\mathbf{v}$ cannot be applied immediately, because $\mathbf{v}$ is not an nontwisted vector, and only such vectors can describe stretching. Therefore, in expression (6.46), we multiply $\mathbf{h}$ and $\mathbf{v}$ by the right-handed unit twisted scalar 1^* in order to make them nontwisted quantities:

$$W = \frac{1}{2}\, \mathbf{E}[[\mathbf{l}]]\, \mathbf{g}[[\mathbf{u}]]\, (1^*\mathbf{h})[[1^*\mathbf{v}]]. \tag{6.51}$$

The energy W is now a function of the nontwisted vector $1^*\mathbf{v} = \mathbf{r}$ determining the size of the capacitor in the direction of stretching:

$$W_D(\mathbf{r}) = \frac{1}{2}\, \mathbf{E}[[\mathbf{l}]]\, \mathbf{g}[[\mathbf{u}]]\, 1^*\mathbf{h}[[\mathbf{r}]].$$

We presently consider a force acting on a hypothetical wall D determined by the exterior product $\mathbf{l} \wedge \mathbf{u}$. Unfortunately, this product is the nontwisted bivector, which is not appropriate for the present problem. A better quantity is the twisted bivector $\mathbf{P} = \mathbf{l} \wedge 1^*\mathbf{u}$ (see Fig. 6.81). We may write down $\mathbf{g}[[\mathbf{u}]] = 1^*\mathbf{g}[[1^*\mathbf{u}]]$, hence

$$W_D(\mathbf{r}) = \frac{1}{2}\, \mathbf{E}[[\mathbf{l}]]\, 1^*\mathbf{g}[[1^*\mathbf{u}]]\, 1^*\mathbf{h}[[\mathbf{r}]] = \frac{1}{2}\, \mathbf{E}[[\mathbf{l}]]\, \mathbf{g}[[1^*\mathbf{u}]]\, \mathbf{h}[[\mathbf{r}]]. \tag{6.52}$$

Fig. 6.81 Exterior product
$P = l \wedge 1^*u$

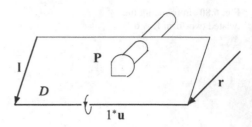

Fig. 6.82 Introducing the
nontwisted two-form $E \wedge g$

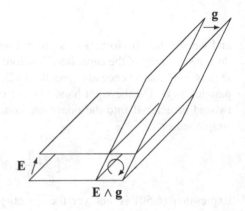

Then the force acting on the wall D described by is (with the minus) the exterior
derivative of (6.52):

$$F_D = -d \wedge W_D(\mathbf{r}) = -\frac{1}{2}\, \mathbf{h}\, E[[l]]\, g[[1^*u]].$$

After using the parallelism $E \,||\, \mathbf{u}, \ g \,||\, l$ and (2.12) we write:

$$F_D = -\frac{1}{2}\, \mathbf{h}\, (E \wedge g)[[l \wedge 1^*u]] = -\frac{1}{2}\, \mathbf{h}\, (E \wedge g)[[\mathbf{P}]]. \tag{6.53}$$

This is a linear mapping of the twisted bivector $\mathbf{P}$ into the nontwisted one-form
proportional to $\mathbf{h}$. This mapping can be written as the simple tensor:

$$\kappa'_D = -\frac{1}{2}\, \mathbf{h} \otimes (E \wedge g). \tag{6.54}$$

The two-form $E \wedge g$ is depicted in Fig. 6.82.

The two simple tensors (6.50) and (6.54) have been obtained under the assumption of constant charge density σ on the plates. This is rather non-physical
assumption, since it demands adding charges on the stretched plates. So we reject
this assumption and replace it by the assumption of constant charges on the plates.
We accordingly change the derivation.

During stretching parallel to $\mathbf{u}$, the surface density of charges should be multiplied by the factor $\|\mathbf{u}\|/\|\mathbf{r}\|$, where $\mathbf{u}$ is the initial vector, $\mathbf{r}$ is the variable vector, increasing parallelly to $\mathbf{u}$, and $\|\cdot\|$ is any measure of length. The electric induction, according to the formula $\mathbf{D}_A = \sigma$ ($\mathbf{D}_A$ denotes restriction of $\mathbf{D}$ to A), should be multiplied by the same factor. If the capacitor is filled by a linear medium, the same factor multiplies the electric field strength. Thus, starting from (6.46), we write instead of (6.47) the following

$$W_C(\mathbf{r}) = \frac{1}{2} \left(\frac{\|\mathbf{u}\|}{\|\mathbf{r}\|} \right)^2 \mathbf{E}[[\mathbf{l}]] \, \mathbf{g}[[\mathbf{r}]] \, \mathbf{h}[[\mathbf{v}]]. \tag{6.55}$$

Since the measure of length is arbitrary, we choose $\|\mathbf{r}\| = \mathbf{g}[[\mathbf{r}]]$ and there remains

$$W_C(\mathbf{r}) = \frac{1}{2} \left(\frac{\mathbf{g}[[\mathbf{u}]]}{\mathbf{g}[[\mathbf{r}]]} \right)^2 \mathbf{E}[[\mathbf{l}]] \, \mathbf{g}[[\mathbf{r}]] \, \mathbf{h}[[\mathbf{v}]] = \frac{1}{2} \frac{(\mathbf{g}[[\mathbf{u}]])^2 \mathbf{E}[[\mathbf{l}]] \, \mathbf{h}[[\mathbf{v}]]}{\mathbf{g}[[\mathbf{r}]]}. \tag{6.56}$$

The exterior derivative (with minus sign) of this function is

$$\mathbf{F}_C(\mathbf{r}) = -\mathbf{d} \wedge W_C = \frac{1}{2} \, \mathbf{g} \, \frac{(\mathbf{g}[[\mathbf{u}]])^2 \mathbf{E}[[\mathbf{l}]] \, \mathbf{h}[[\mathbf{v}]]}{(\mathbf{g}[[\mathbf{r}]])^2}.$$

Its value for the initial vector $\mathbf{r} = \mathbf{u}$ is

$$\mathbf{F}_C(\mathbf{u}) = \frac{1}{2} \, \mathbf{g} \, \mathbf{E}[[\mathbf{l}]] \, \mathbf{h}[[\mathbf{v}]]. \tag{6.57}$$

This result is similar to (6.48), differing only by the sign. After expressing this by the twisted bivector $\mathbf{R} = -\mathbf{l} \wedge \mathbf{v}$, we arrive at an expression analogous to (6.49):

$$\mathbf{F}_C = -\frac{1}{2} \, \mathbf{g} \, (\mathbf{E} \wedge \mathbf{h})[[\mathbf{R}]].$$

Thus the result (6.50) has to be replaced by the simple tensor

$$\kappa_C = -\frac{1}{2} \, \mathbf{g} \otimes (\mathbf{E} \wedge \mathbf{h}). \tag{6.58}$$

We recall that the one-forms $\mathbf{g}$ and $\mathbf{h}$ are factors of the electric induction $\mathbf{D} = \mathbf{g} \wedge \mathbf{h}$. By comparing Figs. 6.78 and 6.79 we notice that the nontwisted one-form $\mathbf{g}$ has the same direction as that of the twisted bivector $\mathbf{R}$. The minus sign means that this time the electric field repels the hypothetical wall C. In order to keep the size of plates constant in the direction of $\mathbf{v}$, one has to press on them from the outside; therefore (6.58) describes the *repulsive stress*.

Fig. 6.83 Magnetic capacitor

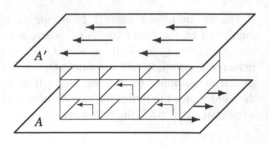

Analogously, by considering a stretching parallel to the vector **v** we would arrive to the formula

$$\kappa_D = \frac{1}{2}\,\mathbf{h} \otimes (\mathbf{E} \wedge \mathbf{g}). \tag{6.59}$$

in place of (6.54). This also describes the *repulsive stress*. That the electric field "elbows" in directions parallel to the plates, is testified by the well-known image of the field lines leaning out from the boundaries of the capacitor.

The three simple tensors (6.44), (6.58) and (6.59) can be joined into a *stress tensor of the electric field* described by field strength **E** and the electric induction **D** = **g** ∧ **h**:

$$\kappa_e = \frac{1}{2}\,\{-\mathbf{E} \otimes \mathbf{D} - \mathbf{g} \otimes (\mathbf{E} \wedge \mathbf{h}) + \mathbf{h} \otimes (\mathbf{E} \wedge \mathbf{g})\}. \tag{6.60}$$

The first term gives a non-zero value for components (of bivectors) parallel to the one-form **E** (components such as the bivector **Q** shown in Fig. 6.83), the two last terms act on components parallel to the two-form **D** (such as **P** and **R**). Due to the identity (3.22)

$$(\mathbf{E} \wedge \mathbf{D})\lfloor \mathbf{S} = -\mathbf{E}\,\mathbf{D}[[\mathbf{S}]] + \mathbf{g}\,(\mathbf{E} \wedge \mathbf{h})[[\mathbf{S}]] - \mathbf{h}\,(\mathbf{E} \wedge \mathbf{g})[[\mathbf{S}]], \tag{6.61}$$

valid for any bivector **S** and the factorization **D** = **g** ∧ **h**, we obtain for the two last terms in (6.60)

$$-\mathbf{g} \otimes (\mathbf{E} \wedge \mathbf{h}) + \mathbf{h} \otimes (\mathbf{E} \wedge \mathbf{g}) = -(\mathbf{E} \wedge \mathbf{D})\lfloor (\cdot) - \mathbf{E} \otimes \mathbf{D}.$$

Therefore, we can replace (6.60) by

$$\kappa_e = -\mathbf{E} \otimes \mathbf{D} - \frac{1}{2}(\mathbf{E} \wedge \mathbf{D})\lfloor (\cdot). \tag{6.62}$$

In the second term the energy density of the electric field $w_e = \frac{1}{2}(\mathbf{E} \wedge \mathbf{D})$ is present; hence we rewrite (6.62) as

$$\kappa_e = -\mathbf{E} \otimes \mathbf{D} - w_e\lfloor (\cdot). \tag{6.63}$$

We recall considerations in Sect. 3.2 about the contraction formula (3.38) $\mathbf{F} = -p \lfloor \mathbf{S}$ in which the twisted three-form p of pressure relates a surface area twisted bivector $\mathbf{S}$ with a one-form $\mathbf{F}$ of force exerted on it. We noticed that this relation is *isotropic*. The second term in (6.63) is also a contraction of the twisted three-form $w_e = w'_e \mathbf{f}_*^{123}$ (with the real number w'_e) by a twisted bivector $\mathbf{S}$. Therefore, we are allowed to claim that the second term in (6.63) is an isotropic part of the electric stress tensor. The first term in (6.63) is anisotropic because it yields zero for any bivector parallel to $\mathbf{D}$.

It is worth mentioning that in the traditional language, in which all directed quantities are represented as vectors, the effect of κ_e on surface area $\mathbf{S}$ can be written as

$$\mathbf{F} = \kappa_e[[\mathbf{S}]] = -\mathbf{E}\,(\mathbf{D}\cdot\mathbf{S}) + w'_e\,\mathbf{S}.$$

If one wants to use a matrix form of κ_e, acting on the column vector $\mathbf{S} = (S_1, S_2, S_3)^T$, one may write down

$$\kappa_e = - \begin{pmatrix} E_1 D^1 & E_1 D^2 & E_1 D^3 \\ E_2 D^1 & E_2 D^2 & E_2 D^3 \\ E_3 D^1 & E_3 D^2 & E_3 D^3 \end{pmatrix} + \begin{pmatrix} w'_e & 0 & 0 \\ 0 & w'_e & 0 \\ 0 & 0 & w'_e \end{pmatrix}, \tag{6.64}$$

or for its elements

$$(\kappa_e)_i{}^j = -E_i D^j + w'_e \delta_i{}^j. \tag{6.65}$$

If the electric field is only a mathematical construction, it would be difficult to imagine that it may be subject to mechanical stresses. Therefore we accept the notion, that the electric field—and generally: the electromagnetic field—is the physical system, and the found stresses are present not only in contacts of the field with other objects (such as the capacitor plates), but also inside the field itself. This is by analogy with a rubber band tensed between two hooks. The stresses are present in the whole volume of the rubber, not only in the contacts with the hooks.

6.4.2 Magnetic Stress Tensor

We now pass to the mechanical properties of the physical system "magnetic field". We shall find them on an example of a homogeneous field inside a *plane magnetic capacitor* (see Fig. 6.83), that is, a pair of two parallel plates on which electric currents flow in parallel with the same density in opposite directions. Let us consider a fragment of the plate A described by the twisted bivector $\mathbf{Q}$ with orientation to the outside of the capacitor. We factorize it as $\mathbf{Q} = \mathbf{u} \wedge \mathbf{v}$, where the nontwisted vector

Fig. 6.84 Exterior products
$Q = u \wedge v$ and $B = f \wedge k$

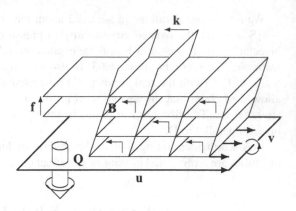

u has the direction of the surface currents on A, and the twisted vector v is parallel to B (see Fig. 6.84).

Through this fragment of the plate, the electric current flows with magnitude I_A and the moment of current $I_A u$ (see Eq. (3.39)). The plate A' acts on it by the Lorentz force given by Eq. (3.40):

$$F_A = B_{A'} \lfloor (I_A u),$$

where $B_{A'}$ is the magnetic induction generated by the plate A'. We know that $B_{A'}$ is half the size of B in the capacitor, so

$$F_A = \frac{1}{2} I_A B \lfloor u. \tag{6.66}$$

We factorize $B = f \wedge k$, where the nontwisted one-form f is parallel to $Q = u \wedge v$ (see Fig. 6.84). Then the expression (6.66) assumes the form

$$F_A = \frac{1}{2} I_A (f \wedge k) \lfloor u = \frac{1}{2} I_A (f k[[u]] - k f[[u]]),$$

and, after using the parallelism $f \parallel u$, there remains

$$F_A = \frac{1}{2} I_A f k[[u]]. \tag{6.67}$$

Comparison of the orientations of k and u allows us to notice, that $k[[u]] < 0$; hence the orientation of the force F_A is opposite to that of the one-form f. Thus we conclude that the antiparallel currents *repel* each other.

We want to represent the current I_A by the density of the surface current i: $I_A = i[[v_A]]$, where v_A is the twisted vector v *flattened* to A (see (6.18)). However, the orientation of v_A according to the prescription shown in Figs. 6.16 and 6.17 is opposite to that of i; hence we should rather write down $I_A = -i[[v_A]]$. We insert this into (6.67):

$$\mathbf{F}_A = -\frac{1}{2}\,\mathbf{f}\,\mathbf{i}[[\mathbf{v}_A]]\,\mathbf{k}[[\mathbf{u}]]. \tag{6.68}$$

We use the interface condition (6.22) which we rewrite in the shape:

$$\mathbf{H}_{1A} - \mathbf{H}_{2A} = \mathbf{i},$$

where $\mathbf{H}_{1A}$ is the restriction to A of the magnetic field $\mathbf{H}_1$ on the side of A pointed out by the arrow attached to A. Since $\mathbf{H}_1 = 0$ and $\mathbf{H}_2 = \mathbf{H}$, we obtain

$$-\mathbf{H}_A = \mathbf{i}. \tag{6.69}$$

Thus we rewrite (6.68) as

$$\mathbf{F}_A = \frac{1}{2}\,\mathbf{f}\,\mathbf{H}_A[[\mathbf{v}_A]]\,\mathbf{k}[[\mathbf{u}]].$$

Due to the identity $\mathbf{H}[[\mathbf{v}]] = \mathbf{H}_A[[\mathbf{v}_A]]$ we are allowed to write down

$$\mathbf{F}_A = \frac{1}{2}\,\mathbf{f}\,\mathbf{H}[[\mathbf{v}]]\,\mathbf{k}[[\mathbf{u}]].$$

The image of the magnetic field $\mathbf{H}$ for the magnetic capacitor is depicted in Fig. 6.85 without the plate A'. Vectors $\mathbf{u}, \mathbf{v}$ and the one-form $\mathbf{k}$ are added in Fig. 6.86. Since $\mathbf{k} \parallel \mathbf{v}$ and $\mathbf{H} \parallel \mathbf{u}$, we may add an extra term:

$$\mathbf{F}_A = \frac{1}{2}\,\mathbf{f}\,(\mathbf{H}[[\mathbf{v}]]\,\mathbf{k}[[\mathbf{u}]] - \mathbf{H}[[\mathbf{u}]]\,\mathbf{k}[[\mathbf{v}]])$$

and use the identity (2.12) $(\mathbf{H} \wedge \mathbf{k})[[\mathbf{u} \wedge \mathbf{v}]] = \mathbf{H}[[\mathbf{u}]]\mathbf{k}[[\mathbf{v}]] - \mathbf{H}[[\mathbf{v}]]\mathbf{k}[[\mathbf{u}]]$:

$$\mathbf{F}_A = -\frac{1}{2}\,\mathbf{f}\,(\mathbf{H} \wedge \mathbf{k})[[\mathbf{u} \wedge \mathbf{v}]] \tag{6.70}$$

The orientation of the exterior product $\mathbf{H} \wedge \mathbf{k}$ is established as in Fig. 1.52. It is visible in Fig. 6.86 that the twisted two-form $\mathbf{H} \wedge \mathbf{k}$ has the same orientation as

Fig. 6.85 Field $\mathbf{H}$ inside the magnetic capacitor

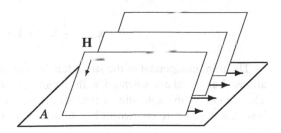

Fig. 6.86 Introducing the
nontwisted two-form $\mathbf{H} \wedge \mathbf{k}$

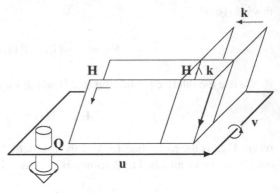

Fig. 6.87 Magnetic field $\mathbf{H}$
in an angular solenoid

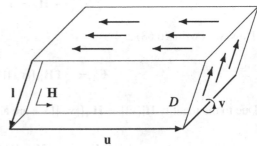

the twisted bivector $\mathbf{Q} = \mathbf{u} \wedge \mathbf{v}$; hence $(\mathbf{H} \wedge \mathbf{k})[[\mathbf{u} \wedge \mathbf{v}]] > 0$, and we once more ascertain that the one-forms $\mathbf{F}_A$ and $\mathbf{f}$ have opposite orientations.

Expression (6.70) is a linear mapping of the twisted bivector $\mathbf{Q} = \mathbf{u} \wedge \mathbf{v}$, parallel to $\mathbf{f}$, into the nontwisted one-form $\mathbf{F}_A$ parallel to A; hence we have a component of the *stress tensor of the magnetic field*:

$$\kappa_A = -\frac{1}{2}\,\mathbf{f} \otimes (\mathbf{H} \wedge \mathbf{k}), \qquad (6.71)$$

acting on the plane A, parallel to the two-form $\mathbf{B} = \mathbf{f} \wedge \mathbf{k}$. In order to keep the plate A at a constant distance from A' one has to press it from the outside; therefore (6.71) describes the *repulsive stress*.

We omit the similar reasoning leading to the second component of the stress tensor of the magnetic field acting on a face C (its attitude is visible in Fig. 6.79) parallel to $\mathbf{k}$ which, thereby, is—like A—also parallel to $\mathbf{B}$:

$$\kappa_C = \frac{1}{2}\,\mathbf{k} \otimes (\mathbf{H} \wedge \mathbf{f}). \qquad (6.72)$$

The third component of the stress tensor will be found from the energy considerations. Let us take a solenoid with a parallelogram cross-section (see Fig. 6.87). The mouth of the solenoid is the parallelogram D described by the nontwisted bivector $\tilde{\mathbf{P}} = \mathbf{l} \wedge \mathbf{u}$, (shown in Fig. 6.88) where the nontwisted vector $\mathbf{l}$ is parallel

Fig. 6.88 Nontwisted
bivector $\tilde{\mathbf{P}} = \mathbf{l} \wedge \mathbf{u}$

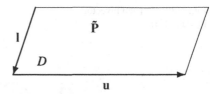

simultaneously to $\mathbf{H}$ and $\mathbf{k}$. The volume of the solenoid is expressed by the twisted trivector $V = \mathbf{v} \wedge \tilde{\mathbf{P}}$. The energy of the magnetic field stored in the solenoid is

$$W = \frac{1}{2}(\mathbf{H} \wedge \mathbf{B})[[V]] = \frac{1}{2}(\mathbf{H} \wedge \mathbf{B})[[\mathbf{v} \wedge \tilde{\mathbf{P}}]].$$

We use the identity (3.35):

$$(\mathbf{H} \wedge \mathbf{B})[[\mathbf{v} \wedge \tilde{\mathbf{P}}]] = \mathbf{H}[[\mathbf{v}]]\,\mathbf{B}[[\tilde{\mathbf{P}}]] + \mathbf{B}[[(\mathbf{H}\lrcorner\tilde{\mathbf{P}}) \wedge \mathbf{v}]].$$

Due to $\mathbf{H} \parallel \tilde{\mathbf{P}}$ the second term on the right-hand side is zero; hence

$$W = \frac{1}{2}\mathbf{H}[[\mathbf{v}]]\,B[[\tilde{\mathbf{P}}]]. \tag{6.73}$$

Both scalar factors $\mathbf{H}[[\mathbf{v}]]$ and $\mathbf{B}[[\tilde{\mathbf{P}}]]$ are positive; hence W is positive.

Let us ponder, what happens during stretching the solenoid in such a manner that $\mathbf{v}$ increases. First of all it has to be changed into the nontwisted vector $\mathbf{r} = 1^*\mathbf{v}$. It follows from the orientation of $\mathbf{v}$ that $\mathbf{r}$ has the appropriate orientation from the inside of the solenoid to the parallelogram D, which is to be moved away. Secondly, the nontwisted bivector $\tilde{\mathbf{P}}$ should be changed into the twisted one $1^*\tilde{\mathbf{P}}$. Fortunately, $1^*\tilde{\mathbf{P}}$ is oriented to outside the solenoid, which is the same as in all considerations up to now. Thus the twisted bivector $\mathbf{P} = 1^*\tilde{\mathbf{P}}$ must be introduced (see Fig. 6.89). We substitute $\mathbf{v} = 1^*\mathbf{r}$ and $\tilde{\mathbf{P}} = 1^*\mathbf{P}$ in (6.73):

$$W = \frac{1}{2}\mathbf{H}[[1^*\mathbf{r}]]\,\mathbf{B}[[1^*\mathbf{P}]] = \frac{1}{2}\mathbf{H}[[\mathbf{r}]]\,\mathbf{B}[[\mathbf{P}]]. \tag{6.74}$$

During stretching parallel to $\mathbf{r}$ (i.e., to $\mathbf{v}$) with the conservation of the current flowing on the solenoid, the density of the surface current is to be multiplied by the factor $\|\mathbf{r}_0\|/\|\mathbf{r}\|$, where $\mathbf{r}_0$ is the initial vector, and $\|\cdot\|$ is any measure of length. If we assume the value of the magnetic field one-form as such a measure, then the factor can be written as $\mathbf{H}_0[[\mathbf{r}_0]]/\mathbf{H}_0[[[\mathbf{r}]]$, where $\mathbf{H}_0$ is the magnetic field in the solenoid for $\mathbf{r} = \mathbf{r}_0$. The magnetic field strength must be multiplied by the same factor by virtue of the formula (6.69) $\mathbf{H}_A = -\mathbf{i}$, and if the magnetic medium filling the solenoid is linear, the magnetic induction must be multiplied too. Therefore instead of (6.74) one should introduce the function

Fig. 6.89 Twisted bivector
$\mathbf{P} = \mathbf{l} \wedge \mathbf{1}^*\mathbf{u}$

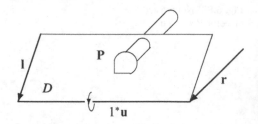

$$W_D(\mathbf{r}) = \frac{1}{2}\left(\frac{\mathbf{H}_0[[\mathbf{r}_0]]}{\mathbf{H}_0[[\mathbf{r}]]}\right)^2 \mathbf{H}_0[[\mathbf{r}]] \, \mathbf{B}_0[[\mathbf{P}]] = \frac{1}{2}\frac{(\mathbf{H}_0[[\mathbf{r}_0]])^2 \, \mathbf{B}_0[[\mathbf{P}]]}{\mathbf{H}_0[[\mathbf{r}]]}. \tag{6.75}$$

The exterior derivative (with the minus sign) of this function is

$$\mathbf{F}_D(\mathbf{r}) = -\mathbf{d} \wedge W_D = -\frac{1}{2}\,\mathbf{H}_0\,\frac{(\mathbf{H}_0[[\mathbf{r}_0]])^2 \, \mathbf{B}_0[[\mathbf{P}]]}{(\mathbf{H}_0[[\mathbf{r}]])^2},$$

and its value for the initial vector $\mathbf{r} = \mathbf{r}_0$ is

$$\mathbf{F}_D(\mathbf{r}_0) = -\frac{1}{2}\,\mathbf{H}_0\,\mathbf{B}_0[[\mathbf{P}]].$$

We now omit the index 0:

$$\mathbf{F}_D = -\frac{1}{2}\,\mathbf{H}\,\mathbf{B}[[\mathbf{P}]]. \tag{6.76}$$

This is a linear mapping of the twisted bivector $\mathbf{P}$ into a nontwisted one-form; hence it can be represented as the simple tensor

$$\kappa_D = -\frac{1}{2}\,\mathbf{H} \otimes \mathbf{B} \tag{6.77}$$

of the *magnetic field stress parallel to* $\mathbf{H}$. The expression $\mathbf{B}[[\mathbf{P}]]$ is a twisted scalar with the right orientation, i.e., 1^*, hence $-\mathbf{H}$ in (6.76) is multiplied by the twisted right scalar 1^*, which yields a nontwisted one-form oriented to the inside of the solenoid. In order to keep the size of the solenoid constant in the direction of $\mathbf{r}$, one must pull them from the outside; therefore (6.77) describes the *attractive stress*.

The three simple tensors (6.71), (6.72) and (6.77) can be united in a single *stress tensor of the magnetic field* expressed by the field strength $\mathbf{H}$ and induction $\mathbf{B} = \mathbf{f} \wedge \mathbf{k}$:

$$\kappa_m = \frac{1}{2}\{-\mathbf{H} \otimes \mathbf{B} - \mathbf{f} \otimes (\mathbf{H} \wedge \mathbf{k}) + \mathbf{k} \otimes (\mathbf{H} \wedge \mathbf{f})\}. \tag{6.78}$$

The first term gives the non-zero contribution for components (of bivectors) parallel to the one-form **H**, and the two last terms act nontrivially on components parallel to the two-form **B**. Due to the identity analogous to (6.61)

$$(\mathbf{B} \wedge \mathbf{H}) \lfloor \mathbf{S} = -\mathbf{H}\,\mathbf{B}[[\mathbf{S}]] + \mathbf{f}\,(\mathbf{H} \wedge \mathbf{k})[[\mathbf{S}]] - \mathbf{k}\,(\mathbf{H} \wedge \mathbf{f})[[\mathbf{S}]], \qquad (6.79)$$

valid for any bivector **S** and $\mathbf{B} = \mathbf{f} \wedge \mathbf{k}$, we can replace two last terms in (6.78) by $-(\mathbf{B} \wedge \mathbf{H})\lfloor (\cdot) - \mathbf{H} \otimes \mathbf{B}$ which yields

$$\kappa_m = -\mathbf{H} \otimes \mathbf{B} - \frac{1}{2}\,(\mathbf{B} \wedge \mathbf{H})\lfloor (\cdot). \qquad (6.80)$$

If both fields, electric and magnetic, are present, we should use a *stress tensor of the electromagnetic field* that unites two expressions (6.62) and (6.80):

$$\kappa = -\mathbf{E} \otimes \mathbf{D} - \mathbf{H} \otimes \mathbf{B} - \frac{1}{2}\,(\mathbf{E} \wedge \mathbf{D} + \mathbf{B} \wedge \mathbf{H})\lfloor (\cdot). \qquad (6.81)$$

We recognize in the second term the energy density (4.13) of the electromagnetic field, so we write down

$$\kappa = -\mathbf{E} \otimes \mathbf{D} - \mathbf{H} \otimes \mathbf{B} - w\lfloor (\cdot). \qquad (6.82)$$

6.5   Reflection and Refraction of Plane Waves

Let a plane electromagnetic wave impede on plane Σ separating two different anisotropic dielectrics. Assume that the wave comes from the side of medium 1 and that the *incident wave* is the eigenwave:

$$\mathbf{E}(\mathbf{r}, t) = \psi(\mathbf{k}[[\mathbf{r}]] - \omega t)\mathbf{E}_0,$$

$$\mathbf{B}(\mathbf{r}, t) = \psi(\mathbf{k}[[\mathbf{r}]] - \omega t)\mathbf{u} \wedge \mathbf{E}_0,$$

$$\mathbf{H}(\mathbf{r}, t) = \psi(\mathbf{k}[[\mathbf{r}]] - \omega t)\mathbf{H}_0,$$

$$\mathbf{D}(\mathbf{r}, t) = -\psi(\mathbf{k}[[\mathbf{r}]] - \omega t)\mathbf{u} \wedge \mathbf{H}_0,$$

where $\psi(\cdot)$ is a scalar function of a scalar argument, ω plays the role of frequency, $\mathbf{k}[[\mathbf{r}]]$ is the value of the linear form **k** of wave covector on the radius vector **r**, and $\mathbf{u} = \omega^{-1}\mathbf{k}$ is the phase slowness. We expect that a *reflected wave* and a *wave penetrating* the other medium will arise. The latter is also called a *refracted wave*. If they are eigenwaves, they can be written down in similar formulas. Both new waves oscillate in the same rhythm as the incident wave, so we expect the same scalar function ψ and the same frequency ω in their time and space dependence. They can, however, differ by wave covectors and constants standing after ψ:

reflected wave

$$\mathbf{E}^{(r)}(\mathbf{r}, t) = \psi(\mathbf{k}^{(r)}[[\mathbf{r}]] - \omega t)\mathbf{E}_0^{(r)},$$

$$\mathbf{B}^{(r)}(\mathbf{r}, t) = \psi(\mathbf{k}^{(r)}[[\mathbf{r}]] - \omega t)\mathbf{u}^{(r)} \wedge \mathbf{E}_0^{(r)},$$

$$\mathbf{H}^{(r)}(\mathbf{r}, t) = \psi(\mathbf{k}^{(r)}[[\mathbf{r}]] - \omega t)\mathbf{H}_0^{(r)},$$

$$\mathbf{D}^{(r)}(\mathbf{r}, t) = -\psi(\mathbf{k}^{(r)}[[\mathbf{r}]] - \omega t)\mathbf{u}^{(r)} \wedge \mathbf{H}_0^{(r)},$$

penetrating wave

$$\mathbf{E}^{(p)}(\mathbf{r}, t) = \psi(\mathbf{k}^{(p)}[[\mathbf{r}]] - \omega t)\mathbf{E}_0^{(p)},$$

$$\mathbf{B}^{(p)}(\mathbf{r}, t) = \psi(\mathbf{k}^{(p)}[[\mathbf{r}]] - \omega t)\mathbf{u}^{(p)} \wedge \mathbf{E}_0^{(p)},$$

$$\mathbf{H}^{(p)}(\mathbf{r}, t) = \psi(\mathbf{k}^{(p)}[[\mathbf{r}]] - \omega t)\mathbf{H}_0^{(p)}.$$

$$\mathbf{D}^{(p)}(\mathbf{r}, t) = -\psi(\mathbf{k}^{(p)}[[\mathbf{r}]] - \omega t)\mathbf{u}^{(p)} \wedge \mathbf{H}_0^{(p)},$$

If they are not eigenwaves, they would be superpositions of two such waves:

$$\mathbf{E}^{(r)}(\mathbf{r}, t) = \psi(\mathbf{k}_{(1)}^{(r)}[[\mathbf{r}]] - \omega t)\mathbf{E}_{01}^{(r)} + \psi(\mathbf{k}_{(2)}^{(r)}[[\mathbf{r}]] - \omega t)\mathbf{E}_{02}^{(r)}$$

and

$$\mathbf{E}^{(p)}(\mathbf{r}, t) = \psi(\mathbf{k}_{(1)}^{(p)}[[\mathbf{r}]] - \omega t)\mathbf{E}_{01}^{(p)} + \psi(\mathbf{k}_{(2)}^{(p)}[\mathbf{r}] - \omega t)\mathbf{E}_{02}^{(p)}.$$

Similar formulas should then be written for the other fields $\mathbf{B}$, $\mathbf{D}$ and $\mathbf{H}$.

Return to the simpler case when the reflected and penetrating waves are eigenwaves. We assume that the wave covectors of all three waves restricted to Σ are equal, since the spatial changes of the fields of the two new waves are forced by changes of the incident wave. Hence we write the condition

$$\mathbf{k}_\Sigma = \mathbf{k}_\Sigma^{(r)} = \mathbf{k}_\Sigma^{(p)}. \tag{6.83}$$

Fig. 6.90 Wave covectors for
the incident, reflected and
refracted waves

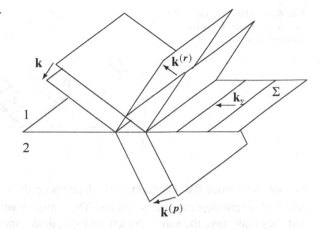

This implies that the restrictions of all **E**-fields to Σ have the same scalar factor:

$$\mathbf{E}(\mathbf{r}, t)|_{\mathbf{r} \in \Sigma} = \psi(\mathbf{k}_\Sigma[[\mathbf{r}]] - \omega t)\mathbf{E}_0 \,,$$

$$\mathbf{E}^{(r)}(\mathbf{r}, t)|_{\mathbf{r} \in \Sigma} = \psi(\mathbf{k}_\Sigma[[\mathbf{r}]] - \omega t)\mathbf{E}_0^{(r)},$$

$$\mathbf{E}^{(p)}(\mathbf{r}, t)|_{\mathbf{r} \in \Sigma} = \psi(\mathbf{k}_\Sigma[[\mathbf{r}]] - \omega t)\mathbf{E}_0^{(p)}$$

and similarly for **D**, **H** and **B**. Condition (6.83) is illustrated in Fig. 6.90. The angle between the planes of **k** and that of Σ is equal to the incidence angle, and similarly the angle between $\mathbf{k}^{(r)}$ and Σ is the refraction angle, whichever metric is used to measure angles.

A question arises: what is the attitude of the wave covector for the reflected wave?[2] There is no easy answer, be cause we know that two reflected waves can appear as two eigenwaves. Each of them has its own phase slowness $\mathbf{u}_{(i)}$ determined by the properties of the medium and its own wave covector given by $\mathbf{k}_{(i)}^{(r)} = \omega\mathbf{u}_{(i)}$. The index $i \in \{1, 2\}$ enumerates the eigenwaves. In this case the reflection law is

---

[2] In the traditional vector language condition (6.83) means equality of the components parallel to Σ:

$$\mathbf{k}_{\|} = \mathbf{k}_{\|}^{(r)} = \mathbf{k}_{\|}^{(p)}.$$

The first of them assumes the following form in the incidence plane:

$$k \sin\theta = k^{(r)} \sin\theta^{(r)},$$

where θ and $\theta^{(r)}$ are the *incidence* and *reflection angles* respectively. In the isotropic medium there is $k^{(r)} = k$; hence we obtain $\sin\theta^{(r)} = \sin\theta$ and $\theta^{(r)} = \theta$, which is the *reflection law*: the reflection angle is equal to the incidence angle. In the anisotropic medium generally $k^{(r)} \neq k$, and thus the reflection law may be not satisfied.

Fig. 6.91 The assumed two-forms **D** and **B** for the incident wave

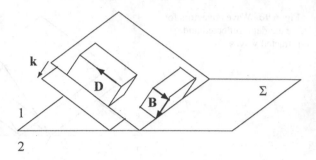

not satisfied, since the reflection angle depends on the magnitude of the slowness, and this in turn depends on its attitude. The wave covectors $\mathbf{k}_{(i)}^{(r)}$ may be different, but they must have the same restrictions to Σ, determined by the wave covector of the incident wave:

$$\mathbf{k}_{1\Sigma}^{(r)} = \mathbf{k}_{2\Sigma}^{(r)} = \mathbf{k}_{\Sigma}.$$

If we cannot find $\mathbf{k}_{(i)}^{(r)}$, let us, at least, ponder what can be said about the attitudes of the electromagnetic quantities. Consider a particular case. Assume that the incident wave has its magnetic induction **B** parallel to the intersection edge of the plane of **k** with Σ, i.e., to $\mathbf{k}_{\Sigma}$. This corresponds to the *transverse magnetic wave*, since then, for the isotropic medium, the magnetic induction in all three waves (as we shall see in a moment) is perpendicular to the incidence plane. For such an eigenwave the electric induction **D** is perpendicular to **B**, but still parallel to **k** (see Fig. 6.91).

Two-form **B** is parallel to Σ, hence we expect that the magnetic induction $\mathbf{B}^{(r)}$ of the reflected wave has the same attitude. As concerns the orientation, two possibilities are admissible depending on the properties of the media. (I have chosen compatible orientations of $\mathbf{B}^{(r)}$ and **B** in the next figures.) The sum $\mathbf{B} + \mathbf{B}^{(r)}$ still has the same attitude, so the net induction in the first medium close to the interface has zero restriction to Σ:

$$(\mathbf{B} + \mathbf{B}^{(r)})_{\Sigma} = 0.$$

In this situation the continuity condition (6.27) for the magnetic induction gives only the requirement in the second medium, that **B** must be parallel to Σ.

The electric induction $\mathbf{D}^{(r)}$ of the reflected wave is perpendicular to the magnetic induction in both metrics and simultaneously parallel to $\mathbf{k}^{(r)}$; hence we already know its attitude. The orientation, however must fit to relation (5.114), that is $\mathbf{D} = -\mathbf{u} \wedge \mathbf{H}$. We know now the whole direction of $\mathbf{D}^{(r)}$, shown in Fig. 6.92.

We know that the magnetic induction in the second medium must be parallel to Σ. It must, moreover, be parallel to the planes of $\mathbf{k}^{(p)}$. The combination of these two conditions gives uniquely the attitude of $\mathbf{B}^{(p)}$—parallel to **B** and $\mathbf{B}^{(r)}$.

Fig. 6.92 The two-forms **D**
and **B** for the reflected wave

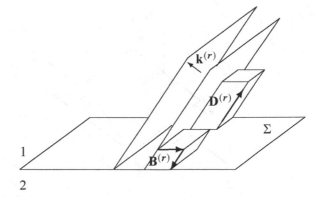

The orientation is determined by the sum $\mathbf{B} + \mathbf{B}^{(r)}$, in which the first term is greater; hence the whole direction of $\mathbf{B}^{(p)}$ is the same as that of $\mathbf{B}$. But all of this corresponds to a wave running in the second medium only when this direction fits to eigenwave. If this is not so, the magnetic induction obtained in this way is only the sum of the magnetic inductions of two eigenwaves:

$$\mathbf{B}^{(p)} = \mathbf{B}^{(p)}_{(1)} + \mathbf{B}^{(p)}_{(2)}.$$

Either of the fields $\mathbf{B}^{(p)}_{(i)}$ are parallel to $\mathbf{k}^{(p)}$, but not necessarily to Σ.

The magnetic field strengths $\mathbf{H}$ and $\mathbf{H}^{(r)}$ are perpendicular simultaneously to $\mathbf{B}$ and $\mathbf{B}^{(r)}$ in the metric of the first medium. Their sum $\mathbf{H} + \mathbf{H}^{(r)}$ as the net field strength in the first medium is perpendicular to $\mathbf{B} + \mathbf{B}^{(r)}$. The continuity condition (6.23) does not yet determine yet the direction of $\mathbf{H}^{(p)}$ uniquely. Only the perpendicularity condition of $\mathbf{B}^{(p)}$ and $\mathbf{H}^{(p)}$ in the metric of the second medium does this. Anyway, the magnetic metrics of the two media are the same as that of the vacuum, since such was the assumption at the beginning of the present section. The observations about the magnetic field strengths are illustrated in Fig. 6.93 when the wave in the second medium is the eigenwave.

Let us consider the second particular case. Assume that the initial eigenwave impinges on Σ in such a way that its electric induction $\mathbf{D}$ is parallel to the wedge of intersection of the planes of $\mathbf{k}$ with Σ, that is, to $\mathbf{k}_\Sigma$. This is called a *transverse electric wave*. For such an eigenwave $\mathbf{B}$ is perpendicular to $\mathbf{D}$, and simultaneously parallel to $\mathbf{k}$ (see Fig. 6.94).

The two-form $\mathbf{D}$ is parallel to Σ; hence we expect that $\mathbf{D}^{(r)}$ has the same attitude. For the orientation two possibilities are admissible, depending on the properties of the two media. (I have chosen the orientation of $\mathbf{D}^{(r)}$ compatible with that of $\mathbf{D}$ in the next figures.) The sum $\mathbf{D} + \mathbf{D}^{(r)}$ still has the same attitude; hence the net induction in the first medium close to the interface has zero restriction to Σ:

$$(\mathbf{D} + \mathbf{D}^{(r)})_\Sigma = 0.$$

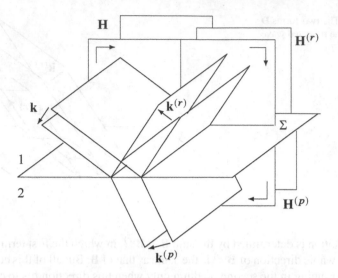

Fig. 6.93 Magnetic field one-forms for the three waves

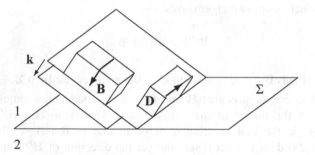

Fig. 6.94 The assumed two-forms **D** and **B** for the second incident wave

In this situation the continuity condition (6.26) imposes on the electric induction in the second medium only the condition that it must be parallel to Σ.

The magnetic induction $\mathbf{B}^{(r)}$ of the reflected wave is perpendicular to the electric induction in both metrics and parallel to $\mathbf{k}^{(r)}$; hence we know its attitude. But the orientation must fit the relation (5.16), that is $\mathbf{B} = \mathbf{u} \wedge \mathbf{E}$, and thus we know the whole direction of $\mathbf{B}^{(r)}$. We show this in Fig. 6.95.

We know already that the electric induction in the second medium must be parallel to Σ. It must also be parallel to the plane of $\mathbf{k}^{(p)}$. The composition of these two conditions determines uniquely the attitude of $\mathbf{D}^{(p)}$—parallel to **D** and $\mathbf{D}^{(r)}$. The orientation is determined by the sum $\mathbf{D} + \mathbf{D}^{(r)}$, in which the first term is greater; hence the whole direction of $\mathbf{D}^{(p)}$ is the same as that of **D**. All this, however, corresponds to the second running wave only when this direction corresponds to the eigenwave. If this is not so, the magnetic induction obtained is only the sum of two

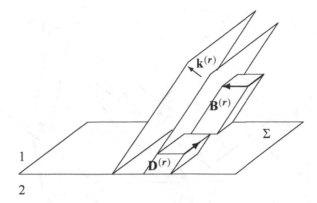

Fig. 6.95 The two-forms **D** and **B** for the second reflected wave

inductions corresponding to the two eigenwaves:

$$\mathbf{D}^{(p)} = \mathbf{D}^{(p)}_{(1)} + \mathbf{D}^{(p)}_{(2)},$$

where both fields $\mathbf{D}^{(p)}_{(i)}$ are parallel to $\mathbf{k}^{(p)}$, but not necessarily to Σ.

The electric field stengths **E** and $\mathbf{E}^{(r)}$ are perpendicular simultaneously to **D** and $\mathbf{D}^{(r)}$ in the first medium metric. Their sum $\mathbf{E} + \mathbf{E}^{(r)}$ as the net strength in the first medium is perpendicular to $\mathbf{D} + \mathbf{D}^{(r)}$. The continuity condition (6.21) does not yet determine the direction of $\mathbf{E}^{(p)}$ uniquely. Only the perpendicularity of $\mathbf{D}^{(p)}$ and $\mathbf{E}^{(p)}$ in the second medium metric does this. The observations about the electric field strengths are illustrated on Fig. 6.96 if the wave in the second medium is the eigenwave.

The next special case is interesting when the incident wave has its wave covector parallel to interface Σ. In the traditional vectorial language we say that the wave falls perpendicular to the interface. Then the restriction $\mathbf{k}_\Sigma$ is zero and by virtue of (6.83) the same occurs for the reflected and penetrating waves. Thus the wave covectors $\mathbf{k}^{(r)}$ and $\mathbf{k}^{(p)}$ are also parallel to Σ. This is depicted in Fig. 6.97.

In this situation the reflected wave has the same attitude of its wave covector as for the incident wave; hence both can be eigenwaves with parallel forms **B** and **D**, and hence both have the same value of slowness and thus by virtue of $\mathbf{k} = \omega\mathbf{u}$, $\mathbf{k}^{(r)} = -\omega\mathbf{u}$ they have the same wave covector. This is depicted in Fig. 6.97 by the same separations between the planes of **k** and $\mathbf{k}^{(r)}$. The penetrating wave generally decomposes into two eigenwaves with different slownesses $\mathbf{u}_{(1)}$ and $\mathbf{u}_{(2)}$; hence they may have different wave covectors $\mathbf{k}^{(p)}_{(i)} = \omega\mathbf{u}_{(i)}$. I have tried to show this fact in Fig. 6.96, because the distances between the planes of $\mathbf{k}^{(p)}_{(1)}$ and $\mathbf{k}^{(p)}_{(2)}$ are distinct.

In Fig. 6.98, the electric and magnetic inductions of the incident and reflected waves are shown. I have chosen both magnetic inductions with the same direction,

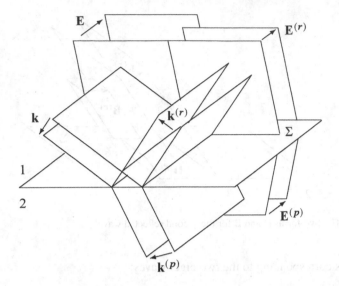

Fig. 6.96 Electric field one-forms for the three waves

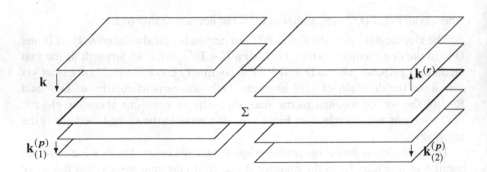

Fig. 6.97 Wave covectors in the particular case

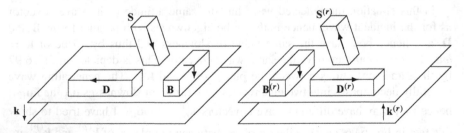

Fig. 6.98 Electric and magnetic inductions, and Poynting forms of the incident and reflected waves

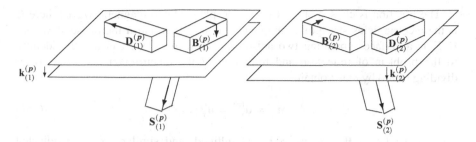

Fig. 6.99 Electric and magnetic inductions, and Poynting forms of the penetrating eigenwaves

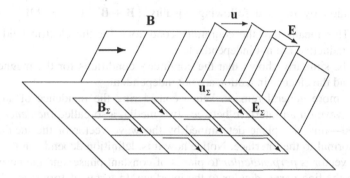

Fig. 6.100 Image of the relation $\mathbf{B} = \mathbf{u} \wedge \mathbf{E}$ and its restriction to Σ

which implies the same directions of the magnetic field strengths. Then the electric inductions must have opposite directions, in order to satisfy the relations $\mathbf{D} = \mathbf{u} \wedge \mathbf{H}$ and $\mathbf{D}^{(r)} = -\mathbf{u} \wedge \mathbf{H}^{(r)}$. There is a minus sign in the second formula because the slowness of the reflected wave is opposite to that of the incident wave. The Poynting forms $\mathbf{S} = \mathbf{E} \wedge \mathbf{H}$ for the incident wave and $\mathbf{S}^{(r)} = \mathbf{E}^{(r)} \wedge \mathbf{H}^{(r)}$ for the reflected wave are also marked—they have the same attitudes, but opposite orientations, since the directions of $\mathbf{H}$ and $\mathbf{H}^{(r)}$ are the same, and those of $\mathbf{E}$ and $\mathbf{E}^{(r)}$ are opposite.

For the penetrating eigenwaves the magnetic inductions $\mathbf{B}_{(i)}^{(p)}$ do not need to be parallel to the magnetic inductions $\mathbf{B}$ of the incident wave; only their sum $\mathbf{B}_{(1)}^{(p)} + \mathbf{B}_{(2)}^{(p)}$ must satisfy this condition, and similarly with the electric induction (see Fig. 6.99). Distinct Poynting forms correspond to these two penetrating waves—this means that the energy is transported in different directions. This is observed as the presence of two rays propagating differently and is called *birefringence*.

Let us take up restrictions of the two-forms to the plane. We show in Fig. 6.100 how the two-form $\mathbf{B}$ is obtained in the external product $\mathbf{B} = \mathbf{u} \wedge \mathbf{E}$. It is clear from this that the following identity holds: $\mathbf{B}_\Sigma = \mathbf{u}_\Sigma \wedge \mathbf{E}_\Sigma$, which can be expressed in words: restriction of a product is a product of restrictions. A similar identity is fulfilled for an improper two-form: then one of the factors is an improper one-form.

The notation is chosen specially to fit to plane electromagnetic waves: here $\mathbf{E}$ is the electric fields strength, $\mathbf{B}$ the magnetic induction and $\mathbf{u}$ the phase slowness. If Σ is a surface separating two anisotropic dielectrics, we can use the identity to the problem of reflection and refraction of the electromagnetic wave. After dividing (6.83) by ω we obtain

$$\mathbf{u}_\Sigma = \mathbf{u}_\Sigma^{(r)} = \mathbf{u}_\Sigma^{(p)}. \tag{6.84}$$

Since the relation $\mathbf{B}_\Sigma = \mathbf{u}_\Sigma \wedge \dot{\mathbf{E}}_\Sigma$ is fulfilled, and similarly for the reflected and penetrating waves, from the equality $\left(\mathbf{E} + \mathbf{E}^{(r)}\right)_\Sigma = \left(\mathbf{E}_{(1)}^{(p)} + \mathbf{E}_{(2)}^{(p)}\right)_\Sigma$ after multiplication by $\mathbf{u}_\Sigma$, the following equality $\left(\mathbf{B} + \mathbf{B}^{(r)}\right)_\Sigma = \left(\mathbf{B}_{(1)}^{(p)} + \mathbf{B}_{(2)}^{(p)}\right)_\Sigma$ follows. This means that the continuity conditions for the electric field and the magnetic induction are not independent.

It can be similarly shown that the continuity conditions for the magnetic field strength and the electric induction are not independent.

Now a more general remark. When considering the incidence of an electromagnetic wave on an interface between two media, a so-called *incidence plane* is introduced—this is a plane determined by the wave vector of the incident wave and the normal to the interface. Notice how this definition depends on the metrics: the wave vector is *perpendicular* to planes of constant phase, and the normal is by definition the line *perpendicular* to the interface. In terms of forms, another thing becomes important, namely a line of intersection of some plane of constant phase with the interface—it may be called an *equal-phase line*. It is a notion independent of metric, because both considered planes are independent of the scalar product.

Chapter 7
Electromagnetism in Space-Time

This chapter is devoted to considerations in four-dimensional space-time. It starts in Sect. 7.1 with a description of directed quantities in a four-dimensional vector space. The transition from three-dimensional space to a plane described in a previous chapter is helpful to perform the passage in opposite direction, namely from three to four dimensions, in Sect. 7.2. A distinguished vector protruding in the fourth dimension is necessary for this—its direction determines time axis. In space-time, two field quantities, namely a *Faraday two-form* **F** joining the **B** and **E** fields, and a *Gauss two-from* uniting the **D** and **H** fields, are sufficient to write down two of Maxwell's equations with the use of exterior derivatives of differential forms. They are independent of the metric of space-time.

Momentum, energy and the energy-stress tensor are considered in Sect. 7.3 as a mapping of a volume trivector into a momentum one-form. It has four components, depending on direction of the three-volume in four-dimensional space-time. Solving Maxwell's equations in the form of plane waves is quite quick in a four-dimensional formulation, which is shown in Sect. 7.4. The distinguished role of electromagnetic waves is essential for introducing, in Sect. 7.5, the scalar product in space-time, which changes it into Minkowski space. A reason is found to favour one of two, seemingly equivalent, scalar products. This section contains elements of particle mechanics. As shown in Sect. 7.6, the metric allows us to find the constitutive relation between **F** and **G**, containing the Hodge map. Owing to this, the two Maxwell equations can be expressed by the coordinates of the Faraday field only.

7.1 Directed Quantities in Four Dimensions

An impediment exists in illustrating considerations in four dimensions by figures for an obvious reason: it is difficult to imagine the fourth dimension, and impossible for us to see it. One can reach the concepts needed only by means of reasoning,

© Springer Nature Switzerland AG 2021
B. Jancewicz, *Directed Quantities in Electrodynamics*,
https://doi.org/10.1007/978-3-030-90471-5_7

Fig. 7.1 Two straight lines
nonparallel to a plane R. Line
B is not lying in our
three-dimensional space

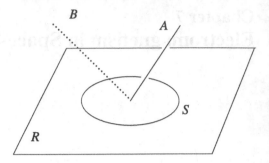

mostly by analogy and sometimes independently. Only by argumentation can we, for instance, accept that among all the planes passing through the origin of coordinates two planes exist with only one common point—just the origin. All linear combinations of vectors from these two planes span the whole four-dimensional space. If two planes are chosen (both passing through the origin) with a common straight line, the linear combinations of vectors from them would have spanned only a three-dimensional subspace.

The next problem also belongs to the mental gymnastics over four dimensions. In three dimensions we are capable of imagining a ring S threaded on a straight line A—this line cannot lie in the plane R of the ring: it only passes through its center O. In four dimensions, however, we may admit the existence of another straight line B passing through the centre O of the same ring, but without any other common point with the three-dimensional space spanned by R and A (see Fig. 7.1). (Line B "is not from this world", so it is drawn dotted.)

Ring S is threaded also on B. The same can be said about an arbitrary straight line from plane P spanned by A and B. Therefore, we have to state that S is threaded on the whole plane P. The word *threaded* means here that S may be translated parallel to P and will always be at the same distance from it, and never intersects it. The two planes P and R have only one common point O and span the whole four-dimensional space. When we become reconciled with this, it is worthwhile to say that on the plane P a ring can also be considered with its centre in O. This ring, in turn, is threaded on R. In this manner we have two rings threaded on each other and having a common centre. Instead of rings one may consider parallelograms and repeat all these statements about their threading. The rings and parallelograms can be endowed with arrows to make them directed. The parallelograms are better in this respect in that they are determined by pairs of vectors, the order of which is important.

We discuss now nontwisted multivectors in four-dimensional space (without repeating the adjective "nontwisted"). For multivectors of grades from zero to three there are no problems—their definitions are the same, only there is more room for their attitudes. Therefore, the set of vectors is four-dimensional, the set of bivectors is six-dimensional (because equal to the number of combinations of two from the four basic vectors), and the set of trivectors is four-dimensional (the number of combinations of three from the four).

There is a problem with bivectors, namely the definition of their sum. Not each pair of bivectors has factorizations onto exterior products with a common factor. This is always possible in three-dimensional space because two nonparallel planes have a common edge. This is no longer true in four dimensions. If the planes representing the attitudes of two bivectors have only one point in common, merely the pair of bivectors written with a plus sign is considered to be their *sum*, because nothing more can be done with them. Such a sum does not have an attitude as a single plane. As things are, bivectors with attitudes composed of one plane are called *simple bivectors* or *volutors* (from Latin *voluto* = roll, rotate). Volutors can be represented as the exterior product of vectors. A natural question arises: whether a sum of two non-simple bivectors, i.e., when each of them is a pair of volutors, must be represented as a quadruple of volutors, or if a smaller number is sufficient? The answer is in the following.

Lemma 7.1 *Each bivector can be represented as a sum of two volutors.*

Proof Let $\{e_1, e_2, e_3, e_4\}$ be the vector basis. Then the bivector basis consists of six elements: $\{e_{12}, e_{23}, e_{31}, e_{14}, e_{24}, e_{34}\}$. An arbitrary bivector can be developed on these elements:

$$\mathbf{S} = \frac{1}{2} \sum_{i,j=1}^{4} S^{ij} \mathbf{e}_{ij},$$

where $S^{ij} = -S^{ji}$. The sum may be divided into two terms $\mathbf{S} = \mathbf{S}_I + \mathbf{S}_{II}$, where

$$\mathbf{S}_I = S^{12} \mathbf{e}_{12} + S^{23} \mathbf{e}_{23} + S^{31} \mathbf{e}_{31},$$

$$\mathbf{S}_{II} = S^{14} \mathbf{e}_{14} + S^{24} \mathbf{e}_{24} + S^{34} \mathbf{e}_{34} = (S^{14} \mathbf{e}_1 + S^{24} \mathbf{e}_2 + S^{34} \mathbf{e}_3) \wedge \mathbf{e}_4.$$

The second term is the exterior product of vectors, so is a volutor, and the first one is contained in three-dimensional space, and hence also is a volutor. ∎

The lemma, unfortunately, says nothing about the uniqueness of the decomposition of a given bivector, but shows that there are at most as many decompositions as different three-dimensional subspaces.

Lemma 7.2 (Schouten [46], p. 24) *Bivector S is simple if and only if* $\mathbf{S} \wedge \mathbf{S} = 0$.

Proof $\Longrightarrow$. If $\mathbf{S}$ is simple, i.e., if $\mathbf{S} = \mathbf{a} \wedge \mathbf{b}$ for two vectors $\mathbf{a}$, $\mathbf{b}$, then by virtue of the associativity and anticommutativity of the exterior product of vectors

$$\mathbf{S} \wedge \mathbf{S} = (\mathbf{a} \wedge \mathbf{b}) \wedge (\mathbf{a} \wedge \mathbf{b}) = \mathbf{a} \wedge \mathbf{b} \wedge \mathbf{a} \wedge \mathbf{b} = -\mathbf{a} \wedge \mathbf{a} \wedge \mathbf{b} \wedge \mathbf{b} = 0.$$

$\Longleftarrow$. Let $\mathbf{S} = \frac{1}{2} \sum_{ij=1}^{4} S^{ij} \mathbf{e}_{ij}$ where $S^{12} \neq 0$ and $\mathbf{S} \wedge \mathbf{S} = 0$. One can check by explicit calculation that

$$\mathbf{S} \wedge \mathbf{S} = 2 \left(S^{12} S^{34} + S^{31} S^{24} + S^{23} S^{14} \right) \mathbf{e}_{1234}.$$

The condition $S \wedge S = 0$ implies

$$S^{12}S^{34} + S^{31}S^{24} + S^{23}S^{14} = 0. \tag{7.1}$$

Take two vectors

$$\mathbf{a} = S^{12}\mathbf{e}_1 + S^{32}\mathbf{e}_3 + S^{42}\mathbf{e}_4, \quad \mathbf{b} = S^{12}\mathbf{e}_2 + S^{13}\mathbf{e}_3 + S^{14}\mathbf{e}_4$$

and calculate their exterior product:

$$\mathbf{a} \wedge \mathbf{b} = S^{12}S^{12}\mathbf{e}_{12} + S^{12}S^{13}\mathbf{e}_{13} + S^{12}S^{14}\mathbf{e}_{14} + S^{32}S^{12}\mathbf{e}_{32} + S^{42}S^{12}\mathbf{e}_{42}$$

$$+(S^{32}S^{14} - S^{42}S^{13})\,\mathbf{e}_{34}.$$

It is seen from (7.1) that $S^{32}S^{14} - S^{42}S^{13} = S^{12}S^{34}$; hence

$$\mathbf{a} \wedge \mathbf{b} = S^{12}(S^{12}\mathbf{e}_{12} + S^{13}\mathbf{e}_{13} + S^{14}\mathbf{e}_{14} + S^{32}\mathbf{e}_{32} + S^{42}\mathbf{e}_{42} + S^{34}\mathbf{e}_{34}) = S^{12}\mathbf{S},$$

or

$$\mathbf{S} = \frac{1}{S^{12}}\,\mathbf{a} \wedge \mathbf{b}.$$

We see that $\mathbf{S}$ is the exterior product of two vectors; hence it is simple. ∎

There are no such problems with the sums of trivectors because their attitudes are three-dimensional spaces, and two such spaces have at least a common plane. Therefore, for two *simple trivectors* (i.e., being represented as the exterior product of three vectors) $\mathbf{R}$ and $\mathbf{T}$ one may find a volutor $\mathbf{S}$ as the common factor: $\mathbf{R} = \mathbf{a} \wedge \mathbf{S}$, $\mathbf{T} = \mathbf{b} \wedge \mathbf{S}$, where $\mathbf{a}$ and $\mathbf{b}$ are vectors. Then their *sum* is defined as $\mathbf{R} + \mathbf{T} = (\mathbf{a} + \mathbf{b}) \wedge \mathbf{S}$—this is also a simple trivector.

The exterior product of a vector with a non-simple bivector also gives a simple trivector because others are absent. Let us investigate this in a specific example. Let $\{\mathbf{e}_1, \mathbf{e}_2, \mathbf{e}_3, \mathbf{e}_4\}$ be the vector basis. Take the non-simple bivector $\mathbf{S} = \mathbf{e}_{12} + \mathbf{e}_{34}$. Calculate its exterior product with vector $\mathbf{a} = \mathbf{e}_1 + \mathbf{e}_3$:

$$\mathbf{a} \wedge \mathbf{S} = (\mathbf{e}_1 + \mathbf{e}_3) \wedge (\mathbf{e}_{12} + \mathbf{e}_{34}) = (\mathbf{e}_1 + \mathbf{e}_3) \wedge \mathbf{e}_{12} + (\mathbf{e}_1 + \mathbf{e}_3) \wedge \mathbf{e}_{34} = \mathbf{e}_{312} + \mathbf{e}_{134}.$$

We notice now the common bivector factor $\mathbf{e}_1 \wedge \mathbf{e}_3$; hence we may write

$$\mathbf{a} \wedge \mathbf{S} = -\mathbf{e}_{132} + \mathbf{e}_{134} = \mathbf{e}_{13} \wedge (\mathbf{e}_4 - \mathbf{e}_2).$$

Fig. 7.2 A parallelogram
whose three sides are not in
our three-dimensional space

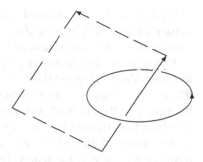

Being the exterior product of three vectors, this is a simple trivector. The following basic elements are in the linear space of trivectors: e_{123}, e_{234}, e_{341}, e_{412}. Therefore, this linear space is four-dimensional.

To define a *quadrivector*[1] one must first define its orientation. As the four-dimensional orientation, we take a pair of directed rings lying on two planes with only one common point; this point is the common centre of the two rings. All other pairs of such rings obtained from this one by arbitrary rotations in space are considered to be equivalent. The family of equivalent pairs of such directed rings will be called *four-handedness*. Penrose and Rindler [42] call it *four-screw*. Only two distinct four-handednesses are possible and they are called *opposite*. One of them can be defined with the aid of three-dimensional right handedness as follows.

As we know, the handedness is a composition of rotary with translatory motion. Leave the ring of rotary motion invariant and join the spatial arrow of the translatory motion with the arrow of the forward time flow (see Fig. 7.2). The two arrows determine a parallelogram (marked by dashed lines on the figure). The plane of this parallelogram has only one common point with the plane of the ring. If we replace now the parallelogram by a ring with the same attitude and orientation, we have already two directed rings lying on two planes with only one common point and hence have four-handedness, which can be called *four-right*.

We are now ready for the next definition: a *quadrivector* is a geometrical object with the following relevant features: *attitude*—the four-dimensional space, *orientation*—four-handedness, and *magnitude*—four-dimensional volume. The set of quadrivectors is a one-dimensional linear space.

If we accepted earlier that the ring can be replaced by a pair of vectors, then a pair of rings may be replaced by an ordered quadruple of vectors. If the order of rings is not important, then in the named quadruple of vectors only pairs of vectors can be interchanged: the first two with the last two. A similarity to the three-dimensional orientation is the following—triples of vectors can be used there instead of the ring plus the straight arrow, and the first vector may go to the end or the last one to the beginning.

---

[1] This name is coined for multivectors of grade four to distinguish them from 'four-vectors', meaning vectors in four-dimensional space.

That is all about nontwisted multivectors in four dimensions. We pass now to twisted multivectors. A *twisted scalar* has the following features: *attitude*—point, *orientation*—four-handedness, and *magnitude*—absolute value.

Our next exercise in mental gymnastics over four dimensions comes now. One may consider an ellipsoid threaded on a straight line. Admit the existence (I would like to write "imagine", but I can not demand this) of a straight line passing through the centre of the ellipsoid, but outgoing from the whole three-dimensional space in which the ellipsoid is contained. There exists such a metric that the ellipsoid is the sphere in it, and the named straight line is perpendicular to the three-dimensional space of the sphere. In that metric an arbitrary point on this line has equal distance from all points of the sphere—similar to the situation in three-dimensional Euclidean space for a line perpendicular to a circle. If such a circle is called threaded on the line, we may equally well call that *ellipsoid threaded on the line*. The ellipsoid may have an orientation marked by curved arrows on it, as on Fig. 1.26. In this manner a three-dimensional outer orientation is ascribed to the straight line.

After successfully performing this exercise we may accept the following definition. A *twisted vector* is an object with the following features: *attitude*—straight line, *orientation*—ellipsoid threaded on the line and endowed with three-dimensional orientation, and *magnitude*—length. One may introduce a basis in the linear space of twisted vectors by the follwing prescription. A twisted vector $\mathbf{e}_1^*$ has a segment of nontwisted vector $\mathbf{e}_1$ and the orientation of the complementary nontwisted trivector $\mathbf{e}_{234}$. Similarly, $\mathbf{e}_2^*$ has the orientation of $\mathbf{e}_{341}$, $\mathbf{e}_3^*$—of $\mathbf{e}_{412}$, and $\mathbf{e}_4^*$—of $\mathbf{e}_{123}$. It is possible to introduce the mapping of nontwisted trivectors into twisted vectors:

$$\mathbf{T} = T^{123}\mathbf{e}_{123} + T^{234}\mathbf{e}_{234} + T^{341}\mathbf{e}_{341} + T^{412}\underline{f}e_{412} \to \mathbf{T}$$
$$= T^{123}\mathbf{e}_4^* + T^{234}\mathbf{e}_1^* + T^{341}\mathbf{e}_2^* + T^{412}\mathbf{e}_3^* \tag{7.2}$$

which is an isomorphism between two four-dimensional linear spaces. This mapping is analogous to (2.73).

Twisted vectors can be added according to the triangle rule by juxtaposing them such that the ellipsoid passing through the junction preserves the orientation. Of course, a directed parallelepiped can also be considered instead of the ellipsoid, with the parallelepiped threaded on the line.

The exterior product of two twisted vectors yields a nontwisted volutor according to the prescription given in Fig. 1.21 (left), which is repeated here as Fig. 7.3. The orientations are marked on it as parallelepipeds with several edges outside of our three-dimensional space drawn as dotted lines. Now the prescription is: two twisted vectors $\mathbf{c}$ and $\mathbf{d}$ are juxtaposed such that the parallelepiped of one segment threaded though the junction coincides with that of the other segment. Then two other sides of the parallelogram are drawn, and this is the exterior product $\mathbf{c} \wedge \mathbf{d}$ with orientation from $\mathbf{c}$ to $\mathbf{d}$.

If we have already accepted a ring threaded on a plane, we may agree on the next definition. A *simple twisted bivector* is a quantity with the following

Fig. 7.3 Exterior product of two twisted vectors. Their orientations are parallelepipeds with some edges outside of our three-dimensional space

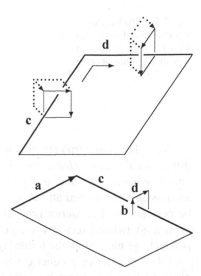

Fig. 7.4 Exterior product $\mathbf{a} \wedge \mathbf{c}$ of an ordinary vector $\mathbf{a}$ with a twisted vector $\mathbf{c}$ in three-dimensional space

features: *attitude*—plane, *orientation*—directed ring threaded on the plane, and *magnitude*—area. Addition of twisted bivectors causes the same problems as nontwisted bivectors, and they are solved in the same manner, i.e., by introducing simple and non-simple quantities. A simple twisted bivector will be called a *twisted volutor*.

The exterior product of a nontwisted vector $\mathbf{a}$ with a twisted vector $\mathbf{c}$ gives a twisted volutor, and the prescription for the orientation of the product is transferred from Sect. 1.2 with appropriate changes. We present it first in Fig. 7.4 with the orientation of the twisted vector $\mathbf{c}$ marked by a parallelogram, as in three-dimensional space. To represent the situation in four dimensions, we add to the parallelogram $\mathbf{b} \wedge \mathbf{d}$ a third vector $\mathbf{e}$ anchored to the tip point of $\mathbf{d}$ but sticking out in the fourth dimension (see Fig. 7.5), and speak about the parallelepiped $\mathbf{b} \wedge \mathbf{d} \wedge \mathbf{e}$. Now the prescription for the orientation. We juxtapose twisted vector $\mathbf{c}$ to the tip of $\mathbf{a}$. In the parallelepiped $\mathbf{b} \wedge \mathbf{d} \wedge \mathbf{e}$ threaded on $\mathbf{c}$ we choose the edge $\mathbf{d}$ parallel to plane R determined by $\mathbf{a}$ and $\mathbf{c}$. The two other edges determine the volutor $\mathbf{b} \wedge \mathbf{e}$, whose attitude is the plane P having only one common point with R. Therefore, we treat the boundary of parallelogram $\mathbf{b} \wedge \mathbf{e}$ as threaded on parallelogram $\mathbf{a} \wedge \mathbf{c}$, and hence it determines the orientation of product $\mathbf{a} \wedge \mathbf{c}$ as the twisted volutor.[2]

---

[2] This prescription is not yet full, because anticommutativity is not visible from it. Notice that at the beginning of $\mathbf{d}$, there is a parallelogram with an other orientation than on the end. For the orientation we choose that one which pierces the parallelogram determined by $\mathbf{a}$ and $\mathbf{c}$. When we juxtapose $\mathbf{c}$ to the beginning of $\mathbf{a}$, then the other parallelogram will be chosen and we obtain the opposite orientation. The anticommutativity $\mathbf{c} \wedge \mathbf{a} = -\mathbf{a} \wedge \mathbf{c}$ is visible.

Fig. 7.5 Exterior product
a ∧ c of an ordinary vector a
with a twisted vector c in
four-dimensional space

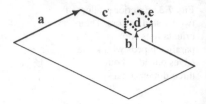

We call a *twisted trivector* a quantity with the following features: *attitude*—three-dimensional space, *orientation*—arrow piercing this space in the fourth dimension, and *magnitude*—volume. It is probably obvious that only two distinct orientations are possible for a fixed attitude—we call them opposite. The twisted trivector can be obtained as the exterior product of a twisted vector with a nontwisted bivector or of a nontwisted vector with an twisted bivector. To define the orientation of the product, we may adopt the following prescriptions.

Take the exterior product **c** ∧ **S** of the twisted vector **c** and volutor **S**. Let the orientation of **c** be given by a parallelepiped **b** ∧ **S** threaded on it. The plane of **S** and the straight lines (that of **b** and of **c**) span the whole four-dimensional space; hence **b** protrudes out of the three-space spanned by **c** and **S**. Thus we take the orientation of **b** as the orientation of the product **c** ∧ **S**.

We consider now the exterior product **a** ∧ **R** of a nontwisted vector **a** with a twisted volutor **R**, whose orientation is given by a parallelogram **b** ∧ **a** threaded on it. The plane of **R** and the plane of volutor **b** ∧ **a** spa the whole four-dimensional space, hence **b** protrudes out of the three-space spanned by **a** and **R**. Thus we take the orientation of **b** as that of the product **a** ∧ **R**.

The *twisted quadrivector* is a quantity with the following features: *attitude*—four-dimensional space, *orientation*—sign, and *magnitude*—a measure of four-dimensional volume. It can be obtained as the exterior product of a twisted vector with a trivector, of a vector with a twisted trivector or of a volutor with a twisted volutor. It is rather fortunate that in those products a twisted quantity has the same type of orientation as that of the other factor, so it is sufficient only to check whether the orientations are compatible or not—then the product has a plus or minus sign. Further thought is needed to decide which products should be anticommutative.

We have already named all the nontwisted and twisted multivectors—there are 10 of them. It is now time for nontwisted and twisted forms. The *zero-form* has the following features: *attitude*—four-dimensional space, *orientation*—sign, and *magnitude*—absolute value, whereas the *twisted zero-form* has: *attitude*—four-dimensional space, *orientation*—four-handedness, and *magnitude*—absolute value.

The algebraic definition of the *one-form* or *linear form* is always the same: the linear mapping of vectors into scalars. In four dimensions, however, its kernel has dimension three. Therefore, the *attitude* of the one-form is the three-space, and its geometric image is a family of parallel and equidistant spaces, with an arrow passing between two neighbouring spaces or piercing one of them. This straight arrow plays the role of the *orientation* of the one-form. The *magnitude* is the linear density.

Fig. 7.6 Exterior product of
two nontwisted one-forms in
four-dimensional space

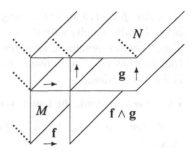

A simplified geometric image of the one-form is a four-dimensional layer
contained between neighbouring three-spaces. Just this image will help us to
understand what a two-form is because we attain it by considering the exterior
product of one-forms. Take two one-forms **f** and **g** with nonparallel attitudes, that
is three-spaces M and N. Their intersection is two-dimensional, i.e., it is a plane,
which we take as the attitude of the exterior product **f** ∧ **g**. The common part of **f**
and **g** is a tube or prism with a two-dimensional "axis" and two-dimensional cross-
section. In this section with the shape of a parallelogram, we place a directed ring
or a broken arrow from the juxtaposed arrows of two factors. This directed ring
can also be threaded on the two-dimensional "axis", being the attitude of **f** ∧ **g**.
We repeat Fig. 1.48 (left) here as Fig. 7.6; the fourth dimension, belonging to the
attitudes of **f**, **g** and **f** ∧ **g**, is marked by the dotted lines.

The described exterior product of one-forms yields a *simple two-form*. Summarize its features: *attitude*—plane, *orientation*—directed ring or parallelogram
threaded on the plane, and *magnitude*—surface density in a plane that has only one
common point with the attitude.

If we have two simple two-forms whose attitudes (as the planes) have a common
edge, their sum may be defined as before in the three-dimensional space cutting that
edge—in this manner a simple two-form is obtained. If, however, the attitudes of
the two-forms have only a common point, their sum is understood as the pair of
two-forms and called a *non-simple two-form*.

The two-forms are, naturally, linear mappings of bivectors into scalars. A value
B[[**S**]] of the two-form **B** on volutor **S** is obtained by counting the two-dimensional
"tubes" of **B** cut by **S** and checking whether the orientations of the two quantities are
compatible. The value of **B** on a non-simple bivector that is the sum of two volutors
$\mathbf{S} = \mathbf{S}_I + \mathbf{S}_{II}$ is taken as the sum $\mathbf{B}[[\mathbf{S}]] = \mathbf{B}[[\mathbf{S}_I]] + \mathbf{B}[[\mathbf{S}_{II}]]$ of the values on the
volutors. The linear space of all two-forms is six-dimensional.

A *three-form* is the linear mapping of trivectors into scalars. Its geometric
image is a family of tubes with one-dimensional axes, all with equal three-
dimensional cross-sections—just on these sections the orientation is marked as
an oriented sphere or parallelepiped. Let us name the features of the three-
form: *attitude*—straight line, *orientation*—directed parallelepiped threaded on the
line, and *magnitude*—spatial density. The linear space of all three-forms is four-
dimensional.

A *four-form* is a linear mapping of quadrivectors into scalars. Its geometric image is the family of cells with the same four-dimensional volume and with given four-dimensional orientation. Features of the four-form: *attitude*—point, *orientation*—four-handedness, and *magnitude*—four-density. The linear space of all four-forms is one-dimensional.

Exercise 7.1 Name the features of all twisted forms in four-dimensional space.

We have altogether 10 kinds of nontwisted and twisted forms in four-dimensional space.

7.2 Premetric Electrodynamics in Space-Time

As the four-dimensional space, we choose *space-time*, which is our three-dimensional space Σ with the added axis T of time flowing in the fourth dimension. Denote it as the cartesian product $\Sigma \times T$. We know from special relativity that the world line of an inertial observer distinguishes a direction out of Σ, which for this observer coincides with the time axis T. We shall start from directed quantities lying in Σ and ponder which of them and in what manner can be transferred to four dimensions. This will be a procedure inverse to the one applied in Chap. 6, where we were passing from the higher dimension to a lower one.

We should notice that not all physical quantities considered as scalars in Σ are spatio-temporal scalars in the sense that they do not depend on the basis (it is said physically that they do not depend on the reference frame). This concerns, for instance, time, because we deliberately take it as one component of the position vector in space-time. Special relativity says that energy and the electric potential are not scalars under the Lorentz transformations. I know of only three physical quantities called scalars in three-dimensional space that remain scalars in space-time—the *action S*, the *electric charge Q* and the *magnetic flux* Φ_m. If a *magnetic charge g* exists, the proportionality between Φ_m and g would occur—it follows from the magnetic Gauss law, in which the right-hand side would not be zero any more. If the electric charge is a scalar and time is not, the electric current as the derivative with respect to time, $I = \frac{dQ}{dt}$, is also not a scalar.

In order to build spatio-temporal quantities a vector is needed, protruding from Σ into the fourth dimension, corresponding to the time flow for an observer resting in Σ. We choose for it vector $\mathbf{e}_T$, showing the directed segment which the observer "passes" in a unit of time, see Fig. 7.7, where the space Σ is drawn as a plane. The attitude of $\mathbf{e}_T$ coincides with the time axis T of the observer and the orientation distinguishes the positive side of the axis. If we admit that this vector has an arbitrary fixing point on Σ, this would mean that we have a set of synchronized clocks "travelling" in the space-time.

Fig. 7.7 Time axis
protruding from
three-dimensional space

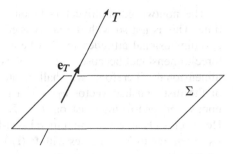

Fig. 7.8 Temporal and
spatial parts of a
spatio-temporal vector

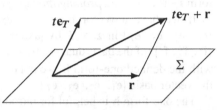

A pair $\{\mathbf{r}, t\}$ describing an event (t is the time of the event and $\mathbf{r}$ is radius vector of the place of the event) can now be replaced by the spatio-temporal vector:

$$\hat{\mathbf{x}} = \mathbf{r} + t\,\mathbf{e}_T. \tag{7.3}$$

The first term in the sum is called the *spatial part* and the second is *temporal part* of $\hat{\mathbf{x}}$. A configuration of the named vectors is depicted in Fig. 7.8. Time is no longer a scalar quantity in (7.3): it became a coordinate of a four-dimensional vector. In this connection, the second becomes an extensional unit, appropriate for particular segments of space-time. In order to contain compatible units, the second term in (7.3) should have the unit of metres; therefore we have to ascribe the unit m/s for the vector $\mathbf{e}_T$:

$$[\mathbf{e}_T] = \frac{\mathrm{m}}{\mathrm{s}}. \tag{7.4}$$

We know from a previous chapter that in addition to the nontwisted vector protruding from subspace Σ a reciprocal form is needed, that is, a nontwisted one-form parallel to σ, such that its value on the protruding vector is one. In this manner $\mathbf{e}_T$ determines uniquely the one-form $\mathbf{h}$ such that

$$\mathbf{h}[[\mathbf{e}_T]] = 1. \tag{7.5}$$

When acting on position vector (7.3), this one-form chooses its time coordinate:

$$\mathbf{h}[[\hat{\mathbf{x}}]] = t. \tag{7.6}$$

The nontwisted multivectors belonging to Σ belong naturally also to space-time. This is not so with the nontwisted one-forms. For instance, one-forms have two-dimensional attitudes in Σ, whereas in space-time, their attitudes should be three-dimensional because a set of points with constant value of the form is three-dimensional. Therefore, we shall create their *prolongations* using the existence of the distinguished vector $\mathbf{e}_T$. For a given nontwisted one-form $\mathbf{f}$ on Σ, many nontwisted one-forms exist on $T \times \Sigma$, which restricted to Σ give the same $\mathbf{f}$. However, only one spatio-temporal one-form $\mathbf{f}_T$ exists that is parallel to $\mathbf{e}_T$ and its *restriction* to Σ coincides with $\mathbf{f}$: $(\mathbf{f}_T)_\Sigma = \mathbf{f}$. Precisely this spatio-temporal one-form $\mathbf{f}_T$ is called the *prolongation of spatial one-form* $\mathbf{f}$.

If we have a vector basis $\{\mathbf{e}_1, \mathbf{e}_2, \mathbf{e}_3\}$ in Σ, we compose a vector basis $\{\mathbf{e}_1, \mathbf{e}_2, \mathbf{e}_3, \mathbf{e}_T\}$ in $\Sigma \times T$ by juxtaposing vector $\mathbf{e}_T$. We prolong the three forms $\{\mathbf{f}^1, \mathbf{f}^2, \mathbf{f}^3\}$ (of the basis dual to $\{\mathbf{e}_1, \mathbf{e}_2, \mathbf{e}_3\}$ in Σ) to $\{\mathbf{f}_T^1, \mathbf{f}_T^2, \mathbf{f}_T^3\}$ in $T \times \Sigma$. Together with the defined one-form $\mathbf{h}$, we obtain the one-form basis $\{\mathbf{f}_T^1, \mathbf{f}_T^2, \mathbf{f}_T^3, \mathbf{h}\}$ dual to the vector basis $\{\mathbf{e}_1, \mathbf{e}_2, \mathbf{e}_3, \mathbf{e}_T\}$.

The one-form $\mathbf{h}$ is helpful for introducing the spatio-temporal wave covector. In this purpose we recall considerations from Sect. 6.1, where formula (6.5) was the decomposition of the one-form $\mathbf{k}$ onto a component $\mathbf{k_a}$ parallel to the protruding vector $\mathbf{a}$ and another component parallel to the form $\mathbf{f}$ dual of $\mathbf{a}$:

$$\mathbf{k} = \mathbf{k_a} + \lambda\mathbf{f}.$$

Afterwards, the one-form $\mathbf{k}$ is replaced by the pair $\{\mathbf{k_{a_\Sigma}}, \lambda\}$. We now proceed conversely; hence we must have the pair: a spatial one-form and a scalar. If that spatial one-form is the wave covector $\mathbf{k}$, the appropriate scalar is the circular frequency ω. We take the prolongation of $\mathbf{k}$ as parallel to the protruding vector $\mathbf{e}_T$, that is $\mathbf{k}_T$. We now write the whole spatio-temporal wave covector:

$$\hat{\mathbf{k}} = \mathbf{k}_T + \omega\mathbf{h}. \tag{7.7}$$

The circular frequency ω is not a zero-form quantity (as named in Table 3.3), it now becomes a coordinate of a four-dimensional wave covector. In order to ensure compatible units, the second term in (7.7) should have the unit m^{-1}; therefore we have to ascribe the unit s/m for the one-form $\mathbf{h}$:

$$[\mathbf{h}] = \frac{\mathrm{s}}{\mathrm{m}}. \tag{7.8}$$

The units in (7.4) and (7.8) fit the duality $\mathbf{h}[[\mathbf{e}_T]] = 1$.

It could seem that the plus sign present in (7.7) is natural. Let us, however, look at the value of (7.7) on some spatio-temporal vector (7.3):

$$\hat{\mathbf{k}}[[\hat{\mathbf{x}}]] = (\mathbf{k}_T + \omega\mathbf{h})[[\mathbf{r} + t\mathbf{e}_T]] = \mathbf{k}_T[[\mathbf{r}]] + t\,\mathbf{k}_T[[\mathbf{e}_T]] + \omega\mathbf{h}[[\mathbf{r}]] + \omega t\mathbf{h}[[\mathbf{e}_T]].$$

Since $\mathbf{k}_T \parallel \mathbf{e}_T$, the second term vanishes, and since $\mathbf{h} \parallel \Sigma$, the third term also vanishes. Moreover, $\mathbf{h}[[\mathbf{e}_T]] = 1$ so there remains:

$$\hat{\mathbf{k}}[[\hat{\mathbf{x}}]] = \mathbf{k}_T[[\mathbf{r}]] + \omega t = \mathbf{k}[[\mathbf{r}]] + \omega t.$$

This expression does not coincide with the phase of the plane wave given by one of formulas (5.5):

$$\phi = \mathbf{k}[[\mathbf{r}]] - \omega t, \tag{7.9}$$

where the signs are different. In order to account for this, instead of (7.7) we should take

$$\hat{\mathbf{k}} = \mathbf{k}_T - \omega \mathbf{h} \tag{7.10}$$

as the *spatio-temporal wave covector*, and the phase can be written as $\phi = \hat{\mathbf{k}}[[\hat{\mathbf{x}}]]$.[3] We should remark that another formula for the phase was also admitted in (5.5), namely $\phi' = \omega t - \mathbf{k}[[\mathbf{r}]]$. The fact that we have chosen (7.9), has an implication for introducing a scalar product in space-time (see Sect. 7.5).

We create a four-dimensional counterpart $\hat{\mathbf{p}}$ of the momentum of any object (e.g., a particle) as the following combination of the three-dimensional scalar of energy $\mathcal{E}$ and the *momentum one-form* $\mathbf{p}$:

$$\hat{\mathbf{p}} = \mathbf{p}_T - \mathcal{E}\mathbf{h} \tag{7.11}$$

We call this an *energy-momentum one-form*. With the use of relation (7.8), one can check that its unit is

$$[\mathbf{p}] = [\mathcal{E}\mathbf{h}] = \frac{\mathrm{Pl}}{\mathrm{m}}. \tag{7.12}$$

It is now time to consider the exterior derivative in space-time. Remind the operator of the form derivative in three dimensions:

$$\mathbf{d} = \mathbf{f}^1 \frac{\partial}{\partial x^1} + \mathbf{f}^2 \frac{\partial}{\partial x^2} + \mathbf{f}^3 \frac{\partial}{\partial x^3}.$$

---

[3] Different signs of the spatial and time components in (7.10) imply that when time flows, the constant phase surfaces $\phi = \phi_0 = \mathrm{const}$ move in Σ in the direction of growing values of $\mathbf{k}$, because when seeking points corresponding to phase ϕ_0 in time t we look for vectors $\mathbf{r}$, fulfilling the equation $\mathbf{k}[\mathbf{r}] = \omega t + \phi_0$.

When we introduce its spatio-temporal prolongation

$$\mathbf{d}_T = \mathbf{f}_T^1 \frac{\partial}{\partial x^1} + \mathbf{f}_T^2 \frac{\partial}{\partial x^2} + \mathbf{f}_T^3 \frac{\partial}{\partial x^3},$$

then the *spatio-temporal form derivative* is the operator

$$\hat{\mathbf{d}} = \mathbf{d}_T + \mathbf{h} \frac{\partial}{\partial t}. \tag{7.13}$$

Notice that both sides have the same unit s/m. By analogy with (3.78) we may write

$$\mathbf{h} = \hat{\mathbf{d}}t. \tag{7.14}$$

Having used (7.14), we can write (7.10) also as

$$\hat{\mathbf{k}} = \mathbf{k}_T - \omega \hat{\mathbf{d}}t.$$

An increment of phase on the spatio-temporal directed segment $d\hat{\mathbf{x}}$ can now be written as

$$\phi = \hat{\mathbf{k}}[[d\hat{\mathbf{x}}]] = \mathbf{k}_T[[d\mathbf{r}]] - \omega dt = \mathbf{k}[[d\mathbf{r}]] - \omega dt. \tag{7.15}$$

We similarly introduce a spatio-temporal one-form uniting the magnetic potential $\mathbf{A}$ with the electric potential Φ:

$$\hat{\mathbf{A}} = \mathbf{A}_T - \Phi \mathbf{h}$$

or equivalently

$$\hat{\mathbf{A}} = \mathbf{A}_T - \Phi \hat{\mathbf{d}}t. \tag{7.16}$$

We call the the *spatio-temporal electromagnetic potential*. As mentioned in Table 3.3, the electric potential in the geometric unit system has the unit Pl/Cs, hence

$$[\Phi \mathbf{h}] = \frac{\text{Pl}}{\text{Cs}} \cdot \frac{\text{s}}{\text{m}} = \frac{\text{Pl}}{\text{Cm}}. \tag{7.17}$$

This is the same unit as in Table 3.3 for the magnetic potential.

Let the operator (7.13) act on the one-form (7.16):

$$\hat{\mathbf{d}} \wedge \hat{\mathbf{A}} = \left(\mathbf{d}_T + \mathbf{h} \frac{\partial}{\partial t} \right) \wedge (\mathbf{A}_T - \Phi \mathbf{h}) = \mathbf{d}_T \wedge \mathbf{A}_T + \mathbf{h} \wedge \frac{\partial \mathbf{A}_T}{\partial t} - \mathbf{d}_T \Phi \wedge \mathbf{h},$$

We use Exercise 6.6 $(M \wedge N)_\Sigma = M_\Sigma \wedge N_\Sigma$, interpreted now the other way around: prolongation of the exterior product is the exterior product of their prolongations:

$$\hat{\mathbf{d}} \wedge \hat{\mathbf{A}} = (\mathbf{d} \wedge \mathbf{A})_T - (\mathbf{d}\Phi)_T \wedge \mathbf{h} + \mathbf{h} \wedge \frac{\partial \mathbf{A}_T}{\partial t} = (\mathbf{d} \wedge \mathbf{A})_T + \left(-\mathbf{d}\Phi - \frac{\partial \mathbf{A}}{\partial t}\right)_T \wedge \mathbf{h}.$$

We recognize the magnetic induction according to formula (4.10) in the first bracket, and the electric field strength (4.11) in the second one.

In this manner we ascertain that the exterior derivative

$$\mathbf{F} = \hat{\mathbf{d}} \wedge \hat{\mathbf{A}} \qquad (7.18)$$

gives a spatio-temporal two-form uniting the electric field strength $\mathbf{E}$ and the magnetic induction $\mathbf{B}$ through their spatio-temporal prolongations:

$$\mathbf{F} = \mathbf{B}_T + \mathbf{E}_T \wedge \mathbf{h} = \mathbf{B}_T + \mathbf{E}_T \wedge \hat{\mathbf{d}}t. \qquad (7.19)$$

This quantity is traditionally called the *electromagnetic field tensor* or *Faraday tensor*; we may call it the *electromagnetic field two-form* or *Faraday two-form*. Its part parallel to Σ (one may call it the *space part*) is connected to the electric field strength, and the part parallel to $\mathbf{e}_T$ (the *time part*) is connected to the magnetic induction. The second term on the right-hand side in (7.19) has the unit

$$[\mathbf{E}\,\mathbf{h}] = \frac{\mathrm{V}}{\mathrm{m}} \cdot \frac{\mathrm{s}}{\mathrm{m}} = \frac{\mathrm{Wb}}{\mathrm{m}^2} = \frac{\mathrm{Pl}}{\mathrm{Cm}^2}. \qquad (7.20)$$

This is the same unit as for the magnetic induction in Table 3.3.

We shall show how to pick up the two parts. Calculate at the beginning the following contraction with the vector $\mathbf{e}_T$:

$$\mathbf{F}\lfloor \mathbf{e}_T = (\mathbf{B}_T + \mathbf{E}_T \wedge \mathbf{h})\lfloor \mathbf{e}_T = \mathbf{B}_T\lfloor \mathbf{e}_T + \mathbf{E}_T\,\mathbf{h}[[\mathbf{e}_T]] - \mathbf{h}\,\mathbf{E}_T[[\mathbf{e}_T]].$$

Since $\mathbf{B}_T \parallel \mathbf{e}_T$ and $\mathbf{E}_T \parallel \mathbf{e}_T$, the two last terms are zero; moreover $\mathbf{h}[[\mathbf{e}_T]] = 1$, and there remains

$$\mathbf{F}\lfloor \mathbf{e}_T = \mathbf{E}_T. \qquad (7.21)$$

Calculate now the exterior product

$$\mathbf{F} \wedge \mathbf{h} = \mathbf{B}_T \wedge \mathbf{h} + \mathbf{E}_T \wedge \mathbf{h} \wedge \mathbf{h} = \mathbf{B}_T \wedge \mathbf{h}$$

and the contraction

$$(\mathbf{F} \wedge \mathbf{h})\lfloor \mathbf{e}_T = (\mathbf{B}_T \wedge \mathbf{h})\lfloor \mathbf{e}_T = -(\mathbf{B}_T\lfloor \mathbf{e}_T) \wedge \mathbf{h} + \mathbf{B}_T\,\mathbf{h}[[\mathbf{e}_T]] = \mathbf{B}_T.$$

Now the restrictions to Σ give the spatial forms

$$(\mathbf{F}\lfloor\mathbf{e}_T)_\Sigma = \mathbf{E}, \qquad \{(\mathbf{F}\wedge\mathbf{h})\lfloor\mathbf{e}_T\}_\Sigma = \mathbf{B}.$$

Calculate the spatio-temporal exterior derivative of (7.19):

$$\hat{\mathbf{d}}\wedge\mathbf{F} = \left(\mathbf{d}_T + \mathbf{h}\frac{\partial}{\partial t}\right)\wedge(\mathbf{B}_T + \mathbf{E}_T\wedge\mathbf{h}) = \mathbf{d}_T\wedge\mathbf{B}_T + \mathbf{d}_T\wedge\mathbf{E}_T\wedge\mathbf{h} + \mathbf{h}\wedge\frac{\partial\mathbf{B}_T}{\partial t}.$$

$$\hat{\mathbf{d}}\wedge\mathbf{F} = (\mathbf{d}\wedge\mathbf{B})_T + \left(\mathbf{d}\wedge\mathbf{E} + \frac{\partial\mathbf{B}}{\partial t}\right)_T\wedge\mathbf{h}. \tag{7.22}$$

Both brackets are zero by virtue of Maxwell's equations (4.6) and (4.7); hence we obtain

$$\hat{\mathbf{d}}\wedge\mathbf{F} = 0. \tag{7.23}$$

In this manner the spatio-temporal equation (7.23) unites two spatial Maxwell equations. Merely vanishing of the time component (the first three-form on the right-hand side of (7.22)) gives $\mathbf{d}\wedge\mathbf{B} = 0$, and vanishing of the spatial component yields $\mathbf{d}\wedge\mathbf{E} + \partial\mathbf{B}/\partial t = 0$. Identity (7.23) is called the *homogeneous Maxwell equation*. We obtain it also directly from (7.18), since the double exterior derivative is zero. If the Faraday two-form is treated as a primary object, the existence of the electromagnetic potential follows from (7.23) due to the Poincaré lemma.

The two other electromagnetic fields $\mathbf{H}$ and $\mathbf{D}$ are twisted forms; thus by analogy with (7.19) we build the spatio-temporal twisted two-form

$$\mathbf{G} = \mathbf{D}_T - \mathbf{H}_T\wedge\mathbf{h} = \mathbf{D}_T - \mathbf{H}_T\wedge\hat{\mathbf{d}}t. \tag{7.24}$$

The need for the minus sign will become clear in a while. This quantity has no separate name; I propose *Gauss two-form*. Its space part (parallel to Σ) is connected with the magnetic field strength, and the time part (parallel to $\mathbf{e}_T$)—with the electric induction. Both terms on the right-hand side in (7.24) have the unit

$$[\mathbf{D}] = [\mathbf{H}\,\mathbf{h}] = \frac{\mathrm{C}}{\mathrm{m}^2}. \tag{7.25}$$

Calculate its spatio-temporal exterior derivative:

$$\hat{\mathbf{d}}\wedge\mathbf{G} = \left(\mathbf{d}_T + \mathbf{h}\frac{\partial}{\partial t}\right)\wedge(\mathbf{D}_T - \mathbf{H}_T\wedge\mathbf{h}) = \mathbf{d}_T\wedge\mathbf{D}_T - \mathbf{d}_T\wedge\mathbf{H}_T\wedge\mathbf{h} + \mathbf{h}\wedge\frac{\partial\mathbf{D}_T}{\partial t},$$

$$\hat{\mathbf{d}}\wedge\mathbf{G} = (\mathbf{d}\wedge\mathbf{D})_T - \left(\mathbf{d}\wedge\mathbf{H} - \frac{\partial\mathbf{D}}{\partial t}\right)_T\wedge\mathbf{h}.$$

Maxwell's equations (4.5) and (4.8) (just the minus sign in (7.24) assures fitting of the second bracket to (4.8)) allow us to change either bracket:

$$\hat{\mathbf{d}} \wedge \mathbf{G} = \rho_T - \mathbf{j}_T \wedge \mathbf{h}. \tag{7.26}$$

In this manner the spatio-temporal equation (7.26) unites two spatial Maxwell equations: simply equating the time components of the two sides gives the equation $\mathbf{d} \wedge \mathbf{D} = \rho$, and equating the spatial components yields $\mathbf{d} \wedge \mathbf{H} - \partial \mathbf{D}/\partial t = \mathbf{j}$.

Equation (7.26) also gives a prescription for creating, from the pair $\{\rho, \mathbf{j}\}$, the spatio-temporal twisted three-form:

$$\hat{\mathbf{j}} = \rho_T - \mathbf{j}_T \wedge \mathbf{h} = \rho_T - \mathbf{j}_T \wedge \hat{\mathbf{d}}t, \tag{7.27}$$

which could be called the *spatio-temporal density of the electric charge and current* or the *electric charge-current density three-form*. This quantity has the unit

$$[\rho] = [\mathbf{j}\,\mathbf{h}] = \frac{C}{m^3}. \tag{7.28}$$

Equation (7.26) can be written as

$$\hat{\mathbf{d}} \wedge \mathbf{G} = \hat{\mathbf{j}}. \tag{7.29}$$

This is called the *nonhomogeneous Maxwell equation*.

Let us check how the exterior derivative of (7.27) looks like:

$$\hat{\mathbf{d}} \wedge \hat{\mathbf{j}} = \left(\mathbf{d}_T + \mathbf{h}\frac{\partial}{\partial t} \right) \wedge (\rho_T - \mathbf{j}_T \wedge \mathbf{h}) = (\mathbf{d} \wedge \rho)_T + \mathbf{h} \wedge \frac{\partial \rho_T}{\partial t} - (\mathbf{d} \wedge \mathbf{j})_T \wedge \mathbf{h}.$$

The first term is zero, because there are no quadrivectors in Σ. So there remains

$$\hat{\mathbf{d}} \wedge \hat{\mathbf{j}} = -\frac{\partial \rho_c}{\partial t} \wedge \mathbf{h} - (\mathbf{d} \wedge \mathbf{j})_T \wedge \mathbf{h} = -\left(\frac{\partial \rho}{\partial t} + \mathbf{d} \wedge \mathbf{j} \right)_T \wedge \mathbf{h}.$$

The last bracket is zero by dint of the continuity equation (4.9). In this manner the law of conservation of electric charge assumes the simple form

$$\hat{\mathbf{d}} \wedge \hat{\mathbf{j}} = 0. \tag{7.30}$$

The same result is obtained directly from (7.29), because the double exterior derivative is zero. It is worth mentioning that the condition (7.30) means that the twisted three-form $\hat{\mathbf{j}}$ is closed in the whole space-time, and this, due to the Poincaré lemma, implies the existence of a twisted two-form $\mathbf{G}$ such that $\hat{\mathbf{j}} = \hat{\mathbf{d}} \wedge \mathbf{G}$. One can say that the very existence of the Gauss form follows from the conservation of electric charge.

Calculate now the following exterior product:

$$\mathbf{F} \wedge \mathbf{G} = (\mathbf{B}_T + \mathbf{E}_T \wedge \mathbf{h}) \wedge (\mathbf{D}_T - \mathbf{H}_T \wedge \mathbf{h}) = \mathbf{B}_T \wedge \mathbf{D}_T + \mathbf{E}_T \wedge \mathbf{h} \wedge \mathbf{D}_T - \mathbf{B}_T \wedge \mathbf{H}_T \wedge \mathbf{h}$$

$$\mathbf{F} \wedge \mathbf{G} = (\mathbf{B} \wedge \mathbf{D})_T + (\mathbf{E} \wedge \mathbf{D})_T \wedge \mathbf{h} - (\mathbf{B} \wedge \mathbf{H})_T \wedge \mathbf{h}.$$

The first term on the right-hand side is zero, because the exterior product of two two-forms is zero in the three-dimensional space; hence there remains

$$\mathbf{F} \wedge \mathbf{G} = (\mathbf{E} \wedge \mathbf{D} - \mathbf{B} \wedge \mathbf{H})_T \wedge \mathbf{h}. \tag{7.31}$$

The expression in the bracket is the doubled Lagrangian density L_{int}, Eq. (4.15) from Sect. 4.1. Therefore $\frac{1}{2} \mathbf{F} \wedge \mathbf{G}$ is the spatio-temporal twisted four-form corresponding to the spatial Lagrangian density of the electromagnetic field. The fourth (with the minus sign) grade of this twisted form corresponds to the fact that the triple integral gives the Lagrangian itself and the fourth integral with respect to time yields the action; thus only a fourfold integral gives a spatio-temporal scalar. Of course, one should take as the integration variable the twisted quadrivector of the four-volume of a spatio-temporal region.

Let us look at the unit of (7.31)

$$[\mathbf{F}\,\mathbf{G}] = \frac{\mathrm{Pl}}{\mathrm{Cm}^2} \cdot \frac{\mathrm{C}}{\mathrm{m}^3} = \frac{\mathrm{Pl}}{\mathrm{m}^4}. \tag{7.32}$$

This is confirmation that fourfold integration over a space-time region gives the scalar quantity, action.

We calculate another exterior product

$$\hat{\mathbf{j}} \wedge \hat{\mathbf{A}} = (\rho_T - \mathbf{j}_T \wedge \mathbf{h}) \wedge (\mathbf{A}_T - \Phi \mathbf{h}) = \rho_T \wedge \mathbf{A}_T - \mathbf{j}_T \wedge \mathbf{h} \wedge \mathbf{A}_T - \rho_T \Phi \wedge \mathbf{h}$$

$$= (\rho_T \wedge \mathbf{A}_T + (\mathbf{j}_T \wedge \mathbf{A}_T - \rho_T \Phi) \wedge \mathbf{h} = (\rho \wedge \mathbf{A})_T + (\mathbf{j} \wedge \mathbf{A} - \rho \Phi)_T \wedge \mathbf{h}.$$

The expression $\rho \wedge \mathbf{A}$, as the external product of the three-form ρ with the one-form $\mathbf{A}$ in the three-dimensional hyperplane Σ, is zero. Therefore, we obtain

$$\hat{\mathbf{j}} \wedge \hat{\mathbf{A}} = (\mathbf{j} \wedge \mathbf{A} - \rho \Phi)_T \wedge \mathbf{h}. \tag{7.33}$$

We recognize in the bracket the Lagrangian density L_{int}, Eq. (4.16) from Sect. 4.1. Thus (7.33) is the spatio-temporal twisted four-form corresponding to the Lagrangian density of the interaction of electric charges with electromagnetic field. In this manner, we arrive at the spatio-temporal total Lagrangian density:

$$L = \frac{1}{2}(\mathbf{F} \wedge \mathbf{G}) + \hat{\mathbf{j}} \wedge \hat{\mathbf{A}} \tag{7.34}$$

Two other forms of grade four exist, which can be built from the electromagnetic field quantities, namely $\mathbf{F} \wedge \mathbf{F}$ and $\mathbf{G} \wedge \mathbf{G}$; they are

$$\mathbf{F} \wedge \mathbf{F} = (\mathbf{B}_T + \mathbf{E}_T \wedge \mathbf{h}) \wedge (\mathbf{B}_T + \mathbf{E}_T \wedge \mathbf{h}) = \mathbf{B}_T \wedge \mathbf{E}_T \wedge \mathbf{h} + \mathbf{E}_T \wedge \mathbf{h} \wedge \mathbf{B}_T$$

$$\mathbf{F} \wedge \mathbf{F} = 2\,\mathbf{B}_T \wedge \mathbf{E}_T \wedge \mathbf{h} = 2\,(\mathbf{B} \wedge \mathbf{E})_T \wedge \mathbf{h}, \tag{7.35}$$

$$\mathbf{G} \wedge \mathbf{G} = (\mathbf{D}_T - \mathbf{H}_T \wedge \mathbf{h}) \wedge (\mathbf{D}_T - \mathbf{H}_T \wedge \mathbf{h}) = -2\,(\mathbf{D} \wedge \mathbf{H})_T \wedge \mathbf{h}. \tag{7.36}$$

Both expressions can be non-zero in spite of being the exterior products of a form with itself—this can happen only for non-simple two-forms. Since the two quantities are nontwisted four-forms, their integrals with the twisted quadrivector of the four-volume give twisted scalars. The geometric unit system ascibes to them units:

$$[\mathbf{F}^2] = \frac{\mathrm{Pl}^2}{\mathrm{C}^2\mathrm{m}^4}, \qquad [\mathbf{G}^2] = \frac{\mathrm{C}^2}{\mathrm{m}^4}. \tag{7.37}$$

Fourfold integration yields scalar quantities. The first is the square of some magnetic flux (or magnetic charge), and the second is the square of some electric charge. What are these quantities?

Premetric electrodynamics uses differential forms; hence the quantities and equations are generally covariant. This means that they preserve their form under arbitrary transformations of coordinates. Loosely speaking, we can claim that they are *covariant* under Galilei, Lorentz and any other transformation. The three quantities (7.31), (7.35) and (7.36), however, are related to Lorentz *invariants*. Before ascertaining this, some comments are necessary. When considering the basis transformations in Sect. 2.2, we assumed that the basis is transformed simultaneously with the coordinates such that the directed quantities remain the same. These are so-called *passive transformations*: the quantities do not change, only their representation by coordinates changes. If one uses this interpretation, the named three quantities can not change at all. But there exist also *active transformations*: the basis is left invariant, only the coordinates are changed, or vice versa; the two kinds of active transformations are related by the fact that the respective matrices are transposed to each other.

All nontwisted quadrivectors are proportional and they transform by the determinant of the matrix changing vector basis; similarly twisted quadrivectors transform by the modulus of the determinant. The nontwisted four-forms transform by the inverse determinant and the twisted four-forms by the inverse modulus of the determinant. Since the proper Lorentz transformations have determinant one, expressions (7.31), (7.35) and (7.36) are invariant under these transformations.

The reader may be interested in the question: why we do not present, analogously to Table 3.2, a table of directed quantities in four-dimensional space-time? The answer is that there are not so many of them. For the multivectors we can mention the directed segments, areas, three-volumes and four-volumes necessary for the appropriate integrals. For the exterior forms we can mention the following:

1. wave covector, energy-momentum and electromagnetic potential as nontwisted one-forms,
2. Faraday field as nontwisted two-form, Gauss field as a twisted two-form,
3. electric charge-current density as a twisted three-form,
4. Lagrangian density as a twisted four-form.

7.3   Energy and Momentum of the Electromagnetic Field

We wish now to find a spatio-temporal quantity that unites the energy density and energy flux density. Calculate the following product using (7.21):

$$\mathbf{G} \wedge (\mathbf{F} \lfloor \mathbf{e}_T) = (\mathbf{D}_T - \mathbf{H}_T \wedge \mathbf{h}) \wedge \mathbf{E}_T = \mathbf{E}_T \wedge \mathbf{D}_T - \mathbf{E}_T \wedge \mathbf{H}_T \wedge \mathbf{h},$$

$$\mathbf{G} \wedge (\mathbf{F} \lfloor \mathbf{e}_T) = (\mathbf{E} \wedge \mathbf{D})_T - (\mathbf{E} \wedge \mathbf{H})_T \wedge \mathbf{h}. \tag{7.38}$$

Similar calculations with $\mathbf{F}$ and $\mathbf{G}$ interchanged (with the use of the identity $\mathbf{G} \lfloor \mathbf{e}_T = -\mathbf{H}_T$) give:

$$\mathbf{F} \wedge (\mathbf{G} \lfloor \mathbf{e}_T) = -(\mathbf{B}_T + \mathbf{E}_T \wedge \mathbf{h}) \wedge \mathbf{H}_T = -\mathbf{H}_T \wedge \mathbf{B}_T + \mathbf{E}_T \wedge \mathbf{H}_T \wedge \mathbf{h},$$

$$\mathbf{F} \wedge (\mathbf{G} \lfloor \mathbf{e}_T) = -(\mathbf{H} \wedge \mathbf{B})_T + (\mathbf{E} \wedge \mathbf{H})_T \wedge \mathbf{h}. \tag{7.39}$$

Take now the difference of (7.38) and (7.39):

$$\mathbf{G} \wedge (\mathbf{F} \lfloor \mathbf{e}_T) - \mathbf{F} \wedge (\mathbf{G} \lfloor \mathbf{e}_T) = (\mathbf{E} \wedge \mathbf{D})_T + (\mathbf{H} \wedge \mathbf{B})_T - 2(\mathbf{E} \wedge \mathbf{H})_T$$

$$= (\mathbf{E} \wedge \mathbf{D} + \mathbf{H} \wedge \mathbf{B})_T - 2\,(\mathbf{E} \wedge \mathbf{H})_T \wedge \mathbf{h}.$$

We obtained a twisted three-form, which unites the electromagnetic energy density w Eq. (4.13)] and the Poynting form $\mathbf{S}$ (Eq. (4.14)):

$$\hat{\mathbf{S}} = \frac{1}{2} \{\mathbf{G} \wedge (\mathbf{F} \lfloor \mathbf{e}_T) - \mathbf{F} \wedge (\mathbf{G} \lfloor \mathbf{e}_T)\} = \frac{1}{2}\,(\mathbf{E} \wedge \mathbf{D} + \mathbf{H} \wedge \mathbf{B})_T - (\mathbf{E} \wedge \mathbf{H})_T \wedge \mathbf{h}.$$

$$\hat{\mathbf{S}} = w_T - \mathbf{S}_T \wedge \mathbf{h} = w_T - \mathbf{S}_T \wedge \mathbf{d}t. \tag{7.40}$$

The temporal part (parallel to $\mathbf{e}_T$) corresponds to the energy density, and the spatial part (parallel to Σ) to the energy flux density. The right-hand side is similar to combination (7.27). Expression (7.40) unites the energy density and the energy flux density of the electromagnetic field. This quantity has the unit

$$[w] = [\mathbf{S}\,\mathbf{h}] = \frac{J}{m^3} = \frac{Pl}{s\,m^3}. \tag{7.41}$$

The vector $\mathbf{e}_T$ is present in the contractions $\mathbf{F}\lfloor\mathbf{e}_T$, $\mathbf{G}\lfloor\mathbf{e}_T$; hence the whole expression is observer-dependent.

We calculate now the exterior derivative of $\hat{\mathbf{S}}$:

$$\hat{\mathbf{d}} \wedge \hat{\mathbf{S}} = \left(\mathbf{d}_T + \mathbf{h}\frac{\partial}{\partial t}\right) \wedge (w_T - \mathbf{S}_T \wedge \mathbf{h}) = \mathbf{d}_T \wedge w_T + \mathbf{h} \wedge \frac{\partial w_T}{\partial t} - \mathbf{d}_T \wedge (\mathbf{S}_T \wedge \mathbf{h})$$

$$= (\mathbf{d} \wedge w)_T + \mathbf{h} \wedge \frac{\partial w_T}{\partial t} - \mathbf{d}_T \wedge (\mathbf{h} \wedge \mathbf{S}_T).$$

The energy density w is a three-form, so $\mathbf{d} \wedge w$ would be a four-form in the three-space, where there is no room for this; hence the first term is zero and there remains

$$\hat{\mathbf{d}} \wedge \hat{\mathbf{S}} = \mathbf{h} \wedge \frac{\partial w_T}{\partial t} - (\mathbf{d}_T \wedge \mathbf{h}) \wedge \mathbf{S}_T + \mathbf{h} \wedge (\mathbf{d} \wedge \mathbf{S})_T.$$

The form $\mathbf{h}$ is constant in time and space, so $\mathbf{d}_a \wedge \mathbf{h} = 0$, and from this:

$$\hat{\mathbf{d}} \wedge \hat{\mathbf{S}} = \mathbf{h} \wedge \left(\frac{\partial w_T}{\partial t} + \mathbf{d} \wedge \mathbf{S}\right)_T. \tag{7.42}$$

Assume now that the medium is electrically and magnetically linear and stationary in time. Then we are allowed to use Theorem 4.1:

$$\hat{\mathbf{d}} \wedge \hat{\mathbf{S}} = \mathbf{h} \wedge (\mathbf{E} \wedge \mathbf{j})_T. \tag{7.43}$$

When the electric currents are absent, i.e., $\mathbf{j} = 0$, then the right-hand side is zero, so the bracket in (7.42) is zero, which means that a decline of energy density creates an energy flux and conversely.

We want to find a formula for the energy-momentum of the electromagnetic field contained in a small three-dimensional region in Σ expressed by the field quantities $\mathbf{F}$ and $\mathbf{G}$. Since such a region is represented by the volume trivector V, we need a mapping of trivectors into one-forms, which is built of $\mathbf{F}$ and $\mathbf{G}$. The simplest way is invoking the expression (4.19) for the spatial momentum, i.e.,

$$\mathbf{p} = \mathbf{D}\lfloor(V\lfloor\mathbf{B}), \tag{7.44}$$

and replacing $\mathbf{D}$ and $\mathbf{B}$ in it by $\mathbf{G}$ and $\mathbf{F}$, respectively:

$$\mathbf{G}\lfloor(V\lfloor\mathbf{F}). \tag{7.45}$$

We check what this formula gives for the spatial trivector V. We calculate first the expression in the bracket:

$$V \lfloor \mathbf{F} = V \lfloor (\mathbf{B}_T + \mathbf{E}_T \wedge \mathbf{h}) = V \lfloor \mathbf{B}_T + (V \lfloor \mathbf{E}_T) \lfloor \mathbf{h}.$$

The contraction $V \lfloor \mathbf{E}_T$ is a spatial bivector, its contraction with $\mathbf{h}$ gives zero; hence the following vector remains

$$V \lfloor \mathbf{F} = V \lfloor \mathbf{B}_T. \qquad (7.46)$$

We calculate now the whole expression (7.45)

$$\mathbf{G} \lfloor (V \lfloor \mathbf{F}) = (\mathbf{D}_T - \mathbf{H}_T \wedge \mathbf{h}) \lfloor (V \lfloor \mathbf{B}_T) = \mathbf{D}_T \lfloor (V \lfloor \mathbf{B}_T) - (\mathbf{H}_T \wedge \mathbf{h}) \lfloor (V \lfloor \mathbf{B}_T).$$

Identity (3.5) yields

$$(\mathbf{H}_T \wedge \mathbf{h}) \lfloor (V \lfloor \mathbf{B}_T) = \mathbf{h} [[V \lfloor \mathbf{B}_T]] \mathbf{H}_T - \mathbf{H}_T [[V \lfloor \mathbf{B}_T]] \mathbf{h}.$$

The value of $\mathbf{h}$ on the spatial vector $V \lfloor \mathbf{B}_T$ is zero, hence

$$\mathbf{G} \lfloor (V \lfloor \mathbf{F}) = \mathbf{D}_T \lfloor (V \lfloor \mathbf{B}_T) + \mathbf{H}_T [[V \lfloor \mathbf{B}_T]] \mathbf{h}$$

Due to (3.36) and (3.37), the second term in the sum can be written differently:

$$\mathbf{G} \lfloor (V \lfloor \mathbf{F}) = \mathbf{D}_T \lfloor (V \lfloor \mathbf{B}_T) - (\mathbf{H}_T \wedge \mathbf{B}_T)[[V]] \mathbf{h}. \qquad (7.47)$$

The expression obtained should be compared with the energy-momentum one-form (7.11). The first term in (7.47) is a prolongation of (7.44), but, since V is a pure spatial trivector, it is the spatial momentum $\mathbf{p}$. It would be desirable that the second term in (7.47) corresponds (with the minus sign) to the second term in (7.11), but $\mathbf{H}_T \wedge \mathbf{B}_T$ is not the energy density of the whole electromagnetic field. The energy density of the electric field is missing. Another expression is needed in which $\mathbf{G}$ and $\mathbf{F}$ exchanged places.

We omit analogous calculations, leading to the equality:

$$\mathbf{F} \lfloor (V \lfloor \mathbf{G}) = \mathbf{B}_T \lfloor (V \lfloor \mathbf{D}_T) + (\mathbf{E}_T \wedge \mathbf{D}_T)[[V]] \mathbf{h}. \qquad (7.48)$$

Due to Exercise 3.21 the first term is opposite to that in (7.47), whereas the second is the missing energy density. In this manner the difference of (7.47) and (7.48) is needed and we arrive at the following formula for the energy-momentum one-form of the electromagnetic field contained in the region described by V:

$$\hat{\mathbf{p}}(V) = \frac{1}{2} \{ \mathbf{G} \lfloor (V \lfloor \mathbf{F}) - \mathbf{F} \lfloor (V \lfloor \mathbf{G}) \}. \qquad (7.49)$$

After using (7.47) and (7.48), this may be written as:

$$\hat{\mathbf{p}}(V) = \mathbf{D}_T \lfloor (V \lfloor \mathbf{B}_T) - \frac{1}{2} (\mathbf{E}_T \wedge \mathbf{D}_T + \mathbf{H}_T \wedge \mathbf{B}_T)[[V]] \mathbf{h}. \tag{7.50}$$

In the second term, we recognize the prolongation of the spatial energy density w of the electromagnetic field:

$$\hat{\mathbf{p}}(V) = \mathbf{D}_T \lfloor (V \lfloor \mathbf{B}_T) - w_T [[V]] \mathbf{h}. \tag{7.51}$$

In this manner we see that for the purely spatial region V, the expression (7.49) yields the momentum and energy contained in it, in the combination corresponding to (7.11) The mapping $V \to \hat{\mathbf{p}}$, contained in (7.51), is called the *energy-momentum tensor*.

It is worthwhile checking what the formula (7.49) gives for another trivector, describing a mixed spatio-temporal region in the form:

$$V_1 = \mathbf{R} \wedge \mathbf{e}_T \Delta t, \tag{7.52}$$

where $\mathbf{e}_T$ is the previously mentioned vector in the time direction, Δt is the length of a time interval, and $\mathbf{R}$ is a spatial bivector. In order to find the expression (7.49) for (7.52), we calculate first

$$V_1 \lfloor \mathbf{F} = (\mathbf{R} \wedge \mathbf{e}_T \Delta t) \lfloor (\mathbf{B}_T + \mathbf{E}_T \wedge \mathbf{h}) = (\mathbf{R} \wedge \mathbf{e}_T \Delta t) \lfloor \mathbf{B}_T + (\mathbf{R} \wedge \mathbf{e}_T \Delta t) \lfloor (\mathbf{E}_T \wedge \mathbf{h})$$

During the contraction of the first bracket with $\mathbf{B}_T$, the expression $\mathbf{e}_T \rfloor \mathbf{B}_T$ emerges, which is zero; hence, by virtue of (3.26), there remains

$$V_1 \lfloor \mathbf{F} = -\mathbf{B}_T [[\mathbf{R}]] \mathbf{e}_T \Delta t + \Delta t \{ (\mathbf{R} \wedge \mathbf{e}_T) \lfloor \mathbf{E}_T \} \lfloor \mathbf{h}.$$

We use (3.23)

$$V_1 \lfloor \mathbf{F} = -\Delta t \, \mathbf{B}_T [[\mathbf{R}]] \mathbf{e}_T + \Delta t \{ \mathbf{E}_T [[\mathbf{e}_T]] \mathbf{R} - (\mathbf{R} \lfloor \mathbf{E}_T) \wedge \mathbf{e}_T \} \lfloor \mathbf{h}.$$

We have $\mathbf{E}_T [[\mathbf{e}_T]] = 0$, so we are allowed to write down

$$V_1 \lfloor \mathbf{F} = -\Delta t \, \mathbf{B}_T [[\mathbf{R}]] \mathbf{e}_T - \Delta t \{ \mathbf{h}[[\mathbf{e}_T]] \, (\mathbf{R} \lfloor \mathbf{E}_T) - \mathbf{h}[[\mathbf{R} \lfloor \mathbf{E}_T]] \mathbf{e}_T \}.$$

The vector $\mathbf{R} \lfloor \mathbf{E}_T$ lies in Σ, so $\mathbf{h}[[\mathbf{R} \lfloor \mathbf{E}_T]] = 0$. Moreover, $\mathbf{h}[[\mathbf{e}_T]] = 1$ and we obtain eventually

$$V_1 \lfloor \mathbf{F} = -\Delta t \, \{ \mathbf{B}_T [[\mathbf{R}]] \mathbf{e}_T + \mathbf{R} \lfloor \mathbf{E}_T \} \tag{7.53}$$

The first term in the curly bracket is a temporal vector, and the second one is spatial.

We calculate now the first term in (7.49) for the trivector (7.52):

$$\mathbf{G}\lfloor(V_1\lfloor\mathbf{F}) = -(\mathbf{D}_T - \mathbf{H}_T \wedge \mathbf{h})\lfloor\{\mathbf{B}_T[[\mathbf{R}]]\,\mathbf{e}_T + \mathbf{R}\lfloor\mathbf{E}_T\}\,\Delta t$$

$$= -\{\mathbf{D}_T\lfloor(\mathbf{B}_T[[\mathbf{R}]]\,\mathbf{e}_T) - (\mathbf{H}_T \wedge \mathbf{h})\lfloor(\mathbf{B}_T[[\mathbf{R}]]\,\mathbf{e}_T) + \mathbf{D}_T\lfloor(\mathbf{R}\lfloor\mathbf{E}_T) - (\mathbf{H}_T \wedge \mathbf{h})\lfloor(\mathbf{R}\lfloor\mathbf{E}_T)\}\,\Delta t.$$

The expressions $\mathbf{D}_T\lfloor\mathbf{e}_T$, $\mathbf{H}_T[[\mathbf{e}_T]]$, $\mathbf{h}[[\mathbf{R}\lfloor\mathbf{E}_T]]$ are zeros; hence

$$\mathbf{G}(\lfloor V_1\lfloor\mathbf{F}) = \{\mathbf{B}_T[[\mathbf{R}]]\,\mathbf{H}_T - \mathbf{D}_T\lfloor(\mathbf{R}\lfloor\mathbf{E}_T) + \mathbf{H}_T[[\mathbf{R}\lfloor\mathbf{E}_T]]\,\mathbf{h}\}\,\Delta t.$$

The identity $\mathbf{D}_T\lfloor(\mathbf{R}\lfloor\mathbf{E}_T) = -\mathbf{D}_T[[\mathbf{R}]]\,\mathbf{E}_T - (\mathbf{E}_T \wedge \mathbf{D}_T)\lfloor\mathbf{R}$ follows from (3.21). Due to this we replace the second term:

$$\mathbf{G}\lfloor(V_1\lfloor\mathbf{F}) = \{\mathbf{B}_T[[\mathbf{R}]]\,\mathbf{H}_T + \mathbf{D}_T[[\mathbf{R}]]\,\mathbf{E}_T + (\mathbf{E}_T \wedge \mathbf{D}_T)\lfloor\mathbf{R} + \mathbf{H}_T[[\mathbf{R}\lfloor\mathbf{E}_T]]\,\mathbf{h}\}\,\Delta t.$$

For the last term we use the identity $\mathbf{H}_T[[\mathbf{R}\lfloor\mathbf{E}_T]] = -(\mathbf{E}_T \wedge \mathbf{H}_T)[[\mathbf{R}]]$ following from (3.15) and (3.20):

$$\mathbf{G}\lfloor(V_1\lfloor\mathbf{F}) = \{\mathbf{B}_T[[\mathbf{R}]]\,\mathbf{H}_T + \mathbf{D}_T[[\mathbf{R}]]\,\mathbf{E}_T + (\mathbf{E} \wedge \mathbf{D})_T\lfloor\mathbf{R} - (\mathbf{E} \wedge \mathbf{H})_T[[\mathbf{R}]]\,\mathbf{h}\}\,\Delta t. \tag{7.54}$$

Similar calculations for the second term in (7.49) lead to the result:

$$\mathbf{F}\lfloor(V_1\lfloor\mathbf{G}) = \{-\mathbf{D}_T[[\mathbf{R}]]\,\mathbf{E}_T - \mathbf{B}_T[[\mathbf{R}]]\,\mathbf{H}_T - (\mathbf{H} \wedge \mathbf{B})_T\lfloor\mathbf{R} - (\mathbf{H} \wedge \mathbf{E})_T[[\mathbf{R}]]\,\mathbf{h}\}\,\Delta t. \tag{7.55}$$

Inserting the two last expressions into (7.49) yields the energy-momentum one-form of the field contained in the spatio-temporal volume described by the trivector $V_1 = \mathbf{R} \wedge \mathbf{e}_T\,\Delta t$:

$$\hat{\mathbf{p}}(V_1) = \left\{\mathbf{D}_T[[\mathbf{R}]]\,\mathbf{E}_T + \mathbf{B}_T[[\mathbf{R}]]\,\mathbf{H}_T + \frac{1}{2}\,(\mathbf{E} \wedge \mathbf{D} + \mathbf{H} \wedge \mathbf{B})_T\lfloor\mathbf{R} - (\mathbf{E} \wedge \mathbf{H})_T[[\mathbf{R}]]\,\mathbf{h}\right\}\,\Delta t. \tag{7.56}$$

The presence of the time interval Δt means that this is a flow of the energy-momentum through the surface $\mathbf{R}$ in the chosen time Δt. The third term in the curly bracket describes the energy passing through $\mathbf{R}$. The last term is the value of the Poynting form (i.e., the energy flux) (4.14) on the bivector $\mathbf{R}$, multiplied by the one-form $\mathbf{h}$:

$$\hat{\mathbf{p}}(V_1) = \{\mathbf{D}_T[[\mathbf{R}]]\,\mathbf{E}_T + \mathbf{B}_T[[\mathbf{R}]]\,\mathbf{H}_T + w_T\lfloor\mathbf{R} - \mathbf{S}_T[[\mathbf{R}]]\,\mathbf{h}\}\,\Delta t. \tag{7.57}$$

The first three first terms describe a flow of momentum through $\mathbf{R}$ in time Δt. Thanks to the classical formula $\mathcal{F}\Delta t = \Delta\mathbf{p}$, relating the force impulse with the increase of momentum, we interpret the three first terms in the curly bracket as a force exerted by the electromagnetic field on the surface described by $\mathbf{R}$:

$$\mathcal{F} = \mathbf{D}_T[[\mathbf{R}]]\,\mathbf{E}_T + \mathbf{B}_T[[\mathbf{R}]]\,\mathbf{H}_T + w_T\lfloor\mathbf{R} \tag{7.58}$$

A mapping $\mathbf{R} \to \mathcal{F}$, contained in this formula, can be expressed as the tensor

$$\kappa_1 = \mathbf{E}_T \otimes \mathbf{D}_T + \mathbf{H}_T \otimes \mathbf{B}_T + w_T\lfloor. \tag{7.59}$$

This is a space-time prolongation (with the minus sign) of the tensor (6.81), called the *energy-stress tensor* of the electromagnetic field. It has three components, because the spatial bivector $\mathbf{R}$ present in (7.52) may have three linearly independent directions. Hence the volume trivector (7.52) has as many directions.

Thus we have seen that the formula (7.49)—depending on a type of three-dimensional volume (purely spatial or spatio-temporal) in four-dimensional space-time—yields the energy and momentum or energy and stress of the electromagnetic field. The mapping $V \to \hat{\mathbf{p}}$, contained in it, should be called the stress-energy-momentum tensor, but the alternative names *stress-energy tensor* or *energy-momentum tensor* are used. For the purely spatial volume V, (7.49) yields the energy and momentum of the field contained in V, whereas for the spatio-temporal volume $V_1 = \mathbf{R} \wedge \mathbf{e}_T\,\Delta t$ it gives the energy flux through surface $\mathbf{R}$ in time Δt and the momentum transmitted through $\mathbf{R}$ by the stress forces of the field also in time Δt.

7.4 Plane Wave

We recall formula (7.10) for the spatio-temporal phase density (wave covector) of any wave, which for the space-time position vector $\hat{\mathbf{x}} = \mathbf{r} + t\mathbf{e}_T$ in accordance with (7.9) yields for the phase of a plane wave

$$\varphi = \hat{\mathbf{k}}[[\hat{\mathbf{x}}]] = \mathbf{k}[[\mathbf{r}]] - \omega t. \tag{7.60}$$

The different signs in front of the spatial and temporal parts of (7.60) imply that when time flows the surfaces of constant phase $\phi = \phi_0 = \text{const}$ move in Σ in the direction of increasing values of $\mathbf{k}$ (since when looking for points with phase ϕ_0 in time moment t, we seek vectors $\mathbf{r}$ fulfilling the equation $\mathbf{k}[[\mathbf{r}]] = \omega t + \phi_0$).

We want to find the fields of a plane electromagnetic wave. In a space-time region devoid of charges and currents, we are looking for a solution of Maxwell's equations in the form

$$\mathbf{F}(\hat{\mathbf{x}}) = \psi(\phi)\,\tilde{\mathbf{F}}, \tag{7.61}$$

$$\mathbf{G}(\hat{\mathbf{x}}) = \psi(\phi)\,\tilde{\mathbf{G}}, \tag{7.62}$$

Where $\tilde{\mathbf{F}}$, $\tilde{\mathbf{G}}$ are constant two-forms, and ψ is a scalar function of the scalar variable ϕ treated as the phase. Mostly ψ is taken as a combination of sine and cosine function, which is tantamount to the assumption that the wave is harmonic in its time dependence. We present our reasoning without this assumption. The presence of the same function ψ in front of $\tilde{\mathbf{F}}$ and $\tilde{\mathbf{G}}$ expresses the synchronicity of changes of the Faraday form $\mathbf{F}$ and Gauss form $\mathbf{G}$.

The exterior derivatives of (7.61) and (7.62) are

$$\hat{\mathbf{d}} \wedge \mathbf{F} = \psi'(\phi)\,\mathbf{d}\phi \wedge \tilde{\mathbf{F}},$$

$$\hat{\mathbf{d}} \wedge \mathbf{G} = \psi'(\phi)\,\mathbf{d}\phi \wedge \tilde{\mathbf{G}}.$$

The exterior derivative of ϕ may be expressed by the basic one-forms:

$$\hat{\mathbf{d}}\phi = \frac{\partial \phi}{\partial t}\,\mathbf{f}^0 + \frac{\partial \phi}{\partial x^1}\,\mathbf{f}^1 + \frac{\partial \phi}{\partial x^2}\,\mathbf{f}^2 + \frac{\partial \phi}{\partial x^3}\,\mathbf{f}^3. \qquad (7.63)$$

For the time-harmonic wave the coordinate $\frac{\partial \phi}{\partial t} = k_0 = \omega$ should be interpreted as the circular frequency of the plane wave, and the expressions $\frac{\partial \phi}{\partial x^i} = k_i$ as the spatial coordinates of the phase density one-form; therefore $\mathbf{d}\phi$ may be considered as the four-dimensional wave covector. Thus we are allowed to write down:

$$\hat{\mathbf{d}}\phi = \hat{\mathbf{k}}. \qquad (7.64)$$

Since we have assumed that charges and currents are absent, both Maxwell equations (7.23) and (7.29) are homogeneous and yield the conditions

$$\hat{\mathbf{k}} \wedge \tilde{\mathbf{F}} = 0, \qquad (7.65)$$

$$\hat{\mathbf{k}} \wedge \tilde{\mathbf{G}} = 0. \qquad (7.66)$$

Equation (7.65) allows us to look for $\tilde{\mathbf{F}}$ in the form of an exterior product

$$\tilde{\mathbf{F}} = \hat{\mathbf{k}} \wedge \mathbf{l}, \qquad (7.67)$$

where the one-form $\mathbf{l}$ is constant and plays the role of polarization in the ordinary description of electromagnetic waves. Let us take it parallel to vector $\mathbf{e}_T$, which implies $\mathbf{l}[[\mathbf{e}_T]] = 0$ and means that $\mathbf{l}$ is a prolongation of a spatial one-form.

A similar substitution

$$\tilde{\mathbf{G}} = \hat{\mathbf{k}} \wedge \mathbf{m} \qquad (7.68)$$

with a constant one-form $\mathbf{m}$ fulfilling condition $\mathbf{m}[[\mathbf{e}_T]] = 0$ gives a solution of (7.66). The one-form $\mathbf{m}$, however, can not be chosen arbitrarily, because also the constitutive relations between $\mathbf{F}$ and $\mathbf{G}$ should be taken into account. This is not important for our present considerations; therefore we accept the field quantities of the electromagnetic wave in the form

$$\mathbf{F} = \psi(\phi)\,\hat{\mathbf{k}} \wedge \mathbf{l}, \tag{7.69}$$

$$\mathbf{G} = \psi(\phi)\,\hat{\mathbf{k}} \wedge \mathbf{m}, \tag{7.70}$$

where $\hat{\mathbf{k}} = \hat{\mathbf{d}}\phi$. Because of the presence of the common factor $\hat{\mathbf{k}}$, all three external products $\mathbf{F} \wedge \mathbf{G}$, $\mathbf{F} \wedge \mathbf{F}$, $\mathbf{G} \wedge \mathbf{G}$ are zero. This means that both Lorentz invariants vanish for the plane electromagnetic wave.

We proceed to calculate the energy-momentum of this wave contained in a region represented by trivector $V = \mathbf{e}_{123}$ with linearly independent vectors $\mathbf{e}_i$. We have by virtue of (3.29)

$$V \lfloor \mathbf{F} = \psi\,\mathbf{e}_{123} \lfloor (\hat{\mathbf{k}} \wedge \mathbf{l}) = \psi\{(k_3 l_2 - k_2 l_3)\mathbf{e}_1 + (k_1 l_3 - k_3 l_1)\mathbf{e}_2 + (k_2 l_1 - k_1 l_2)\mathbf{e}_3\},$$

where $k_i = \hat{\mathbf{k}}[\mathbf{e}_i]$ and similarly with l_j. We calculate the next quantity

$$\mathbf{G} \lfloor (V \lfloor \mathbf{F}) = \psi^2(\hat{\mathbf{k}} \wedge \mathbf{m}) \lfloor \{(k_3 l_2 - k_2 l_3)\mathbf{e}_1 + (k_1 l_3 - k_3 l_1)\mathbf{e}_2 + (k_2 l_1 - k_1 l_2)\mathbf{e}_3\}$$

$$= (k_3 l_2 - k_2 l_3)(m_1\hat{\mathbf{k}} - k_1\mathbf{m}) + (k_1 l_3 - k_3 l_1)(m_2\hat{\mathbf{k}} - k_2\mathbf{m}) + (k_2 l_1 - k_1 l_2)(m_3\hat{\mathbf{k}} - k_3\mathbf{m})$$

$$= \psi^2\{m_1(k_3 l_2 - k_2 l_3) + m_2(k_1 l_3 - k_3 l_1) + m_3(k_2 l_1 - k_1 l_2)\}\hat{\mathbf{k}}.$$

We write the expression in the curly bracket with the notation of space vectors replacing the one-forms

$$\mathbf{G} \lfloor (V \lfloor \mathbf{F}) = \psi^2\{\mathbf{m} \cdot (\mathbf{k} \times \mathbf{h})\}\hat{\mathbf{k}}.$$

Analogous calculations lead to

$$\mathbf{F} \lfloor (V \lfloor \mathbf{G}) = -\psi^2\{\mathbf{m} \cdot (\mathbf{k} \times \mathbf{h})\}\hat{\mathbf{k}}.$$

Inserting the two last results to (7.49) gives

$$\hat{\mathbf{p}}(V) = \psi^2\,\{\mathbf{m} \cdot (\mathbf{k} \times \mathbf{h})\}\hat{\mathbf{k}}. \tag{7.71}$$

The energy-momentum one-form of the plane electromagnetic wave is proportional to the wave density one-form. This property belongs to the premetric electrodynamics, and hence is rather fundamental. It should be viewed in the light of the de Broglie relations, which are the basis of quantum mechanics.

Let us see also how the two-form (7.69) decomposes into magnetic and electric parts:

$$\mathbf{F} = \psi(\phi)(\mathbf{k_T} - \omega\mathbf{h}) \wedge \mathbf{l} = \psi(\phi)\mathbf{k_T} \wedge \mathbf{l} + \psi(\phi)\omega\mathbf{l} \wedge \mathbf{h}. \tag{7.72}$$

After comparing with (7.19) we notice that

$$\mathbf{B} = \psi(\phi)\mathbf{k} \wedge \mathbf{l_\Sigma}, \quad \mathbf{E} = \omega\psi(\phi)\mathbf{l_\Sigma}, \tag{7.73}$$

where $\mathbf{l_\Sigma}$ is the restriction of $\mathbf{l}$ to the three-space Σ. We see from this that the relation (5.16)

$$\mathbf{B} = \omega^{-1}\mathbf{k} \wedge \mathbf{E}$$

is fulfilled, which has been obtained in consideration of the plane electromagnetic wave in three-dimensional space.

7.5 Scalar Product in Space-Time

The considerations of this chapter up to now belong to the *premetric electrodynamics*, because we did not use any metric in three-dimensional space or in the space-time. In special relativity theory, a discussion about Lorentz transformations occupies a lot of space. These are linear transformations preserving a bilinear form in space-time, which creates from it the so-called *Minkowski space*. We are going to present now this subject assuming that the reader is acquainted with special relativity.

Space-time is being built by a specific observer in our approach. In a four-dimensional linear space M of events a three-dimensional subspace Σ consisting of simultaneous events is connected to the moment of time $t = 0$. We call the vectors belonging to it *spatial vectors*. We distinguish a straight line protruding in the fourth dimension—this corresponds to "motion" of the observer placed at the spatial centre of the coordinate system. We call this line the *time axis* T and the vectors belonging to it *temporal vectors*. We are able to present this graphically only in three dimensions. We draw a subspace of constant time Σ as a plane, and the time axis T as a straight line puncturing Σ obliquely, because a perpendicularity is not yet defined (see Fig. 7.7). We have marked the vector $\mathbf{e}_T$ on T, which corresponds to a segment of one second. We interpret this as a unit of the observer's "movement" in space-time but resting in space Σ. The observer has prepared a collection of clocks deployed in Σ and synchronized with his or her own clock. Hence each vector representing an event can be uniquely decomposed into temporal and spatial parts—this is the decomposition into the cartesian product $M = T \times \Sigma$, shown in Fig. 7.8.

Fig. 7.9 Temporal and
spatial parts of
spatio-temporal phase
velocity

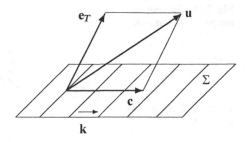

Fig. 7.10 Prolongation $\mathbf{k}_T$ of
a spatial wave covector into
space-time

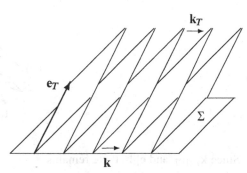

The special relativity theory was created from contemplations about the particular role of light as an electromagnetic wave; therefore we consider the propagation of a plane wave in space-time. By using the background of Fig. 7.7, we draw in Σ the spatial wave covector $\mathbf{k}$ of the plane wave propagating to the right, in the shape of a family of parallel and equidistant lines (see Fig. 7.9). We assume that this occurs in the vacuum, so it is worthwhile to choose in Σ the natural metric that allows us to introduce vector perpendicular to $\mathbf{k}$. In this manner we introduce the vector $\mathbf{c}$ of the phase velocity of the wave. It informs us about how far a specific plane of fixed phase moves during a previously chosen unit of time. Thus the spatio-temporal velocity of such a plane (the phase velocity) is defined by the sum $\mathbf{u} = \mathbf{c} + \mathbf{e}_T$ (see Fig. 7.9). Due to (7.4), both terms of this sum have the same unit.

This was an introduction. We proceed now to represent graphically the spactio-temporal wave covector (7.10), which is the linear combination

$$\hat{\mathbf{k}} = \mathbf{k}_T - \omega\,\mathbf{h}, \tag{7.74}$$

where $\mathbf{h}[[\mathbf{e}_T]] = 1$. The prolongation $\mathbf{k}_T$ of $\mathbf{k}$ looks as in Fig. 7.10. The one-form $\omega\,\mathbf{h}$ is parallel to Σ and looks as in Fig. 7.11. The combination (7.74) is shown in Fig. 7.12. The one-form $\hat{\mathbf{k}}$ has its attitude bent to the right, which corresponds to a wave propagating to the right.

Let us check the value of $\hat{\mathbf{k}}$ on $\mathbf{u}$:

$$\hat{\mathbf{k}}[[\mathbf{u}]] = (\mathbf{k}_T - \omega\mathbf{h})[[\mathbf{c} + \mathbf{e}_T]] = \mathbf{k}_T[[\mathbf{c}]] + \mathbf{k}_T[[\mathbf{e}_T]] - \omega\,\mathbf{h}[[\mathbf{c}]] - \omega\,\mathbf{h}[[\mathbf{e}_T]].$$

Fig. 7.11 Forming the
temporal one-form $\omega\mathbf{h}$

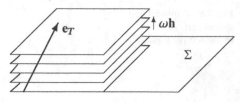

Fig. 7.12 Combination
$\mathbf{k}_T - \omega\,\mathbf{h}$

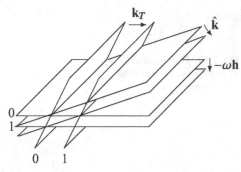

Since $\mathbf{k}_T\|\mathbf{e}_T$ and $\mathbf{c}\|\mathbf{h}$, there remains

$$\hat{\mathbf{k}}[[\mathbf{u}]] = \mathbf{k}_T[[\mathbf{c}]] - \omega = \mathbf{k}[[\mathbf{c}]] - \omega.$$

The expression $\mathbf{k}[[\mathbf{c}]]$ can be written in the traditional language as $\mathbf{k}[[\mathbf{c}]] = kc$, where k is the magnitude of the wave covector. It is, moreover, known that $\omega/k = c$; hence we obtain

$$\hat{\mathbf{k}}[[\mathbf{u}]] = kc - \omega = \omega - \omega = 0.$$

This result should not be a surprise, because we mentioned before Fig. 7.9 that $\mathbf{u}$ describes the movement of the plane of a fixed phase, so the increment of the phase should be zero on this vector. This means geometrically that $\mathbf{u}$ is parallel to $\hat{\mathbf{k}}$; hence the world line of a light signal is parallel to the spatio-temporal wave covector of the light wave.

In the special relativity theory, the role of light is distinguished, hence, correspondingly, we choose a basis and a metric in space-time. If we have the basis $\{\mathbf{e}_1, \mathbf{e}_2, \mathbf{e}_3\}$ in Σ, orthonormal with respect to the natural metric, we choose the unit vector $\mathbf{e}_0$ on the time axis T as such a segment of time, in which the light passes at a distance corresponding to the vector $\mathbf{e}_1$. It follows from the isotropy of space that equally good are the vectors $\mathbf{e}_2$, $\mathbf{e}_3$ or any other vector in Σ corresponding to a unit of length: let it be one metre. In order to compare two temporal vectors $\mathbf{e}_0$ and $\mathbf{e}_T$ (the latter represents one second) we should realize that time needed by light to travel one metre, is c times shorter; hence

$$\mathbf{e}_T = c\,\mathbf{e}_0 \tag{7.75}$$

Fig. 7.13 Normalization of
the temporal vector

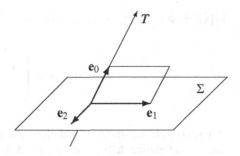

(here c is the speed of light in a vacuum), and we are allowed to write down

$$\hat{\mathbf{x}} = t\mathbf{e}_T + \mathbf{r} = ct\,\mathbf{e}_0 + \mathbf{r}; \tag{7.76}$$

thus $x^0 = ct$. In this manner we create the basis $\{\mathbf{e}_0, \mathbf{e}_1, \mathbf{e}_2, \mathbf{e}_3\}$ in the space-time M (see Fig. 7.13).[4] This should be an orthonormal basis of Minkowski space; therefore the scalar-product matrix with elements $g_{\mu\nu} = g(\mathbf{e}_\mu, \mathbf{e}_\nu)$ is diagonal.

One should, however, note that the Minkowskian scalar product does not satisfy all the axioms named in Sect. 1.1: namely, it is not positive. Strictly speaking one should reject the name "scalar product" and use for instance the name "bilinear form". But the tradition is so strong, that we shall use the commonly accepted name. We enumerate now the properties of the *scalar product in space-time*:

1. *linearity for the first factor:* $g(\mathbf{u} + \mathbf{v}, \mathbf{w}) = g(\mathbf{u}, \mathbf{w}) + g(\mathbf{v}, \mathbf{w})$, $g(\lambda\mathbf{v}, \mathbf{w}) = \lambda g(\mathbf{v}, \mathbf{w})$ for any $\lambda \in \mathbf{R}$,
2. *symmetry:* $g(\mathbf{v}, \mathbf{w}) = g(\mathbf{w}, \mathbf{v})$,
3. *nondegeneracy:* $[g(\mathbf{v}, \mathbf{u}) = 0 \,\forall \mathbf{v} \in M] \Rightarrow \mathbf{u} = 0$.

The question of signs on the diagonal of the scalar product matrix (also called the *metric matrix*) $\mathcal{G} = \{g_{\mu\nu}\}$ has to be decided. We have noticed in Sect. 5.1 that there are two possibilities for introducing the phase of the plane wave propagating in space, namely

$$\phi = \mathbf{k}[[\mathbf{r}]] - \omega t \quad \text{or} \quad \phi' = \omega t - \mathbf{k}[[\mathbf{r}]], \tag{7.77}$$

(see Eq. (5.5)). The inverse metric matrix serves to change the one-form $\hat{\mathbf{k}}$ into the vector $\hat{\mathbf{k}}$ so as to assure the equality $\hat{\mathbf{k}}[[\hat{\mathbf{x}}]] = g(\hat{\mathbf{k}}, \hat{\mathbf{x}})$; hence when we choose

---

[4] Of course, there is no place for the vector $\mathbf{e}_3$ in this figure.

$\hat{\mathbf{k}}[[\hat{\mathbf{x}}]] = \phi = \mathbf{k}[[\mathbf{r}]] - \omega t$ according to (7.74), then the metric matrix has the shape

$$\mathcal{G} = \begin{pmatrix} -1\ 0\ 0\ 0 \\ 0\ 1\ 0\ 0 \\ 0\ 0\ 1\ 0 \\ 0\ 0\ 0\ 1 \end{pmatrix}. \tag{7.78}$$

For such a choice, the diagonal elements are $g_{00} = -1$, $g_{11} = g_{22} = g_{33} = 1$. For the second choice $\hat{\mathbf{k}}[[\hat{\mathbf{x}}] = \phi' = \omega t - \mathbf{k}[[\mathbf{r}]]$, the scalar product matrix is

$$\mathcal{G}' = \begin{pmatrix} 1\ \ 0\ \ 0\ \ 0 \\ 0\ -1\ \ 0\ \ 0 \\ 0\ \ 0\ -1\ \ 0 \\ 0\ \ 0\ \ 0\ -1 \end{pmatrix}, \tag{7.79}$$

and the diagonal elements are $g_{00} = 1$, $g_{11} = g_{22} = g_{33} = -1$. Both choices are equally good. As a consequence of choosing the first option in (7.77) we are obliged to use the scalar product with the matrix (7.78).

Let $\{\mathbf{e}_\mu\}$, $\mu \in \{0, 1, 2, 3\}$ be the *orthonormal basis for space-time*, such that

$$g(\mathbf{e}_\mu, \mathbf{e}_\nu) = g_{\mu\nu}, \tag{7.80}$$

where $g_{\mu\nu}$ are elements of the matrix (7.78). Because of (7.80), the basic elements $\mathbf{e}_\mu$ do not contain any unit, in accordance with (7.4) and (7.75). For a space-time vector expressed in this basis as $\hat{\mathbf{x}} = x^\mu \mathbf{e}_\mu = x^0 \mathbf{e}_0 + \mathbf{r}$ with the summation over μ from 0 to 3, its scalar square is

$$g(\hat{\mathbf{x}}, \hat{\mathbf{x}}) = -(x^0)^2 + (x^1)^2 + (x^2)^2 + (x^3)^2 = -(x^0)^2 + \mathbf{r}^2. \tag{7.81}$$

Due to (7.76), this scalar square can be written differently:

$$g(\hat{\mathbf{x}}, \hat{\mathbf{x}}) = -c^2 t^2 + \mathbf{r}^2.$$

A set of points with fixed scalar square satisfies the equation

$$g(\hat{\mathbf{x}}, \hat{\mathbf{x}}) = (x^1)^2 + (x^2)^2 + (x^3)^2 - (x^0)^2 = \lambda = \text{const.}$$

Depending on the sign of λ, they are the following sets:
   one-sheet hyperboloid for $\lambda > 0$ (Fig. 7.14);
   oblique cone for $\lambda = 0$ (only the upper half is shown in Fig. 7.15);
   two-sheet hyperboloid for $\lambda < 0$ (only the upper part is shown in Fig. 7.16).
   Let us devote some attention to the vector $\mathbf{e}_T$, which influences the prescription of prolongation of forms defined on Σ to the space-time. We claimed earlier that the direction of $\mathbf{e}_T$ is arbitrary with the only condition being that it is nonparallel

Fig. 7.14 Tip points of vectors with positive square (7.81)

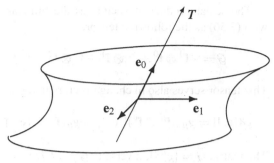

Fig. 7.15 Tip points of vectors with zero square (7.81)

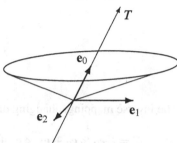

Fig. 7.16 Tip points of vectors with negative square (7.81)

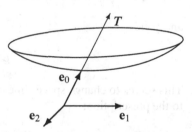

to Σ. In this manner the prescription of prolongation seems to be non-unique. We should explain that this does not matter. In fact, $\mathbf{e}_T$ lies on the world line of the observer resting with respect to the reference system, in which Σ denotes the three-dimensional space of constant time zero. The presented construction of the Minkowskian metric ensures that $\mathbf{e}_T$ parallel to $\mathbf{e}_0$ is orthogonal to Σ. Thus drawing $\mathbf{e}_T$ oblique to Σ has no effect on metrical properties—anyway it must be called perpendicular to the three-space of constant time.

We introduce the dual basis $\mathbf{f}^\nu$ by the condition $\mathbf{f}^\nu[[\mathbf{e}_\mu]] = \delta^\nu_\mu$; then due to (7.75), the parallel one-forms $\mathbf{h}$ and $\mathbf{f}^0$ must satisfy the relation

$$\mathbf{h} = c^{-1}\mathbf{f}^0. \tag{7.82}$$

In this case formula (7.74) can be written alternatively:

$$\hat{\mathbf{k}} = \mathbf{k}_T - \frac{\omega}{c}\mathbf{f}^0.$$

The scalar product of vectors as the bilinear form can be written by analogy with (3.56) as the following tensor:[5]

$$\mathcal{G} = -\mathbf{f}^0 \otimes \mathbf{f}^0 + \mathbf{f}^1 \otimes \mathbf{f}^1 + \mathbf{f}^2 \otimes \mathbf{f}^2 + \mathbf{f}^3 \otimes \mathbf{f}^3 = g_{\mu\nu}\mathbf{f}^\mu \otimes \mathbf{f}^\nu. \tag{7.83}$$

This tensor serves also to change vectors into one-forms:

$$\mathcal{G}[[\mathbf{e}_\lambda]] = g_{\mu\nu}(\mathbf{f}^\mu \otimes \mathbf{f}^\nu)[[\mathbf{e}_\lambda]] = g_{\mu\nu}\mathbf{f}^\mu \, \mathbf{f}^\nu[[\mathbf{e}_\lambda]] = g_{\mu\nu}\mathbf{f}^\mu \, \delta^\nu_\lambda = g_{\mu\lambda}\mathbf{f}^\mu. \tag{7.84}$$

The matrix $\tilde{\mathcal{G}} = \{g^{\mu\nu}\}$, inverse to $\mathcal{G}$ of (7.78), is

$$\tilde{\mathcal{G}} = \begin{pmatrix} -1 & 0 & 0 & 0 \\ 0 & 1 & 0 & 0 \\ 0 & 0 & 1 & 0 \\ 0 & 0 & 0 & 1 \end{pmatrix}. \tag{7.85}$$

The inverse mapping, changing one-forms into vectors, is expressed by the tensor

$$\tilde{\mathcal{G}} = -\mathbf{e}_0 \otimes \mathbf{e}_0 + \mathbf{e}_1 \otimes \mathbf{e}_1 + \mathbf{e}_2 \otimes \mathbf{e}_2 + \mathbf{e}_3 \otimes \mathbf{e}_3 = g^{\mu\nu}\mathbf{e}_\mu \otimes \mathbf{e}_\nu \tag{7.86}$$

and yields

$$\tilde{\mathcal{G}}[[\mathbf{f}^\lambda]] = g^{\mu\lambda}\mathbf{e}_\mu. \tag{7.87}$$

This serves to change space-time one-forms into spatio-temporal vectors according to the prescription:

$$\hat{\mathbf{k}} = k_\nu\mathbf{f}^\nu \;\mapsto\; \hat{\mathbf{k}} = \tilde{\mathcal{G}}[[\hat{\mathbf{k}}]] = k_\nu\tilde{\mathcal{G}}[[\mathbf{f}^\nu]] = k_\nu g^{\mu\nu}\mathbf{e}_\mu = k^\mu\mathbf{e}_\mu, \quad \text{where} \quad k^\mu = g^{\mu\nu}k_\nu.$$

In this mapping, the one-form $\hat{\mathbf{k}}$ with coordinates $(k_0, k_1, k_2, k_3) = (-c^{-1}\omega, k_1, k_2, k_3)$ is changed into the vector $\hat{\mathbf{k}}$ with coordinates $(k^0, k^1, k^2, k^3) = (c^{-1}\omega, k_1, k_2, k_3)$. This vector has a positive time component, and hence may be parallel to the vector $\mathbf{u}$ depicted in Fig. 7.9. The three pluses in matrix (7.85) ensure that the spatial part of vector $\hat{\mathbf{k}}$ has the same orientation as the spatial part of the one-form $\mathbf{k}$. Matrix (7.79) can not ascertain this.

Let us look closely at this. The form $\hat{\mathbf{k}}$, shown in Fig. 7.12, in the orthonormal basis from Fig. 7.13, has coordinates $(-c^{-1}\omega, k, 0, 0) = (-k, k, 0, 0)$. Hence the respective vector $\hat{\mathbf{k}}$ has coordinates $(k, k, 0, 0)$, whereas from comparison of Figs. 7.9 and 7.13 and from (7.75), it is seen that $\mathbf{u}$ has coordinates $(c, c, 0, 0)$. We ascertain that—indeed—vectors $\hat{\mathbf{k}}$ and $\mathbf{u}$ are parallel. Both vectors have zero scalar squares according to (7.81); moreover, they are orthogonal to each other. We may

---

[5] With the summation from 0 to 3 over indices repeated on different levels.

also notice that the vector $\hat{\mathbf{k}}$ is orthogonal to planes of $\hat{\mathbf{k}}$—this should not be strange for us, because the change of a one-form into a vector according to a scalar product gives a vector with its attitude orthogonal to (hyper)planes of the initial form. We are not accustomed, however, to the fact that a vector is simultaneously parallel and orthogonal to a plane.

Exercise 7.2 Verify that the value of the one-form $\hat{\mathbf{k}}$ on an arbitrary vector $\hat{\mathbf{x}}$ is equal to the scalar product of vectors $\hat{\mathbf{k}}$ and $\hat{\mathbf{x}}$:

$$\hat{\mathbf{k}}[[\hat{\mathbf{x}}]] = g(\hat{\mathbf{k}}, \hat{\mathbf{x}}).$$

It is worth noting that the spatio-temporal wave vector $\hat{\mathbf{k}}$ has the same direction as the space-time vector $\mathbf{u}$ of the phase velocity of the light signal only due to the chosen shape (7.78) of the scalar product matrix with one minus and three pluses. If the matrix had elements with opposite signs, the wave vector would have orientation opposite to that of the velocity vector, which would be less natural. In this manner **we have found a reason to favour one of two—seemingly equivalent—scalar products in the Minkowski space**.

A generalization of the mapping (7.87) changing two-forms into bivectors can be introduced. We give it as the following formula

$$\tilde{\mathcal{G}}[[\mathbf{f}^\lambda \wedge \mathbf{f}^\kappa]] = \tilde{\mathcal{G}}[[\mathbf{f}^\lambda]] \wedge \tilde{\mathcal{G}}[[\mathbf{f}^\kappa]] = g^{\mu\lambda}\mathbf{e}_\mu \wedge g^{\nu\kappa}\mathbf{e}_\nu.$$

When acting on arbitrary two-form $\mathbf{F} = \frac{1}{2}F_{\lambda\kappa}\mathbf{f}^{\lambda\kappa}$ this yields

$$\tilde{\mathcal{G}}[[\mathbf{F}]] = \frac{1}{2}F_{\lambda\kappa}\tilde{\mathcal{G}}[[\mathbf{f}^{\lambda\kappa}]] = \frac{1}{2}F_{\lambda\kappa}g^{\mu\lambda}g^{\nu\kappa}\mathbf{e}_{\mu\nu} = \frac{1}{2}F^{\mu\nu}\mathbf{e}_{\mu\nu}. \tag{7.88}$$

In this manner coordinates (with upper indices) of the obtained bivector are related to coordinates (with lower indices) of the two-form by

$$F^{\mu\nu} = g^{\mu\lambda}g^{\nu\kappa}F_{\lambda\kappa}. \tag{7.89}$$

The Faraday form (7.19) with the substitution (7.82) has the shape:[6]

$$\mathbf{F} = \frac{1}{2}B_{ij}\,\mathbf{f}^i \wedge \mathbf{f}^j + E_j\,\mathbf{f}^j \wedge c^{-1}\mathbf{f}^0. \tag{7.90}$$

This means that its coordinates can be written as

$$F_{ij} = B_{ij}, \quad F_{i0} = c^{-1}E_j \text{ for } i, j \in \{1, 2, 3\}. \tag{7.91}$$

---

[6] We omit the subscript T for the prolongated forms.

Apply (7.88) to (7.90):

$$\tilde{\mathcal{G}}[[\mathbf{F}]] = \frac{1}{2} B_{ij}\, g^{mi} g^{nj} \mathbf{e}_m \wedge \mathbf{e}_n + c^{-1} E_j\, g^{mj} \mathbf{e}_m \wedge g^{00} \mathbf{e}_0.$$

The element $g^{00} = -1$, so

$$\tilde{\mathcal{G}}[[\mathbf{F}]] = \frac{1}{2} B_{ij}\, g^{mi} g^{nj} \mathbf{e}_m \wedge \mathbf{e}_n - c^{-1} E_j\, g^{mj} \mathbf{e}_m \wedge \mathbf{e}_0. \tag{7.92}$$

The coordinates of this bivector can be written as

$$F^{ij} = B_{ij}, \qquad F^{j0} = -c^{-1} E_j. \tag{7.93}$$

An obvious generalization of (7.87), changing three-forms into trivectors, can be introduced:

$$\tilde{\mathcal{G}}[[\mathbf{f}^\lambda \wedge \mathbf{f}^\kappa \wedge \mathbf{f}^\sigma]] = \tilde{\mathcal{G}}[[\mathbf{f}^\lambda]] \wedge \tilde{\mathcal{G}}[[\mathbf{f}^\kappa]] \wedge \tilde{\mathcal{G}}[[\mathbf{f}^\sigma]] = g^{\mu\lambda} \mathbf{e}_\mu \wedge g^{\nu\kappa} \mathbf{e}_\nu \wedge g^{\alpha\sigma} \mathbf{e}_\alpha. \tag{7.94}$$

The charge-current density twisted three-form (7.27) has the form

$$\hat{\mathbf{j}} = \rho - \mathbf{j} \wedge \mathbf{h} = \rho - \mathbf{j} \wedge c^{-1} \mathbf{f}^0 = \langle \rho \rangle\, \mathbf{f}_*^{123} - \frac{1}{2c}\, j_{kl}\, \mathbf{f}_*^{kl0} \tag{7.95}$$

where $\langle \rho \rangle$ is the numerical value of the charge density. This three-form has the coordinates

$$\hat{j}_{123} = \langle \rho \rangle, \qquad \hat{j}_{kl0} = -c^{-1} j_{kl}. \tag{7.96}$$

Apply (7.94) to (7.96):

$$\tilde{\mathcal{G}}[[\hat{j}]] = \langle \rho \rangle\, \mathbf{e}_{123}^* + \frac{1}{2c}\, j_{kl}\, \mathbf{e}_{kl0}^*. \tag{7.97}$$

The coordinates of this twisted trivector are

$$\hat{j}^{123} = \langle \rho \rangle, \qquad \hat{j}^{kl0} = c^{-1} j_{kl}. \tag{7.98}$$

The four-dimensional linear space, endowed with the scalar product

$$g(\mathbf{x}, \mathbf{y}) = x^\mu g_{\mu\nu} y^\nu \tag{7.99}$$

with matrix $\mathcal{G} = \{g_{\mu\nu}\}$ expressed by (7.78), is called *Minkowski space-time* or in brief *Minkowski space*. The linear transformations $\Lambda : \mathbf{x} \mapsto \mathbf{x}'$ preserving the scalar product:

$$g(\Lambda \mathbf{x}, \Lambda \mathbf{y}) = g(\mathbf{x}, \mathbf{y}) \tag{7.100}$$

are called *Lorentz transformations*. In typical presentations of special relativity plenty of space is devoted to covariant and contravariant vectors treated as collections of four coordinates—covariant with lower indices, contravariant with upper ones. Which of them are connected with a specific directions in space-time is not mentioned. We say in this presentation about vectors and one-forms that coordinates with upper indices are for vectors, and those with lower indices are for forms. Each type of the quantities has its separate type of direction: for vectors direction is one-dimensional, for one-forms it is three-dimensional. Formulas written traditionally by coordinates can be carried also to our presentation, because they keep their importance; only by multiplying them with basic forms or vectors do we obtain an appropriate directed quantity.

Some authors introduce basis vectors with upper indices: $\mathbf{e}^\nu = g^{\nu\mu}\mathbf{e}_\mu$. Their scalar product with the initial basis vectors is

$$g(\mathbf{e}^\nu, \mathbf{e}_\lambda) = g^{\nu\mu}g(\mathbf{e}_\mu, \mathbf{e}_\lambda) = g^{\nu\mu}g_{\mu\lambda} = \delta^\nu_\lambda.$$

For the scalar product chosen by us with matrix (7.78), this implies that the following relations are true:

$$\mathbf{e}^0 = -\mathbf{e}_0, \quad \mathbf{e}^1 = \mathbf{e}_1, \quad \mathbf{e}^2 = \mathbf{e}_2, \quad \mathbf{e}^3 = \mathbf{e}_3.$$

The same vector $\hat{\mathbf{x}}$ can be written in both bases:

$$\hat{\mathbf{x}} = x^\mu \mathbf{e}_\mu = x_\nu \mathbf{e}^\nu. \tag{7.101}$$

It can be seen from this that introducing two types of vectors (co- and contravariant) is superfluous: the two types of coordinates serve to express one vector by equality (7.101).

Let us express three quantities (7.31), (7.35) and (7.36) by the coordinates. After using expressions $\mathbf{E} \wedge \mathbf{D}$ from Sect. 4.2 and $\mathbf{H} \wedge \mathbf{B}$ from Sect. 4.3, we write

$$\mathbf{F} \wedge \mathbf{G} = [\varepsilon_0(E_i \varepsilon^{ij} E_j) - \mu_0(H_i \mu^{ij} H_j)] \, \mathbf{f}^{123}_{*T} \wedge \mathbf{h}.$$

Similar calculations can be done for the two other quantities:

$$\mathbf{F} \wedge \mathbf{F} = 2\mu_0(E_i \mu^{ij} H_j) \, \mathbf{f}^{123}_T \wedge \mathbf{h},$$

$$\mathbf{G} \wedge \mathbf{G} = 2\varepsilon_0(H_i \varepsilon^{ij} E_j) \, \mathbf{f}^{123}_T \wedge \mathbf{h}.$$

When the medium is such that the permittivity and permeability tensors are proportional (Ismo Lindell [34] calls such a medium *affinely isotropic*), the last two four-forms are proportional to each other. In particular, we obtain for the vacuum:

$$\mathbf{F} \wedge \mathbf{G} = (\varepsilon_0 \mathbf{E}^2 - \mu_0 \mathbf{H}^2) \mathbf{f}_{*T}^{123} \wedge \mathbf{h} = \varepsilon_0 (\mathbf{E}^2 - c^2 \mathbf{B}^2) \, \mathbf{f}_{*T}^{123} \wedge \mathbf{h},$$

$$\mathbf{F} \wedge \mathbf{F} = 2 \, \mathbf{E} \cdot \mathbf{B} \, \mathbf{f}_T^{123} \wedge \mathbf{h},$$

$$\mathbf{G} \wedge \mathbf{G} = 2 \, \mathbf{D} \cdot \mathbf{H} \, \mathbf{f}_T^{123} \wedge \mathbf{h} = 2 \, \frac{\varepsilon_0}{\mu_0} \, \mathbf{E} \cdot \mathbf{B} \, \mathbf{f}_T^{123} \wedge \mathbf{h}.$$

Expressions $\mathbf{E}^2 - c^2 \mathbf{B}^2$ and $\mathbf{E} \cdot \mathbf{B}$ are called *Lorentz invariants of the electromagnetic field* in physics textbooks. Two final quadrivectors are proportional:

$$\mathbf{F} \wedge \mathbf{F} = \frac{\mu_0}{\varepsilon_0} \, \mathbf{G} \wedge \mathbf{G} = Z^2 \, \mathbf{G} \wedge \mathbf{G}, \qquad (7.102)$$

where we have used the quantity $Z = \sqrt{\mu_0/\varepsilon_0}$, known as the *wave impedance of vacuum*. Since $[\frac{\mu_0}{\varepsilon_0}] = \frac{\text{Wb}}{\text{C}^2} = \frac{\text{Pl}^2}{\text{C}^4}$, the unit for Z is Pl/C^2, the same as for the electric resistance (see remark below Table 3.3). It is natural that the coefficient between two quadrivectors is a scalar.

7.5.1 Elements of Particle Mechanics

We know from previous chapters what the geometric character of a physical quantity is. For instance, position is a vector and momentum is a one-form. Velocity, as the derivative of position with respect to time, is not a vector in space-time because time is not invariant under changes of reference system—time is not a space-time scalar. Only a derivative of position with respect to an invariant scalar can be a vector. Such a scalar, invariant under Lorentz transformations, is the *proper time* of a massive particle, that is, time in a reference frame in which the particle is at momentary rest. If the motion of a particle is described by the function $\hat{\mathbf{x}}(\tau)$, where τ is the proper time of the particle, the derivative of this function is called the *four-velocity* or *space-time velocity*:

$$\mathbf{u} = \frac{d\hat{\mathbf{x}}}{d\tau} = \frac{d}{d\tau}(t \, \mathbf{e}_T + \mathbf{r}) = \frac{dt}{d\tau} \, \mathbf{e}_T + \frac{d\mathbf{r}}{dt} \frac{dt}{d\tau} = \frac{dt}{d\tau} \, (\mathbf{e}_T + \mathbf{v}) = \gamma \, (\mathbf{e}_T + \mathbf{v}),$$

$$(7.103)$$

where $\mathbf{v} = \frac{d\mathbf{r}}{dt}$ is the ordinary spatial velocity, and $\gamma = \frac{dt}{d\tau} = (1 - v^2/c^2)^{-1/2}$ is the so-called *Lorentz factor*.

One unites the energy $\mathcal{E}$ and momentum $\mathbf{p}$ of the particle into a space-time *energy-momentum one-form*. We perform this in the combination (7.11)

$$\hat{\mathbf{p}} = \mathbf{p}_T - \mathcal{E} \mathbf{h},$$

where $\mathbf{p}_T$ is the prolongation of the spatial momentum $\mathbf{p}$ to space-time, similar to what is shown in Fig. 7.10 for the wave covector. The value of this one-form on the position vector is

$$\hat{\mathbf{p}}[[\hat{\mathbf{x}}]] = (\mathbf{p}_T - \mathcal{E}\mathbf{h})[[\mathbf{r} + t\mathbf{e}_T]] = \mathbf{p}_T[[\mathbf{r}]] + t\,\mathbf{p}_T[[\mathbf{e}_T]] - \mathcal{E}\mathbf{h}[[\mathbf{r}]] - t\mathcal{E}\mathbf{h}[[\mathbf{e}_T]].$$

Now $\mathbf{p}_T \| \mathbf{e}_T$ implies $\mathbf{p}_T[[\mathbf{e}_T]] = 0$; $\mathbf{h} \| \mathbf{r}$, implies $\mathbf{h}[[\mathbf{r}]] = 0$ and there remains

$$\hat{\mathbf{p}}[[\hat{\mathbf{x}}]] = \mathbf{p}[[\mathbf{r}]] - \mathcal{E}t.$$

This corresponds to the scalar product $g(\hat{\mathbf{p}}, \hat{\mathbf{x}})$ of the momentum and position vectors in the Minkowski space. The value of the energy-momentum on the four-velocity $\mathbf{u}$ is

$$\hat{\mathbf{p}}[[\mathbf{u}]] = (\mathbf{p}_T - \mathcal{E}\mathbf{h})[[\gamma(\mathbf{e}_T + \mathbf{v})]] = \gamma\mathbf{p}_T[[\mathbf{e}_T]] + \gamma\mathbf{p}_T[[\mathbf{v}]] - \mathcal{E}\gamma\mathbf{h}[[\mathbf{e}_T]] - \mathcal{E}\gamma\mathbf{h}[[\mathbf{v}]].$$

Again two terms are zero and there remains

$$\hat{\mathbf{p}}[[\mathbf{u}]] = \gamma\,\mathbf{p}[[\mathbf{v}]] - \gamma\mathcal{E}.$$

After substitution of formulas known from relativistic kinematics: $p^j = m_0\gamma v^j$, $\mathcal{E} = m_0\gamma c^2$, with the rest mass m_0, we obtain $\mathbf{p}[[\mathbf{v}]] = m_0\gamma v^j v^j = m_0\gamma v^2$ and

$$\hat{\mathbf{p}}[[\mathbf{u}]] = m_0\gamma^2\left(v^2 - c^2\right) = \frac{m_0(v^2 - c^2)}{1 - v^2/c^2} = m_0c^2,$$

i.e., minus the rest energy of the particle.

For the description of the dynamics of a massive particle, a *four-force* or *Minkowski force* is also introduced:

$$\mathcal{F} = \gamma\left(\frac{d\mathbf{p}}{dt} - c^{-1}\frac{d\mathcal{E}}{dt}\mathbf{f}^0\right) = \frac{d\mathbf{p}}{d\tau} - \frac{d\mathcal{E}}{d\tau}\mathbf{h} = \frac{d\hat{\mathbf{p}}}{d\tau}. \tag{7.104}$$

This unites the power $\frac{d\mathcal{E}}{dt}$ accelerating the particle and the spatial force $\frac{d\mathbf{p}}{dt}$; therefore this quantity may also be called the *space-time form of power-force*.

A counterpart of the Lorentz force was missing in the premetric part. We now know why. This was because we did not have a space-time covariant velocity vector. Let us calculate the contraction of the Faraday two-form $\mathbf{F}$ with the space-time velocity $\mathbf{u}$ of a charged particle:

$$\mathbf{F}\lfloor\mathbf{u} = (\mathbf{B}_T + \mathbf{E}_T \wedge \mathbf{h})\lfloor\gamma(\mathbf{e}_T + \mathbf{v}) = \gamma\{\mathbf{B}_T\lfloor\mathbf{e}_T + \mathbf{B}_T\lfloor\mathbf{v} + (\mathbf{E}_T \wedge \mathbf{h})\lfloor\mathbf{e}_T + (\mathbf{E}_T \wedge \mathbf{h})\lfloor\mathbf{v}\}.$$

The first term vanishes because of $\mathbf{B}_T \,\|\, \mathbf{e}_T$, so

$$\mathbf{F}\lfloor\mathbf{u} = \gamma\{\mathbf{B}_T\lfloor\mathbf{v} + \mathbf{E}_T\,\mathbf{h}[[\mathbf{e}_T]] - \mathbf{h}\,\mathbf{E}_T[[\mathbf{e}_T]] + \mathbf{E}_T\,\mathbf{h}[[\mathbf{v}]] - \mathbf{h}\,\mathbf{E}_T[[\mathbf{v}]]\}.$$

We have $\mathbf{E}_T[[\mathbf{e}_T]] = 0$ for $\mathbf{E}_T \,\|\, \mathbf{e}_T$ and $\mathbf{h}[[\mathbf{v}]] = 0$ for $\mathbf{h} \,\|\, \mathbf{v}$; moreover $\mathbf{h}[[\mathbf{e}_T]] = 1$, therefore we are left with

$$\mathbf{F}\lfloor\mathbf{u} = \gamma\{\mathbf{E}_T + \mathbf{B}_T\lfloor\mathbf{v} - \mathbf{E}[[\mathbf{v}]]\,\mathbf{h}\}.$$

The identity $\mathbf{B}_T\lfloor\mathbf{v} = (\mathbf{B}\lfloor\mathbf{v})_T$ is valid for $\mathbf{v} \in \Sigma$ Hence we are allowed to write down

$$\mathbf{F}\lfloor\mathbf{u} = \gamma(\mathbf{E} + \mathbf{B}\lfloor\mathbf{v})_T - \gamma\,\mathbf{E}[[\mathbf{v}]]\,\mathbf{h}.$$

We multiply this equality by the charge q of the particle:

$$q\mathbf{F}\lfloor\mathbf{u} = \gamma(q\mathbf{E} + q\mathbf{B}\lfloor\mathbf{v})_T - \gamma q\mathbf{E}[[\mathbf{v}]]\,\mathbf{h} \tag{7.105}$$

The expression in the bracket is the Lorentz force according to (4.20); hence we claim that the nontwisted one-form (7.105) is the *space-time Lorentz force*. Its component parallel to the time axis is $(q\mathbf{F}\lfloor\mathbf{u})_T = \gamma\frac{d\mathbf{p}}{dt}$, which is the appropriate component of the Minkowski force (7.104). Its component parallel to Σ is proportional to the nontwisted spatial scalar $q\mathbf{E}[[\mathbf{v}]]$, which is the power gained by the charged particle due to acceleration by the electric field. Hence whole expression (3.10) is equal to the *power-force* one-form acting on charged particle in the electromagnetic field:

$$\mathcal{F} = q\mathbf{F}\lfloor\mathbf{u}.$$

7.6 Constitutive Relation

It was claimed in Chap. 4 that when considering an electromagnetic field in a medium, a constitutive equation is necessary. We are going to find such a relation between the Faraday form $\mathbf{F}$ and the Gauss form $\mathbf{G}$ in the four-dimensional formulation. Recall the constitutive relation (4.39) in a dielectric medium:

$$\mathbf{D} = \varepsilon_0\mathbf{f}_*^{123}\lfloor\tilde{\mathcal{G}}[[\mathbf{E}]], \tag{7.106}$$

where $\tilde{\mathcal{G}}$ is the mapping of one-forms into vectors, defined by the scalar product.[7] It explicitly shows that the constitutive relation needs a metric. We now have the

---

[7] The reader should distinguish the bold $\mathbf{G}$ denoting the Gauss form from the italic $\mathcal{G}$ denoting the metric tensor.

scalar product (7.99) in the four-dimensional space-time and the mapping (7.87) of one-forms into vectors; hence we can build a mapping of **F** into **G**.

Exercise 7.3 Show the identity with the contraction

$$\mathbf{f}^{0123} \lfloor (\mathbf{e}_m \wedge \mathbf{e}_n) = \epsilon_{mnk} \mathbf{f}^{k0}$$

where $\mathbf{f}^{0123} = \mathbf{f}^0 \wedge \mathbf{f}^1 \wedge \mathbf{f}^2 \wedge \mathbf{f}^3$ is the basic nontwisted four-form and ϵ_{mnk} is the totally antisymmetric symbol.

Exercise 7.4 Show the identity

$$\mathbf{f}^{0123} \lfloor (\mathbf{e}_m \wedge \mathbf{e}_0) = \frac{1}{2} \epsilon_{mkl} \mathbf{f}^{kl}.$$

Similar formulas are valid for the twisted forms:

$$\mathbf{f}_*^{0123} \lfloor (\mathbf{e}_m \wedge \mathbf{e}_n) = \epsilon_{mnk} \mathbf{f}_*^{k0} \tag{7.107}$$

$$\mathbf{f}_*^{0123} \lfloor (\mathbf{e}_m \wedge \mathbf{e}_0) = \frac{1}{2} \epsilon_{mkl} \mathbf{f}_*^{kl}. \tag{7.108}$$

Exercise 7.5 Show that the last two relations can be put in one formula:

$$\mathbf{f}_*^{0123} \lfloor \mathbf{e}_{\mu\nu} = -\frac{1}{2} \epsilon_{\mu\nu\lambda\kappa} \mathbf{f}_*^{\lambda\kappa} \tag{7.109}$$

We expect that the mapping of **F** into **G** has the following form similar to (7.106):

$$\mathbf{F} \mapsto \alpha \, \mathbf{f}_*^{0123} \lfloor \tilde{\mathcal{G}}[[\mathbf{F}]], \tag{7.110}$$

where $\tilde{\mathcal{G}}[[\mathbf{F}]]$ is given by (7.88), with the coefficient α to be determined. The first step has been already done in (7.92). We now perform the next step:

$$\mathbf{f}_*^{0123} \lfloor \tilde{\mathcal{G}}[[\mathbf{F}]] = \frac{1}{2} B_{ij} \, g^{mi} \, g^{nj} \, \mathbf{f}_*^{0123} \lfloor (\mathbf{e}_m \wedge \mathbf{e}_n) - c^{-1} E_j \, g^{mj} \, \mathbf{f}_*^{0123} \lfloor (\mathbf{e}_m \wedge \mathbf{e}_0). \tag{7.111}$$

We calculate the first term on the right-hand side with the substitution $g^{ij} = \delta^{ij}$ and the use of (7.107):

$$\frac{1}{2} B_{ij} \, g^{mi} \, g^{nj} \, \mathbf{f}_*^{0123} \lfloor (\mathbf{e}_m \wedge \mathbf{e}_n) = \frac{1}{2} B_{ij} \delta^{mi} \delta^{nj} \epsilon_{mnk} \mathbf{f}_*^{k0} = B_{12} \mathbf{f}_*^{30} + B_{23} \mathbf{f}_*^{01} + B_{31} \mathbf{f}_*^{20}.$$

Recall (4.47):

$$B_{ij} = \mu_0 \, \epsilon_{ijk} \, \mu^{kl} \, H_l. \tag{7.112}$$

We consider electrodynamics in vacuum, because only in this medium does light have velocity c for any reference frame. Therefore, we put the relative magnetic permeability $\mu^{kl} = \delta^{kl}$, and obtain

$$B_{12} = \mu_0 H_3, \quad B_{23} = \mu_0 H_1, \quad B_{31} = \mu_0 H_2.$$

In this manner we obtain for the first term in (7.111)

$$\frac{1}{2} B_{ij} \, g^{mi} g^{nj} \, \mathbf{f}_*^{0123} \lfloor (\mathbf{e}_m \wedge \mathbf{e}_n) = \mu_0 H_j \mathbf{f}_*^j \wedge \mathbf{f}^0 = \mu_0 \mathbf{H} \wedge \mathbf{f}^0. \tag{7.113}$$

We calculate the second term in (7.111) using (7.108):

$$-c^{-1} g^{mj} E_j \, \mathbf{f}_*^{0123} \lfloor (\mathbf{e}_m \wedge \mathbf{e}_0) = -\frac{1}{2} c^{-1} \delta^{mj} \, E_j \epsilon_{mkl} \mathbf{f}_*^{kl} = -\frac{1}{c} (E_1 \mathbf{f}_*^{23} + E_2 \mathbf{f}_*^{31} + E_3 \mathbf{f}_*^{12}).$$

In relation (4.29), $\varepsilon_0 \epsilon_{ijk} \varepsilon^{kl} E_l = D_{ij}$, we put relative permittivity of the vacuum $\varepsilon^{kl} = \delta^{kl}$, which leads to $\varepsilon_0 E_1 = D_{23}$, $\varepsilon_0 E_2 = D_{31}$, $\varepsilon_0 E_3 = D_{12}$; therefore

$$-c^{-1} g^{mj} E_j \, \mathbf{f}_*^{0123} \lfloor (\mathbf{e}_m \wedge \mathbf{e}_0) = -\frac{1}{c\varepsilon_0} (D_{23} \mathbf{f}_*^{23} + D_{31} \mathbf{f}_*^{31} + D_{12} \mathbf{f}_*^{12}) = -\frac{1}{c\varepsilon_0} \mathbf{D}. \tag{7.114}$$

We insert (7.113) and (7.114) into (7.111):

$$\mathbf{f}_*^{0123} \lfloor \tilde{\mathcal{G}}[[\mathbf{F}]] = \mu_0 \, \mathbf{H} \wedge \mathbf{f}^0 - \frac{1}{c\varepsilon_0} \mathbf{D} = \mu_0 \left(\mathbf{H} \wedge \mathbf{f}^0 - \frac{1}{c\varepsilon_0\mu_0} \mathbf{D} \right).$$

Since $c = 1/\sqrt{\varepsilon_0\mu_0}$, one may replace $\frac{1}{c\varepsilon_0\mu_0} = \frac{c}{c^2\varepsilon_0\mu_0} = c$ and

$$\mathbf{f}_*^{0123} \lfloor \tilde{\mathcal{G}}[[\mathbf{F}]] = \mu_0 (\mathbf{H} \wedge \mathbf{f}^0 - c\mathbf{D}).$$

By dint of (7.82), $\mathbf{f}^0 = c\mathbf{h}$, hence

$$\mathbf{f}_*^{0123} \lfloor \tilde{\mathcal{G}}[[\mathbf{F}]] = \mu_0 (\mathbf{H} \wedge c\mathbf{h} - c\mathbf{D}) = \mu_0 c \, (\mathbf{H} \wedge \mathbf{h} - \mathbf{D}).$$

Formula (7.24) allows us to write down

$$\mathbf{f}_*^{0123} \lfloor \tilde{\mathcal{G}}[[\mathbf{F}]] = -\mu_0 c \mathbf{G} = -\sqrt{\frac{\mu_0}{\varepsilon_0}} \, \mathbf{G}. \tag{7.115}$$

Let us introduce the expression $Z = \sqrt{\mu_0/\varepsilon_0}$, called the *intrinsic impedance of vacuum*. It follows from (7.115) that

$$\mathbf{G} = -\sqrt{\frac{\varepsilon_0}{\mu_0}}\, \mathbf{f}_*^{0123}\, \lfloor \tilde{\mathcal{G}}[[\mathbf{F}]] = -Z^{-1}\mathbf{f}_*^{0123}\, \lfloor \tilde{\mathcal{G}}[[\mathbf{F}]]. \tag{7.116}$$

The minus sign can be replaced by the determinant g of the metric matrix (7.78), $g = \det \mathcal{G}$:

$$\mathbf{G} = Z^{-1}\, g\, \mathbf{f}_*^{0123}\, \lfloor \tilde{\mathcal{G}}[[\mathbf{F}]]. \tag{7.117}$$

In this manner we have obtained the *four-dimensional constitutive relation*. The transition

$$\mathbf{F} \mapsto g\, \mathbf{f}_*^{0123}\, \lfloor \tilde{\mathcal{G}}[[\mathbf{F}]]$$

is known as the *Hodge map*. It is invertible— one may check that

$$\mathbf{F} = Zg\, \mathcal{G}[[\, \mathbf{e}_{3210}^* \lfloor \mathbf{G}]] \tag{7.118}$$

is the inverse of (7.117).

Let us look more closely at relation (7.117). Insert (7.88) in it and use (7.109):

$$\mathbf{G} = Z^{-1}g\, \mathbf{f}_*^{0123}\, \lfloor \left(\frac{1}{2} F^{\mu\nu} \mathbf{e}_{\mu\nu} \right) = \frac{1}{2} Z^{-1} g\, F^{\mu\nu} (\mathbf{f}_*^{0123} \lfloor \mathbf{e}_{\mu\nu}) = -\frac{1}{2} Z^{-1} g\, F^{\mu\nu} \frac{1}{2} \epsilon_{\mu\nu\lambda\kappa} \mathbf{f}_*^{\lambda\kappa}.$$

Hence the coordinates of $\mathbf{G}$ are

$$G_{\lambda\kappa} = -\frac{1}{2} Z^{-1} g \epsilon_{\lambda\kappa\mu\nu} F^{\mu\nu} = \frac{1}{2} Z^{-1} \epsilon_{\lambda\kappa\mu\nu} F^{\mu\nu}, \tag{7.119}$$

where $\epsilon_{\lambda\kappa\mu\nu}$ is the *totally antisymmetric (Levi-Civita) symbol* in four dimensions:

$$\epsilon^{\lambda\kappa\mu\nu} = \epsilon_{\lambda\kappa\mu\nu} = \begin{cases} 1 & \text{for even permutations of } \lambda\kappa\mu\nu \text{ with respect to 0, 1, 2, 3,} \\ -1 & \text{for odd permutations of } \lambda\kappa\mu\nu, \\ 0 & \text{when two indices repeat.} \end{cases}$$

The reader may check that the six coordinates of $\mathbf{G}$ are related to the six coordinates of $\tilde{\mathcal{G}}[[\mathbf{F}]]$ as follows

$$G_{01} = Z^{-1} F^{23}, \quad G_{02} = Z^{-1} F^{31}, \quad G_{03} = Z^{-1} F^{12}, \tag{7.120}$$

$$G_{12} = Z^{-1} F^{03}, \quad G_{23} = Z^{-1} F^{01}, \quad G_{31} = Z^{-1} F^{02}, \tag{7.121}$$

Because of the constitutive relation, the fields $\mathbf{F}$ and $\mathbf{G}$ become dependent on each other. It is interesting to consider the two Maxwell equations (7.23) and (7.29) imposed on only one of the fields—let it be the Faraday Field $\mathbf{F}$. The homogeneous equation (7.23) needs not be changed: we shall only express it by coordinates. We calculate explicitly:

$$\hat{\mathbf{d}} \wedge \mathbf{F} = (\mathbf{f}^0 \partial_0 + \mathbf{f}^1 \partial_1 + \mathbf{f}^2 \partial_2 + \mathbf{f}^3 \partial_3) \wedge (F_{01} \mathbf{f}^{01} + F_{02} \mathbf{f}^{02} + F_{03} \mathbf{f}^{03} + F_{12} \mathbf{f}^{12} + F_{23} \mathbf{f}^{23} + F_{31} \mathbf{f}^{31})$$

$$= \mathbf{f}^0 \partial_0 \wedge (F_{12} \mathbf{f}^{12} + F_{23} \mathbf{f}^{23} + F_{31} \mathbf{f}^{31}) + \mathbf{f}^1 \partial_1 \wedge (F_{02} \mathbf{f}^{02} + F_{03} \mathbf{f}^{03} + F_{23} \mathbf{f}^{23}) +$$

$$+ \mathbf{f}^2 \partial_2 \wedge (F_{01} \mathbf{f}^{01} + F_{03} \mathbf{f}^{03} + F_{31} \mathbf{f}^{31}) + \mathbf{f}^3 \partial_3 \wedge (F_{01} \mathbf{f}^{01} + F_{02} \mathbf{f}^{02} + F_{12} \mathbf{f}^{12}).$$

We put together terms with the same indices with permutations:

$$\hat{\mathbf{d}} \wedge \mathbf{F} = (\partial_0 F_{12} \mathbf{f}_{012} + \partial_1 F_{02} \mathbf{f}_{102} + \partial_2 F_{01} \mathbf{f}_{201}) + (\partial_0 F_{23} \mathbf{f}_{023} + \partial_2 F_{23} \mathbf{f}_{023} + \partial_3 F_{02} \mathbf{f}_{302}) +$$

$$+ (\partial_0 F_{31} \mathbf{f}_{031} + \partial_1 F_{03} \mathbf{f}_{103} + \partial_3 F_{01} \mathbf{f}_{301}) + (\partial_1 F_{23} \mathbf{f}_{123} + \partial_2 F_{31} \mathbf{f}_{231} + \partial_3 F_{12} \mathbf{f}_{312}).$$

We use the antisymmetry of symbols $F_{\mu\nu}$ and $\mathbf{f}^{\lambda\mu\nu}$:

$$\hat{\mathbf{d}} \wedge \mathbf{F} = (\partial_0 F_{12} + \partial_1 F_{20} + \partial_2 F_{01}) \mathbf{f}^{012} + (\partial_0 F_{23} + \partial_2 F_{30} + \partial_3 F_{02}) \mathbf{f}^{023} +$$

$$+ (\partial_0 F_{31} + \partial_1 F_{30} + \partial_3 F_{01}) \mathbf{f}^{031} + (\partial_1 F_{23} + \partial_2 F_{31} + \partial_3 F_{12}) \mathbf{f}^{123}. \tag{7.122}$$

The four basic three-forms $\mathbf{f}^{012}$, $\mathbf{f}^{023}$, $\mathbf{f}^{031}$, $\mathbf{f}^{123}$ are linearly independent; hence the equation $\hat{\mathbf{d}} \wedge \mathbf{F} = 0$ implies that four brackets are zero. In each bracket, a sum of three expressions with cyclic permutations of three indices is present. We denote them by square brackets placed in the indices:

$$\partial_{[0} F_{12]} = \partial_0 F_{12} + \partial_1 F_{20} + \partial_2 F_{01} = 0,$$

$$\partial_{[0} F_{23]} = \partial_0 F_{23} + \partial_2 F_{30} + \partial_3 F_{02} = 0,$$

$$\partial_{[0} F_{31]} = \partial_0 F_{31} + \partial_1 F_{30} + \partial_3 F_{01} = 0,$$

$$\partial_{[1} F_{23]} = \partial_1 F_{23} + \partial_2 F_{31} + \partial_3 F_{12} = 0.$$

These are four equations imposed in six coordinates $F_{\mu\nu}$ of the Faraday form stemming from the homogeneous Maxwell equation (7.23). They may be written down generally:

$$\partial_{[\lambda} F_{\mu\nu]} = 0. \tag{7.123}$$

We now proceed to express the nonhomogeneous Maxwell equation (7.29) in terms of **F**. If the charge-current density $\hat{\mathbf{j}}$ is expressed by basic three-forms:

$$\hat{\mathbf{j}} = \hat{j}_{012}\mathbf{f}^{012} + \hat{j}_{023}\mathbf{f}^{023} + \hat{j}_{031}\mathbf{f}^{031} + \hat{j}_{123}\mathbf{f}^{123},$$

the equation $\hat{\mathbf{d}} \wedge \mathbf{G} = \hat{\mathbf{j}}$ is expressed by coordinates as:

$$\partial_{[\lambda} G_{\mu\nu]} = \hat{j}_{\lambda\mu\nu}. \tag{7.124}$$

Any three-form **T** has four independent coordinates. There exists a means to transform it into another directed quantity with four independent coordinates, namely a vector. This is a contraction with the basic quadrivector $\mathbf{e}_{0123}$:

$$\mathbf{T} = \mathbf{e}_{0123}\lfloor \mathbf{T}.$$

For the twisted three-form $\hat{\mathbf{j}}$, the transformation is a contraction with the twisted quadrivector $\mathbf{e}^*_{0123}$:

$$\hat{\mathbf{j}} = \mathbf{e}^*_{0123}\lfloor \hat{\mathbf{j}}. \tag{7.125}$$

Appropriate calculation yields

$$\hat{\mathbf{j}} = \mathbf{e}^*_{0123}\lfloor \left(\frac{1}{6}\,\hat{j}_{\mu\nu\lambda}\mathbf{f}_*^{\mu\nu\lambda} \right) = \frac{1}{6}\,\hat{j}_{\mu\nu\lambda}\mathbf{e}^*_{0123}\lfloor \mathbf{f}_*^{\mu\nu\lambda} = \frac{1}{6}\,\hat{j}_{\mu\nu\lambda}\,\epsilon^{\mu\nu\lambda\kappa}\,\mathbf{e}_\kappa.$$

The expression for $\hat{\mathbf{j}}$ as the combination of four basics vectors $\mathbf{e}_\kappa$ is now visible; hence we write down its coordinates:

$$\hat{j}^\kappa = \frac{1}{6}\,\epsilon^{\mu\nu\lambda\kappa}\,\hat{j}_{\mu\nu\lambda}. \tag{7.126}$$

This denotes four relations:

$$\hat{j}^0 = \epsilon^{1230}\,\hat{j}_{123} = -\epsilon^{0123}\,\hat{j}_{123} = -\hat{j}_{123}, \quad \hat{j}^1 = \epsilon^{0231}\,\hat{j}_{023} = \epsilon^{0123}\,\hat{j}_{123} = \hat{j}_{023},$$
$$\tag{7.127}$$

$$\hat{j}^2 = \epsilon^{0312}\,\hat{j}_{031} = \epsilon^{0123}\,\hat{j}_{031} = \hat{j}_{031}, \quad \hat{j}^3 = \epsilon^{0123}\,\hat{j}_{012} = \hat{j}_{012}.$$
$$\tag{7.128}$$

These will be helpful for expressing equation (7.124) in terms of **F**.

We express left-hand side of (7.124) for specific indices with the use of (7.120,7.121):

$$\partial_{[0}G_{12]} = \partial_0 G_{12} + \partial_1 G_{20} + \partial_2 G_{01} = \partial_0(-Z^{-1}F^{03}) + \partial_1 Z^{-1}F^{31} + \partial_2(-Z^{-1}F^{23})$$

$$= -Z^{-1}(\partial_0 F^{03} + \partial_1 F^{13} + \partial_2 F^{23}).$$

Similarly

$$\partial_{[0}G_{23]} = -Z^{-1}(\partial_0 F^{01} + \partial_2 F^{21} + \partial_3 F^{31}),$$

$$\partial_{[0}G_{31]} = -Z^{-1}(\partial_0 F^{02} + \partial_1 F^{12} + \partial_3 F^{32}),$$

$$\partial_{[1}G_{23]} = Z^{-1}(\partial_1 F^{10} + \partial_2 F^{20} + \partial_3 F^{30}).$$

We insert this and (7.127, 7.128) into (7.124):

$$-Z^{-1}(\partial_0 F^{03} + \partial_1 F^{13} + \partial_2 F^{23}) = \hat{\jmath}^3,$$

$$-Z^{-1}(\partial_0 F^{01} + \partial_2 F^{21} + \partial_3 F^{31}) = \hat{\jmath}^1,$$

$$-Z^{-1}(\partial_0 F^{02} + \partial_1 F^{12} + \partial_3 F^{31}) = \hat{\jmath}^2,$$

$$Z^{-1}(\partial_1 F^{10} + \partial_2 F^{20} + \partial_3 F^{30}) = \hat{\jmath}^0.$$

After adding an appropriate zero term in each bracket (e.g. $\partial_3 F^{33}$ in first one) we use the summation convention:

$$-Z^{-1}\partial_\mu F^{\mu 3} = \hat{\jmath}^3,$$

$$-Z^{-1}\partial_\mu F^{\mu 1} = \hat{\jmath}^1,$$

$$-Z^{-1}\partial_\mu F^{\mu 2} = \hat{\jmath}^2,$$

$$Z^{-1}\partial_\mu F^{\mu 0} = \hat{\jmath}^0,$$

If the minus sign is absent in the first three equations, this could be written down as

$$\partial_\mu F^{\mu\nu} = Z\,\hat{\jmath}^\nu. \tag{7.129}$$

Thus, we have expressed the two Maxwell equations by coordinates of the Faraday field—(7.123) for the homogeneous equation and (7.129) for the nonhomogeneous one. They are revealed (usually without the Z factor) in traditional relativistic presentations of electrodynamics. We should stress that (7.123) is generally covariant, whereas (7.129) is only Lorentz covariant, because the Hodge map with the metric tensor has been used for its derivation.

The theory considered up to now is called *Maxwell-Lorentz electrodynamics*. We now sketch, according to [20], the most universal case. For a general medium, a local and linear constitutive relation can be written in terms of coordinates in the form:

$$G_{ij} = \frac{1}{2} \kappa_{ij}{}^{kl} F_{kl}.$$ (7.130)

The constitutive tensor $\kappa_{ij}{}^{kl}$, antisymmetric in pairs ij and kl separately, has 36 independent coordinates. It is useful to decompose it into irreducible parts. Within the premetric framework, constraction is the only tool for this.

The first contraction

$$\kappa_i{}^k = \kappa_{il}{}^{kl}$$ (7.131)

has 16 components. The second contraction

$$\kappa = \kappa_k{}^k = \kappa_{kl}{}^{kl}$$ (7.132)

is a single coordinate. The traceless piece of $\kappa_i{}^k$:

$$\pi_i{}^k = \kappa_i{}^k - \frac{1}{4} \kappa \, \delta_i^k$$ (7.133)

has 15 components. κ and $\pi^{\cdot}$ can be immersed in the original constitutive tensor as follows:

$$\kappa_{ij}{}^{kl} = \omega_{ij}{}^{kl} + 2 \pi_{[i}{}^{[k} \delta_{j]}^{l]} + \frac{1}{6} \kappa \, \delta_{[i}^k \delta_{j]}^l$$ (7.134)

where [ ] denotes antisymmetrization and $\omega^{\cdot\cdot}$ is the totally traceless part of $\kappa^{\cdot\cdot}$, that is,

$$\omega_{il}{}^{kl} = 0.$$ (7.135)

The right-hand side of (7.134) is the split of $\kappa^{\cdot\cdot}$ according to numbers of coordinates $36 = 20 + 15 + 1$. The symbol $\omega^{\cdot\cdot}$ is called the *principal part* of the constitutive tensor. The quantity $\pi^{\cdot}$, treated as a function of a space-time point, defines a *skewon field*:

$$S_i{}^j = -\frac{1}{2} \pi_i{}^j$$ (7.136)

whereas κ defines an *axion field*:

$$\alpha = \frac{1}{12} \kappa.$$ (7.137)

Maxwell-Lorentz electrodynamics arises when $S = 0$ and $\kappa = 0$ and

$$\omega_{ij}^{\ kl} = \lambda_0 \sqrt{-g} \, \epsilon_{ijmn} \, g^{mk} \, g^{nl}. \tag{7.138}$$

It should be mentioned that the spatial part of the metric is not unique—it can be accomodated to a medium, especially when it is not isotropic, which was considered in Sects. 4.2 and 4.3.

Lindell and Sihvola [36, 49] considered an exotic medium for which the principal and skewon parts of the constitutive tensor are zero and only the axion part remains. In such a case the constitutive relation assumes the simplest form:

$$G = \alpha \, F \tag{7.139}$$

with the pseudoscalar α. The authors called this medium a *perfect electromagnetic conductor (PEMC)*, bacause the limit $\alpha \to 0$ yields the perfect magnetic conductor, and the limit $1/\alpha \to 0$ gives the perfect electric conductor.

Let us insert expressions (7.19) for the Faraday two-form and (7.24) for the Gauss two-form

$$\mathbf{D}_T - \mathbf{H}_T \wedge \hat{\mathbf{d}}t = \alpha \, (\mathbf{B}_T + \mathbf{E}_T \wedge \hat{\mathbf{d}}t). \tag{7.140}$$

Comparing appropriate components gives two equalities:

$$\mathbf{D}_T = \alpha \, \mathbf{B}_T \quad \text{and} \quad \mathbf{H}_T = -\alpha \, \mathbf{E}_T,$$

which, for the spatial parts, lead to

$$\mathbf{D} = \alpha \, \mathbf{B} \quad \text{and} \quad \mathbf{H} = -\alpha \, \mathbf{E}. \tag{7.141}$$

The reader may recognize relations (1.9) mentioned in the context of the directed nature of the axion field.

Since, according to (7.141), $\mathbf{H}$ is proportional to $\mathbf{E}$, the Poynting two-form $\mathbf{S} = \mathbf{E} \wedge \mathbf{H}$ is zero. This means that when the electromagnetic wave enters a PEMC, the wave can not transmit energy. A plane electromagnetic wave incident normally from an ordinary medium on a slab made of PEMC was considered in [31]. The boundary conditions on the front and back interfaces on two sides imply that the electromagnetic wave does not enter the second medium behind the slab.

Appendix: Energy-Momentum Tensor in Coordinates

We start from the formula (7.49):

$$\mathbf{p}(V) = \frac{1}{2}\{\mathbf{G}\lfloor(V\lfloor\mathbf{F}) - \mathbf{F}\lfloor(V\lfloor\mathbf{G})\}. \tag{7.142}$$

We employ identities (3.31)

$$V\lfloor\mathbf{F} = -\frac{1}{2}\,V^{\mu\nu\lambda}F_{\nu\lambda}\mathbf{e}_\mu = \mathbf{v}, \tag{7.143}$$

and (3.8)

$$\mathbf{B}\lfloor\mathbf{v} = B_{\sigma\mu}v^\mu\mathbf{f}^\sigma \tag{7.144}$$

for the two-form $\mathbf{B}$ and vector $\mathbf{v}$, Thanks to this the first term in (7.142) can be expressed in the basis of one-forms:

$$\mathbf{G}\lfloor(V\lfloor\mathbf{F}) = G_{\sigma\mu}v^\mu\mathbf{f}^\sigma = -\frac{1}{2}\,G_{\sigma\mu}F_{\nu\lambda}V^{\mu\nu\lambda}\mathbf{f}^\sigma \tag{7.145}$$

The second term in (7.142) can be written by the exchange $\mathbf{G} \leftrightarrow \mathbf{F}$

$$\mathbf{F}\lfloor(V\lfloor\mathbf{G}) = -\frac{1}{2}\,F_{\sigma\mu}G_{\nu\lambda}V^{\mu\nu\lambda}\mathbf{f}^\sigma. \tag{7.146}$$

Now the whole expression (7.142) in the basis of one-forms is:

$$\mathbf{p}(V) = -\frac{1}{2}\left(-\frac{1}{2}\,G_{\sigma\mu}F_{\nu\lambda} + \frac{1}{2}\,F_{\sigma\mu}G_{\nu\lambda}\right)V^{\mu\nu\lambda}\mathbf{f}^\sigma \tag{7.147}$$

In such a case the coordinates of the one-form $\mathbf{p}$ may be written down as:

$$p_\sigma(V) = -\frac{1}{2}\left(-\frac{1}{2}\,G_{\sigma\mu}F_{\nu\lambda} + \frac{1}{2}\,F_{\sigma\mu}G_{\nu\lambda}\right)V^{\mu\nu\lambda}. \tag{7.148}$$

The energy-momentum tensor is of fourth rank as the mapping of the volume three-form into the momentum one-form. All considerations up to now are independent on metrics.

Usually in special relativity, the energy-momentum tensor is of second rank as a mappping of one-forms into one-forms. It is worth, therefore, replacing the volume trivector by an apropriate one-form. The contraction with the basis four-form f^{0123} is well suited for this. We introduce the one-form η to represent the volume:

$$\eta = f^{0123}\lfloor V = V^{012}\mathbf{f}^3 + V^{031}\mathbf{f}^2 + V^{023}\mathbf{f}^1 - V^{123}\mathbf{f}^0. \tag{7.149}$$

This has the coordinates

$$\eta_\alpha = \frac{1}{6} \epsilon_{\alpha\beta\gamma\lambda} V^{\mu\nu\lambda}. \tag{7.150}$$

The inverse transition has the form

$$V^{\beta\gamma\lambda} = -\epsilon^{\beta\gamma\lambda\nu} \eta_\nu. \tag{7.151}$$

We insert this into (7.148):

$$p_\sigma(V) = \frac{1}{4} \left(G_{\sigma\beta} F_{\gamma\lambda} - F_{\sigma\beta} G_{\gamma\lambda} \right) \epsilon^{\beta\gamma\lambda\nu} \eta_\nu. \tag{7.152}$$

The energy-momentum tensor is here the mapping $T : \eta \to \mathbf{p}$. Its coordinates have only two indices:

$$T_\sigma{}^\nu = \frac{1}{4} \left(G_{\sigma\beta} F_{\gamma\lambda} - F_{\sigma\beta} G_{\gamma\lambda} \right) \epsilon^{\beta\gamma\lambda\nu} \tag{7.153}$$

If the basis is arbitrary, this expression is still premetric.

The Gauss tensor $\mathbf{G}$ is not preesent in typical presentations of special relativity. It should be expressed by the Faraday tensor $\mathbf{F}$. The formula (7.119) allows us to express the coordinates of $\mathbf{G}$:

$$G_{\sigma\beta} = \frac{1}{2} Z^{-1} \epsilon_{\sigma\beta\kappa\pi} F^{\kappa\pi}, \qquad G_{\gamma\lambda} = \frac{1}{2} Z^{-1} \epsilon_{\gamma\lambda\kappa\pi} F^{\kappa\pi}. \tag{7.154}$$

It is necessary to remark that the basis cannot be arbitrary now. The formula (7.119) rests essentially on relation (7.87), where the metric tensor $g^{\mu\lambda}$ is present. Thus the basis must be orthonormal in the chosen metric. In this way, we arrive at the conclusion that energy-momentum can be expressed only by $\mathbf{F}$ after introducing a scalar product in space-time. We insert (7.154) into (7.153):

$$T_\sigma{}^\rho = \frac{1}{8} Z^{-1} \left(\epsilon_{\sigma\beta\kappa\pi} F^{\kappa\pi} F_{\gamma\lambda} - F_{\sigma\beta} \epsilon_{\gamma\lambda\kappa\pi} F^{\kappa\pi} \right) \epsilon^{\beta\gamma\lambda\nu} \tag{7.155}$$

$$= \frac{1}{8} Z^{-1} \left(\epsilon_{\sigma\beta\kappa\pi} \epsilon^{\beta\gamma\lambda\nu} F^{\kappa\pi} F_{\gamma\lambda} - \epsilon_{\gamma\lambda\kappa\pi} \epsilon^{\beta\gamma\lambda\nu} F_{\sigma\beta} F^{\kappa\pi} \right). \tag{7.156}$$

For the products of epsilons with summing over repeating indices identities exist, changing the products into multi-index deltas:

$$\epsilon_{\sigma\beta\kappa\pi} \epsilon^{\beta\gamma\lambda\nu} == -\delta^{\gamma\lambda\nu}_{\sigma\kappa\pi}, \tag{7.157}$$

$$\epsilon_{\gamma\lambda\kappa\pi} \epsilon^{\beta\nu\lambda\nu} = 2\delta^{\beta\nu}_{\kappa\pi}. \tag{7.158}$$

Therefore:

$$T_\sigma{}^\nu = \frac{1}{8} Z^{-1} \left(-\delta^{\gamma\lambda\nu}_{\sigma\kappa\pi} F_{\gamma\lambda} F^{\kappa\pi} - 2\delta^{\beta\nu}_{\kappa\pi} F_{\sigma\beta} F^{\kappa\pi} \right). \tag{7.159}$$

Multi-index deltas can be expressed by deltas with lower numbers of indices:

$$\delta^{\gamma\lambda\nu}_{\sigma\kappa\pi} = \delta^\gamma_\sigma \delta^{\lambda\nu}_{\kappa\pi} - \delta^\lambda_\sigma \delta^{\gamma\nu}_{\kappa\pi} + \delta^\nu_\sigma \delta^{\gamma\lambda}_{\kappa\pi}, \tag{7.160}$$

$$\delta^{\beta\nu}_{\kappa\pi} = \delta^\beta_\kappa \delta^\nu_\pi - \delta^\nu_\kappa \delta^\beta_\pi. \tag{7.161}$$

Thus the product present in the first term in the bracket in (7.159) can be written as

$$\delta^{\gamma\lambda\nu}_{\sigma\kappa\pi} F_{\gamma\lambda} = (\delta^\gamma_\sigma \delta^{\lambda\nu}_{\kappa\pi} - \delta^\lambda_\sigma \delta^{\gamma\nu}_{\kappa\pi} + \delta^\nu_\sigma \delta^{\gamma\lambda}_{\kappa\pi}) F_{\gamma\lambda} \tag{7.162}$$

$$= \delta^{\lambda\nu}_{\kappa\pi} F_{\sigma\lambda} - \delta^{\gamma\nu}_{\kappa\pi} F_{\gamma\sigma} + \delta^\nu_\sigma \delta^{\gamma\lambda}_{\kappa\pi} F_{\gamma\lambda} \tag{7.163}$$

$$= 2\delta^\nu_\pi F_{\sigma\kappa} - 2\delta^\nu_\kappa F_{\sigma\pi} + 2\delta^\nu_\sigma F_{\kappa\pi}, \tag{7.164}$$

and the whole first term in the bracket as

$$-\delta^{\gamma\lambda\nu}_{\sigma\kappa\pi} F_{\gamma\lambda} F^{\kappa\pi} = -2\delta^\nu_\pi F_{\sigma\kappa} F^{\kappa\pi} + 2\delta^\nu_\kappa F_{\sigma\pi} F^{\kappa\pi} - 2\delta^\nu_\sigma F_{\kappa\pi} F^{\kappa\pi} \tag{7.165}$$

$$= -2F_{\sigma\kappa} F^{\kappa\nu} + 2F_{\sigma\pi} F^{\nu\pi} - 2\delta^\nu_\sigma F_{\kappa\pi} F^{\kappa\pi}. \tag{7.166}$$

We change the summing indices $\pi \to \kappa$ in the second term in (7.166). Moreover we simplify the notation $F_{\kappa\pi} F^{\kappa\pi} = F \cdot F$; hence

$$-\delta^{\gamma\lambda\nu}_{\sigma\kappa\pi} F_{\gamma\lambda} F^{\kappa\pi} = 4 F_{\sigma\kappa} F^{\nu\kappa} - 2\delta^\nu_\sigma F \cdot F. \tag{7.167}$$

We transform the second term in the bracket in (7.159):

$$-2\delta^{\beta\nu}_{\kappa\pi} F_{\sigma\beta} F^{\kappa\pi} = -2(\delta^\beta_\kappa \delta^\nu_\pi F_{\sigma\beta} F^{\kappa\pi} - \delta^\nu_\kappa \delta^\beta_\pi F_{\sigma\beta} F^{\kappa\pi}) \tag{7.168}$$

$$= -2(F_{\sigma\kappa} F^{\kappa\nu} - F_{\sigma\pi} F^{\nu\pi}) = -4F_{\sigma\kappa} F^{\kappa\nu}. \tag{7.169}$$

In this way the whole expression (7.159) takes the form

$$T_\sigma{}^\nu = \frac{1}{8} Z^{-1} (4 F_{\sigma\kappa} F^{\nu\kappa} - 2\delta^\nu_\sigma F \cdot F - 4 F_{\sigma\kappa} F^{\kappa\nu}). \tag{7.170}$$

Thus we obtained

$$T_\sigma{}^\nu = Z^{-1} (F_{\sigma\kappa} F^{\nu\kappa} - \frac{1}{4} \delta^\nu_\sigma F \cdot F). \tag{7.171}$$

We raise the lower index σ using the metric tensor $g^{\mu\sigma}$:

$$T^{\mu\nu} = Z^{-1}(g^{\mu\sigma} F_{\sigma\kappa} F^{\nu\kappa} - \frac{1}{4} g^{\mu\nu} F \cdot F), \qquad (7.172)$$

$$= Z^{-1}(F^{\mu}_{\kappa} F^{\nu\kappa} - \frac{1}{4} g^{\mu\nu} F \cdot F). \qquad (7.173)$$

One may check that by raising and lowering indices in the first term, this can be written alternatively:

$$T^{\mu\nu} = Z^{-1}(F^{\mu\alpha} F^{\nu}_{\alpha} - \frac{1}{4} g^{\mu\nu} F \cdot F). \qquad (7.174)$$

This could be compared with formula (3.60) in the book by Misner, Thorne, and Wheeler [39]:

$$T^{\mu\nu} = F^{\mu\alpha} F^{\nu}_{\alpha} - \frac{1}{4} g^{\mu\nu} F_{\alpha\beta} F^{\alpha\beta}. \qquad (7.175)$$

The factor $Z^{-1} = \sqrt{\varepsilon_0/\mu_0}$, is not present in this expression. It can be equal to 1 if one takes the permittivity and permeability of the vacuum as 1.

Summarising: We have shown that the energy-momentum tensor (7.153) with the appropriate choice of the vacuum constants can be reduced to the traditional form written in the special theory of relativity.

References

1. Baldomir, D.: Differential forms and electromagnetism in 3-dimensional Euclidean Space. IEE Proc. **133A**, 139 (1986)
2. Baldomir, D., Hammond, P.: Global geometry of electromagnetic systems. IEE Proc. **140A**, 142 (1993)
3. Baldomir, D., Hammond, P.: Geometry of Electromagnetic Systems. Clarendon Press, Oxford (1996)
4. Bamberg, P., Sternberg, S.: A Course in Mathematics for Students of Physics. Cambridge University Press, Cambridge (1992)
5. Baylis, W.E.: Electrodynamics. A Modern Geometric Approach. Birkhauser, Boston (1998)
6. Burke, W.L.: Space-Time, Geometry, Cosmology. University Science Books, Mill Valley (1980)
7. Burke, W.L.: Manifestly parity invariant electromagnetic theory and twisted tensors. J. Math. Phys. **24**(1), 65 (1983)
8. Burke, W.L.: Applied Differential Geometry. Cambridge University Press, Cambridge (1985)
9. Byron, F.W., Fuller, R.W.: Mathematics in Classical and Quantum Physics. Addison-Wesley, Reading (1969)
10. de Rham, G.: Varietes Differentiables. Hermann, Paris (1955). English edition Differentiable Manifolds. Springer, Berlin (1984)
11. Deschamps, G.A.: Electromagnetics and differential forms. Proc. IEEE **69**, 676 (1981)
12. Dirac, P.M.: Quantised singularities in the electromagnetic field. Proc. R. Soc. Lond. **A133**, 60 (1931)
13. Dirac, P.M.: The monopole concept. Int. J. Theor. Phys. **17**, 235–247 (1978)
14. Doran, C., Lasenby, A.: Geometric Algebra for Physicists. Cambridge University Press, Cambridge (2003)
15. Edelen, D.G.B.: Applied Exterior Calculus. Wiley, New York (1985)
16. Flanders, H.: Differential Forms with Applications to the Physical Sciences. Dover, New York (1963)
17. Frankel, T.: Gravitational Curvature: An Introduction to Einstein's Theory. Freeman, San Francisco (1979)
18. Frankel, T.: The Geometry of Physics. An Introduction. Cambridge University Press, Cambridge (1997)
19. Hammond, P., Baldomir, D.: Dual energy methods in electromagnetism using tubes and slices. IEE Proc. **135A**, 167 (1988)
20. Hehl, F.W., Obukhov, Y.N.: Foundations of Classical Electrodynamics: Charge, Flux, and Metric. Birkhäuser, Boston (2003)

© Springer Nature Switzerland AG 2021
B. Jancewicz, *Directed Quantities in Electrodynamics*,
https://doi.org/10.1007/978-3-030-90471-5

21. Hehl, F.W., Obukhov, Y.N.: Dimensions and units in electrodynamics. Gen. Relativ. Gravit **37**(4), 733–749 (2005)
22. Hehl, F.W., Obukhov, Y.N.: Linear media in classical electrodynamics and the Post constraint. Phys. Lett. **A334**, 249–259 (2005)
23. Hehl, F.W., Obukhov, Y.N.: Spacetime metric from local and linear electrodynamics: a new axiomatic scheme. In: Ehlers, J., Lämmerzahl, C. (eds.) Special Relativity: Will it Survive the Next 101 Years?, pp. 163–187. Springer, Berlin (2006)
24. Hehl, F.W., Obukhov, Y.N., Rivera, J.-P., Schmid, H.: Magnetoelectric Cr_2O_3 and relativity theory. Eur. Phys. J. **B71**(3), 321–329 (2009)
25. Hehl, F.W., Itin, Y., Obukhov, Y.N.: On Kottler's path: origin and evolution of the premetric program in gravity and in electrodynamics. Int. J. Mod. Phys. **D25**(11), 1640016 (2016)
26. Herrmann, F.: Energy density and stress. A new approach to teaching electromagnetism. Am. J. Phys. **57**(8), 707–714 (1989)
27. Hestenes, V.: New Foundations for Classical Mechanics. D. Reidel, Dordrecht (1986)
28. Ingarden, R.S., Jamiołkowski, A.: Classical Electrodynamics. Elsevier, Amsterdam (1985) (Polish edition in 1980)
29. Jancewicz, B.: Multivectors and Clifford Algebra in Electrodynamics. World Scientific, Singapore (1988)
30. Jancewicz, B.: Directed Quantities in Electrodynamics (in Polish). Wyd. Uniwersytetu Wrocławskiego, Wrocław (2000)
31. Jancewicz, B.: Plane electromagnetic wave in PEMC. J. Electromagn. Waves Appl. **20**(5), 547–659 (2006)
32. Kaiser, G.: Energy-momentum conservation in pre-metric electrodynamics with magnetic charges. J. Phys. A **37**, 7163–7168 (2004)
33. Klitzing, K.v., Dorda, G.: New method for high-accuracy determination of the fine-structure constant based on quantized Hall resistance. Phys. Rev. Lett. **45**, 494–497 (1980)
34. Lindell, I.V.: Methods for Electromagnetic Field Analysis. Clarendon Press, Oxford (1992)
35. Lindell, I.V.: Multiforms, Dyadics, and Electromagnetic Media. Wiley, Hoboken (2015)
36. Lindell, I.V., Sihvola, A.H.: Perfect electromagnetic conductor. J. Electromagn. Waves Appl. **19**(7), 861–869 (2005)
37. Lounesto, P.: Clifford Algebras and Spinors, p. 1. Cambridge University Press, Cambridge (1997)
38. Lounesto, P., Mikkola, R., Vierros, V.: Geometric algebra software for teaching complex numbers, vectors and spinors. J. Comput. Math. Sci. Teach. **9**, 93 (1989)
39. Maxwell, J.C.: A Treatise on Electricity and Magnetism, 2nd edn. Clarendon Press, Oxford (1881)
40. Misner, C.W., Thorne, K.S., Wheeler, J.A.: Gravitation. Freeman and Co., San Francisco (1973). Sec. 2.5
41. Obukhov, Y.N., Hehl, F.W.: Measuring a piecewise constant axion field in classical electrodynamics. Phys. Lett. **A341**, 357–365 (2005)
42. Penrose, R., Rindler, W.: Spinors and Space-Time, vol. I. Cambridge University Press, Cambridge (1984)
43. Pérez, J.-P., Carles, R., Fleckinger, R.: Électromagnétisme. Fondements et Applications. Masson, Paris (1996)
44. Post, E.J.: Formal Structue of Electromagnetics. General Covariance and Electromagnetics. North Holland, Amsterdam (1962). Also Dover, Mineola, NY (1997)
45. Rund, H.: Energy-momentum tensors in the theory of electromagnetic fields admitting electric and magnetic charge distributions. J. Math. Phys. **18**(7), 1312–1315 (1977)
46. Schouten, J.A.: Tensor Analysis for Physicists, 2nd edn. Dover, New York (1989) (1st edn.: Clarendon Press, Oxford 1951)
47. Schutz, B.F.: A First Course in General Relativity. Cambridge University Press, Cambridge (1985)
48. Schwinger, J.: Magnetic charge and charge quantization condition. Phys. Rev. D **12**(10), 3105–3111 (1975)

49. Sihvola, A., Lindell, I.V.: Perfect electromagnetic conductor as building block for complex materials. Electromagnetics **26**(3/4), 279–287 (2006)
50. Spivak, M.: Calculus on Manifolds. Benjamin, New York (1965)
51. Tonti, E.: The Mathematical Structure of Classical and Relativistic Physics. Birkhäuser-Springer, New York (2013)
52. van Dantzig, D.: The fundamental equations of electromagnetism independent of metrical geometry. Proc. Camb. Phil. Soc. **20**, 421–427 (1934)
53. von Westenholz, C.: Differential Forms in Mathematical Physics. North Holland, Amsterdam (1978)
54. Warnick, K.F., Selfridge, R.H., Arnold, D.V.: Teaching electromagnetic field theory using differential forms. IEEE Trans. Educ. **40**(1), 53 (1997)
55. Weyl, H.: Space, Time, Matter. Dover, New York (1922)
56. Weyssenhoff, J.: Principles of Classical Electromagnetism and Optics (in Polish). PWN, Warsaw (1956)
57. Wu, L., Salehi, M., Koirala, N., Moon, J., Oh, S., Armitage, N.P.: Quantized Faraday and Kerr rotation and axion electrodynamics of a 3D topolgical insulator. Science **354**(6316), 1124–1127 (2016)

Printed in the United States
by Baker & Taylor Publisher Services